Low Voltage
Wiring
Handbook

Other Electrical Construction Books of Interest

ANTHONY • *Electric Power System Protection and Coordination*

CADICK • *Electrical Safety Handbook*

CROFT & SUMMERS • *American Electricians' Handbook*

DENNO • *Electric Protective Devices*

HANSELMAN • *Brushless Permanent-Magnet Motor Design*

JOHNSON • *Electrical Contracting Business Handbook*

JOHNSON • *Successful Business Operations for Electrical Contractors*

KOLSTAD • *Rapid Electrical Estimating and Pricing*

KUSKO • *Emergency / Standby Power Systems*

LINDEN • *Handbook of Batteries*

LUNDQUIST • *On-Line Electrical Troubleshooting*

McPARTLAND • *McGraw-Hill's National Electrical Code® Handbook*

McPARTLAND • *Handbook of Practical Electrical Design*

RICHTER & SCHWAN • *Practical Electrical Wiring*

SMEATON • *Switchgear and Control Handbook*

TRAISTER • *Design and Application of Security / Fire-Alarm Systems*

Low Voltage Wiring Handbook

Design, Installation, and Maintenance

Harry B. Maybin

McGraw-Hill, Inc.

New York   San Francisco   Washington, D.C.   Auckland   Bogotá
Caracas   Lisbon   London   Madrid   Mexico City   Milan
Montreal   New Delhi   San Juan   Singapore
Sydney   Tokyo   Toronto

Library of Congress Cataloging-in-Publication Data

Maybin, Harry B.
    Low voltage wiring handbook : design, installation, and
maintenance / Harry B. Maybin.
        p.   cm.
    Includes index.
    ISBN 0-07-041083-6
    1. Telecommunication wiring. 2. Low voltage systems. I. Title.
TK5103.12.M39   1995
621.382′3—dc20
                                 94-35054
                                    CIP

1 2 3 4 5 6 7 8 9 0 DOC/DOC 9 0 9 8 7 6 5 4

ISBN 0-07-041083-6

*The sponsoring editor for this book was Harold B. Crawford, the
editing supervisor was Peggy Lamb, and the production supervisor was
Donald F. Schmidt. It was set in Century Schoolbook by McGraw-Hill's
Professional Book Group composition unit.*

Printed and bound by R. R. Donnelley & Sons Company.

This book is printed on recycled, acid-free paper containing
a minimum of 50% recycled de-inked fiber.

Contents

Preface   xi

Chapter 1. Low Voltage Wiring—Introduction   1.1

1.1	Introduction	1.1
1.2	Information Exchange Process	1.1
1.3	Information Industry Terminology	1.3
1.4	Origins of Telecommunication Technology	1.4
1.5	Industry Standards	1.5
	1.5.1 International Organizations	1.6
	1.5.2 European Organizations	1.6
	1.5.3 United States Organizations	1.7
1.6	Regulatory Implications	1.9
1.7	Connectivity Alternatives	1.19
1.8	Book Overview	1.19

Chapter 2. Wire, Cable, and Coax Applications   2.1

2.1	Public and Private Networks	2.1
2.2	Local Distribution	2.2
2.3	Environment and Applications	2.4
	2.3.1 Horizontal Building Systems	2.5
	2.3.2 Vertical Building Systems	2.6
	2.3.3 Campus Distribution Systems	2.7
	2.3.4 Demarcation Points (MPOP)	2.7
	2.3.5 Environmental Specifications	2.10
	2.3.6 Additional MPOP Specifications	2.11
	2.3.7 Service Entrance	2.11
	2.3.8 House and/or Riser Cable Definitions	2.14
	2.3.9 Cable and Wire Specifications	2.14
	2.3.10 Wire and Cable Terminations	2.18
2.4	Entrance/Interbuilding Cable Sizing	2.18
2.5	Station Wire Sizing	2.18
2.6	Riser/Distribution Cable Sizing	2.20
2.7	Intraagency Tie Cable	2.21
2.8	Cable Performance Tables	2.21

Chapter 3. Optical-Fiber Applications   3.1

3.1   Origins of Optical Transmission   3.1
3.2   Fiber-Optic Modes   3.1
3.3   Optical Communication Links   3.2
3.4   Fiber-Optic Cable Systems   3.4
3.5   Types, Specifications, and Capabilities   3.6
    3.5.1   Single-Mode Fiber   3.6
    3.5.2   Multimode and Graded-Index Opitcal Fibers   3.7
3.6   Light Sources and System Components   3.7
    3.6.1   Light Sources for Fiber-Optic Communication Links   3.7
    3.6.2   Optical Detectors   3.8
    3.6.3   Other Components of the Optical-Fiber Link   3.9
    3.6.4   Fiber Connectors and Splices   3.9
    3.6.5   Wavelength Multiplexers   3.11
    3.6.6   Optical Directional Couplers   3.12
3.7   Optical Loop Calculations   3.12
3.8   Practical Integration   3.14

Chapter 4. Support Structure and Connectivity Hardware   4.1

4.1   Component Definition   4.1
4.2   Apparatus and Satellite Closets   4.1
    4.2.1   Apparatus Closet Specifications   4.2
    4.2.2   Satellite Closet Specifications   4.3
    4.2.3   Satellite Cabinets   4.3
4.3   Testing and Patch Panels   4.4
4.4   Demarcation Points   4.4
    4.4.1   Hardware   4.4
    4.4.2   Physical Security   4.6
4.5   Building Environment   4.8
    4.5.1   Horizontal Overhead Distribution Methods   4.8
    4.5.2   Horizontal Floor Distribution   4.11
    4.5.3   Horizontal Distribution in Older Buildings   4.16
4.6   Campus Distribution Systems   4.19
    4.6.1   Outside Plant Alternatives   4.19
    4.6.2   Aerial Construction   4.21
    4.6.3   Buried Cable   4.22
    4.6.4   Underground Conduit   4.22
4.7   Electrical Separation Requirements   4.24
    4.7.1   National Electrical Code® Position   4.24
    4.7.2   General Utility Company Separation Requirements   4.25
4.8   Electrical Protection, Grounding, and Cable Sheath Bonds   4.26
    4.8.1   Electrical Protection   4.26
    4.8.2   Electrical Protectors   4.27
    4.8.3   Electrical Grounds   4.29

Chapter 5. Noise Suppression and AC Power and Lighting Considerations   5.1

5.1   Sources and Types of Signal Interference   5.1
    5.1.1   Systematic and Fortuitous Distortion   5.2

5.2 Power and Lighting Considerations   5.13
     5.2.1 Power Planning Considerations   5.15
     5.2.2 Lighting   5.16
     5.2.3 Grounding   5.16
     5.2.4 AC Power Requirement   5.16
     5.2.5 Equipment and Power Clearance   5.16
5.3 Main Terminal and Equipment Rooms   5.17
5.4 Apparatus and Satellite Closets   5.17
5.5 Station Locations   5.17
     5.5.1 Voice Telephone Equipment   5.18
     5.5.2 Data   5.18
     5.5.3 Accessories   5.18

Chapter 6. Telecommunication Systems and Services   6.1

6.1 Private Languages and the Sale of Telecommunication Services 6.1
6.2 Transmission Methods   6.3
     6.2.1 Nonvoice Traffic   6.5
     6.2.2 Telephone Connections   6.5
     6.2.3 Simplex, Half-duplex, and Full-duplex Lines   6.6
6.3 Public Network Services   6.8
     6.3.1 An Overview of Centrex Services   6.8
     6.3.2 Digital Switching Systems   6.13
     6.3.3 ISDN   6.13
     6.3.4 Data Communications   6.14
     6.3.5 Features   6.14
6.4 Premises Equipment   6.16
     6.4.1 Capital Investment and Recurring Costs   6.17
     6.4.2 Regulatory Considerations   6.19
     6.4.3 Technology   6.19
     6.4.4 An Overview of Key and Hybrid Systems   6.20
6.5 Cellular Radio, Microwave, and Satellite Options   6.26
     6.5.1 Cellular Radio (Telephone)   6.26
     6.5.2 Short-haul Microwave Communication Systems   6.30
     6.5.3 Infrared Communications   6.34
     6.5.4 Satellite Communications   6.39
6.6 System Integration   6.46
     6.6.1 Bandwidth Partitioning   6.46
     6.6.2 Public/Private Interaction   6.47
     6.6.3 Multivendor Interface Options   6.48

Chapter 7. Data Processing Applications   7.1

7.1 Industry Overview   7.1
7.2 ISO and IEEE Models   7.1
7.3 Wide Area Networks   7.4
7.4 Data Speed   7.7
7.5 Topology   7.8
     7.5.1 Star   7.8
     7.5.2 Bus   7.9
     7.5.3 Ring   7.10
     7.5.4 Branching Tree   7.10

7.6   Protocols and Protocol Conversion   7.11
         7.6.1   Cable Access Protocols   7.12
         7.6.2   Protocol Converters   7.15
7.7   Vendor Connectivity Specifications   7.16

Chapter 8. Management and Control Systems   8.1

8.1   System Administration Requirements   8.1
8.2   Facility Inventory System   8.2
         8.2.1   Data Equipment Forms   8.3
         8.2.2   Single-Line Telephone Set Forms   8.3
         8.2.3   Key Telephone System Inventory Forms   8.3
         8.2.4   PBX Equipment Inventory Forms   8.3
8.3   Cable Numbering Records   8.10
         8.3.1   Type of Cable   8.10
         8.3.2   Terminating Location   8.10
         8.3.3   Cable Numbering   8.10
8.4   Building Closet Numbering   8.11
         8.4.1   Serving Closet Code   8.11
         8.4.2   Floor Number   8.11
         8.4.3   Closet Identification   8.11
8.5   Universal Information Outlet Numbering Plan   8.11
         8.5.1   Floor Number   8.12
         8.5.2   Type of Serving Closet   8.12
         8.5.3   Universal Information Outlet Number   8.12
8.6   Building Terminal Records   8.12
         8.6.1   Network Access Point (NAP), MPOP, or Demarc Record   8.13
         8.6.2   Apparatus and Satellite Closet Terminal Records   8.13
         8.6.3   Equipment Room Records   8.16
         8.6.4   Data Patch Panel Records   8.16
8.7   Cable Assignment Records   8.16
8.8   Location Records   8.21
         8.8.1   Cable Location Records   8.21
         8.8.2   Campus Location Records   8.21
         8.8.3   Aerial Facilities   8.21
         8.8.4   Buried Cable   8.22
         8.8.5   Conduit Location Records   8.22
8.9   Facility Modifications   8.22
8.10 Facility and Equipment Diagnostic System   8.24
         8.10.1   Modular and Nonmodular Set Flowchart   8.24
         8.10.2   Line Troubleshooting Procedure   8.24
         8.10.3   Data Troubleshooting Procedure   8.24
         8.10.4   Diagnostic Equipment   8.26
8.11 Trouble Reporting and Repair Analysis Procedure   8.28

Chapter 9. Planning Guidelines   9.1

9.1   A System Approach to Connectivity Planning   9.1
         9.1.1   Study Objectives   9.2
         9.1.2   Alternatives Considered   9.2

9.2 Study Techniques   9.2
     9.2.1 Internal Rate of Return   9.3
     9.2.2 Present Worth of Expenditure   9.3
     9.2.3 Present Worth of Annual Costs   9.3
     9.2.4 Benefit-to-Cost Ratio   9.4
     9.2.5 Payback Period   9.4
     9.2.6 Break-Even Study   9.4
     9.2.7 Study Technique Summary   9.5
9.3 Economic Variables   9.6
9.4 Information Gathering   9.6
     9.4.1 Per-Unit Cost Estimating Process   9.7
     9.4.2 Cost Applications   9.7
     9.4.3 Additional Costs   9.8
     9.4.4 Per-Unit Costs   9.8
9.5 Capital Expenditure Study Technique   9.16
9.6 Maintenance Expense Study Technique   9.16
     9.6.1 Study Assumptions   9.16
     9.6.2 Selection of Study Alternatives   9.17
     9.6.3 Data Collection   9.17
     9.6.4 Program Time Conversion   9.18
     9.6.5 Cable and Support Structure Lives   9.19
     9.6.6 Special Study Considerations   9.19
9.7 Sample Study Analysis   9.20
9.8 Study Analysis and Budget Process   9.23
     9.8.1 Project Presentation   9.25
9.9 Contract Bid Strategies   9.26
9.10 Bid Specifications   9.26
9.11 Contractor Strategies (the Vendor's Point of View)   9.27
9.12 Bid Evaluation   9.29

Chapter 10. Job Scheduling   10.1

10.1 Purpose   10.1
10.2 Coordination Responsibility   10.1
10.3 Scheduler Authority and Responsibility   10.2
10.4 Scheduling Techniques   10.2
     10.4.1 Office/Building Planning Counterparts   10.3
     10.4.2 Equipment and Service Order Origination   10.4
     10.4.3 Facilities Design/Engineering Department or Vendor   10.4
     10.4.4 Procurement Department or Services   10.4
     10.4.5 Public Sector Utility Companies   10.4
     10.4.6 Vendor and Contractor Schedules   10.5
10.5 Performance Control System   10.5
     10.5.1 Cable Placement Activity   10.5
     10.5.2 Cable Termination Activity   10.6
10.6 Inspection Requirements   10.6
10.7 Conformance Testing   10.7
     10.7.1 Types of Defects   10.7
     10.7.2 Data Cable Tests   10.8

x   Contents

10.7.3   Conformance Tests on Interbuilding or Campus
          Distribution Systems   10.9
10.8   System Acceptance   10.9

Glossary of Telecommunications Technologies   10.11
Index   I.1 (follows Glossary)

Preface

The communications industry is currently in a state of rapid expansion. Primary to the evolution of services and networks is the connectivity portion of the industry. For decades, the marketing and public relations emphasis of major players in the industry has been on the innovative features that will enhance the individual's or entity's ability to rapidly transfer information. Central to this theme is the method by which this information passes. While the digitizing of information has become the primary method, the means by which information travels are today, and will probably remain, the physical connectivity facilities.

This book explores the various types of wire, optical fiber, and coax media. Each medium is described with potentials for trafficking described. Nonphysical systems are also described with their application potentials and component equipment.

It is important for the reader to recognize that this work encompasses several years of field experience in the application of voice and data services for customers ranging from the individual subscriber to the largest Fortune 500 users. The depth of description and applications described herein may be applied to the most diverse and exotic equipment configurations. The presentation of material is intended to be both tutorial and practical so that even the beginner may gain substantial knowledge of the application of connectivity theories.

Harry B. Maybin

Low Voltage
Wiring
Handbook

1

Low Voltage Wiring— Introduction

1.1 Introduction

This book is presented to give the neophyte electrical engineer and information system manager an insight into the complexities of the connectivity layer to the systems currently available to process voice, data, and video information.

This work is based on many years of investigation and empirical involvement with telecommunications and data processing operations in a broad range of environments. The impetus for developing this work was that no compendium of information vital to system analysis and design was available for me when I undertook system analysis; I observed that rather simple approaches would have saved my clients considerable time and expense had such information been available.

The topics covered in this work could, on their own, comprise many volumes. To contain the information for practical use, only information vital to explaining the connectivity layer is included. What others will provide to the point of interconnect will be explained but not detailed. From there, it is hoped that the reader will be able to distinguish between true connectivity and product delivery technique.

1.2 Information Exchange Process

The means by which humans transfer information is called *communication*. Communication occurs between a minimum of two persons, and this exchange can take many forms. Over the course of time, increasingly intricate methods of establishing clear meaning for different sounds and signs developed. Eventually standard sounds and body language became accepted as standard representations for verifiable thoughts and meanings between individuals. Long-distance communication methods were developed for communicating when sight or auditory senses were not able to discern communications because of distance between the communicators. At first, smoke signals and drum beats were codified to pass simple messages. Later, more complex sound and sight patterns were developed to allow transfers of more complex information without direct contact with the receiving party.

Eventually, written languages were developed and standard symbols were used to represent certain sounds, which represented like and verifiable thoughts common among many persons.

From the standardization of symbols and pulses came the Morse code. Digital communication evolved from this standard. Voice synthesizing has become a unique separate development of equal and diverse proportions. In this discipline, voice-frequency ranges were originally transmitted on direct low voltages over a range of frequencies common to the human voice range on a one-on-one basis. To improve communication, time-division multiplexing was developed to allow each voice on a particular communication link to be divided by 60 or more samples per second and reassembled at the receiving end. By this method, a time-slot ordering of voice communication was developed which allowed for simultaneous communication over the same transportation medium. From this, digital and carrier electronics developed.

Today, these basic methods have been taken to a highly sophisticated level. The telecommunications and data processing industries have merged into the information technology industry and have the capability and capacity to process information communications at increasingly higher speeds and efficiencies.

The challenge now is how to best use this capacity. We are seeing an increasing tendency to rely on electronics to pass, retain, and collate all human communication. The industry appears to be not only supporting this position, but developing new technologies to make it possible for humans to not have to communicate on a one-on-one basis at all. This being the case (just try to reach a human bureaucrat these days), true communication is eroding. Massive information files are being developed around incredibly complex security and retrieval systems. While beyond the scope of this work, the reader should be cognizant of this tendency and apply realistic logic to wants and needs when developing performance criteria. Sometimes it is best to allow people to communicate directly.

1.3 Information Industry Terminology

The communications industry has been kept a secret, in terms of component and feature mechanics, for practically 100 years. Partially because of complexity, the physical aspects of communications were never described or shared with the public in general. The overlying strategy of the Bell System was to attain monopoly status, which it did. While the Bell System enjoyed this status, other competing interests did not have the incentive to participate, and the point was pretty much moot. When competition was allowed, several vendors began trying to penetrate the system, and they encountered a maze of terminology and proprietary signaling scenarios.

It took several years for inroads to be made into the public network and standards to evolve. Meanwhile, nonpublic systems were being developed by several manufacturers to compete with Bell System devices. A massive campaign was then undertaken to undermine competition by all parties. (A similar campaign caused the formation of the monopoly in the first place.) Feature packages were developed and defined by each manufacturer without a standard point of reference. This approach has led to some bizarre occurrences in the end-user market, with customers having purchased features which are never used. The latest information indicates that 93 percent of the features available in most private systems is used by less than 5 percent of the users.

The same phenomenon is true in the connectivity arena. There are over 2400 different types and configurations of wire, coaxial cable, and optical fiber available to the public. Each equipment manufacturer and service provider has developed a standard for its system and configures the connectivity in a precise and often unique manner.

The advantage to this strategy is one of marketing. Since there is no regulatory mandate in the area of private system compatibility, vendors have developed systems that are incompatible with one another. System expansion, therefore, is usually one of like for like, excluding advantages available in others.

The proliferation of system configurations extends through all layers of the International Standards Organization (ISO) model. This text deals specifically with level 1 of that model, the physical link.

1.4 Origins of Telecommunication Technology*

As originally invented by Bell, telephone communication went from a particular telephone instrument to only one other telephone instrument. This service was soon extended to connect a number of telephones to the same line, a form of party line. Everyone could hear everyone else, hence there was no privacy, and one call would prevent anyone else from using the line. Clearly, there was the need for a telephone to be switched to any other desired telephone.

One way to perform this switching would be to bring lines from all telephones to all other telephones. A switch at each telephone would then make the connection with the appropriate line to the desired telephone. If the universe of telephone stations to be reached is small, such a system would be workable. However, if the universe is large, then the large number of lines that must terminate at each telephone makes such a system of station switching impractical.

The ultimate solution was discovered and implemented only a few years after the invention of the telephone by Bell. The solution was a centralized switching arrangement. All the lines from all the telephone stations were brought to a common place where the electrical connections were made to connect one station to another. The actual connections were made by people. The central place where the lines all came together is called the *central office* or the *exchange*.

As exchanges grew to cover greater geographic areas, it became impractical to bring lines from the more outlying areas to one central office. More central offices were created, each serving a nearby surrounding area. Connections between central offices were made on lines called *trunks*. As growth continued, special switching offices were developed to handle the trunks between a number of central offices. This centralized switching of trunks was performed at a switching office called a *tandem office*.

A new switching need then developed: to serve long-distance or toll circuits between cities. Switching offices were thus devised to perform switching of toll trunks only. These offices were called *toll offices*.

For businesses, a fair amount of telephone traffic was between telephones that were all located on the customers' premises. These telephones, therefore, could be served most efficiently with a private

---

*Datapro Research Corp., "Switching Systems," Delran, N.J., McGraw-Hill, MT 20-050, February 1988, pp. 201–202.

switch located on the premises. This switch was called a *private branch exchange* (PBX), a term that is still in use today. Present PBXs are automatic (PABXs), using electromechanical or electronic technology (EPABXs).

During the first few decades of telephone communications, switching was a manual operation performed by human beings who made the actual connections of circuits. The connections were made at a switchboard utilizing cords with plugs at the ends. The plug had a tip and a ring, which made the actual connection between the lines. The sleeve was used for signaling and the supervisory purposes in common battery exchanges. The terms *tip* and *ring* continue to be used to this day for the two wires between the central office and the actual telephone instrument. The circuits calling for service and the availability of trunks were indicated by small lamps.

The first major innovation in switching came in 1892 with the installation of an automatic switch controlled by the telephone instrument itself. This switch was conceived by Almond B. Strowger. A later modification of this system included the use of a dial and dial pulses to control the operation of the electromechanical switching device. The Strowger switch was adopted by the Bell System in 1919. Bell System engineers later developed improved automatic switching systems using electromechanical technology. The electromechanical technology was somewhat slow, not very flexible in terms of offering new services, and frequently generated electrical noise in the connection.

The current generation of technology for telephone switching is electronic, using either analog or digital switching techniques. Electronic technology is extremely fast and flexible.*

1.5 Industry Standards†

While it is the technology of telecommunications that captures the headlines, it is industrywide standards that make the applications of this technology possible. The roles of the various standards-setting bodies are outlined below.

---

*Datapro Research Corp., "Switching Systems," Delran, N.J., McGraw-Hill, MT 20-050, February 1988, pp. 201–202.

†Datapro Research Corp., "Communications Standards and Standards Organizations," Delran, N.J., McGraw-Hill, MT 70-700, June 1990, pp. 101–106.

1.5.1 International organizations

At the international level, two organizations are important in setting telecommunications standards. These are:

1. International Telecommunications Union (ITU)
2. International Standards Organization (ISO)

The ITU now has 160 member governments. Carriers, postal, telephone, and telegraph agencies (PTTs), and industrial and scientific organizations are also involved in the ITU's work. The ITU's mission is to promote international cooperation and development of worldwide telecommunications services. The ITU contains two standards-setting bodies:

1. The International Radio Consultative Committee (CCIR)
2. The Consultative Committee on International Telegraph and Telephone (CCITT)

As its name suggests, the CCIR is concerned with matters that have to do with over-the-air communications. All other matters are dealt with by the CCITT.

The International Standards Organization is made up of national standards bodies from 89 states. The purpose of the ISO is to promote the development of standardization throughout the world. It carries out its mission primarily by publishing technical specifications called *international standards*. The ISO's Technical Committee 97, however, has been a major international force behind the development of standards for information systems. Technical Committee 97 has subcommittees dealing with the following matters:

- Information coding
- Telecommunications and information exchange between systems
- Interconnection of equipment
- Text and office systems
- Information processing security
- Information retrieval, transfer, and management
- Optical digital data discs

1.5.2 European organizations

At the European level, two organizations are important to the setting of telecommunication standards. These are:

1. The European Conference of Postal and Telecommunications Administrations (ECPT)

2. The European Computer Manufacturers Association (ECMA)

ECPT coordinates the policies of various European PTTs. It operates through a series of committees which deal with:

- Harmonization coordination, covering such matters as service definitions
- Commercial action dealing with tariffs and related matters
- Transatlantic telecommunications
- Radio communications
- Long-term studies

The ECMA currently has 45 members. Its Technical Committee 32 concerns itself with open systems interconnection (OSI) and related matters through the following subcommittees:

- Public data networks
- LANs
- Interfaces to private switching networks
- Transport and network layers

1.5.3 United States organizations

The most important organizations influencing telecommunications standards setting in the United States are:

- The Federal Communications Commission (FCC)
- The Department of State's "public advisory committees" for the CCITT and the CCIR. These are called respectively the U.S. Organization for the CCITT (USCCITT) and the U.S. Organization for the CCIR (USCCIR)
- The Exchange Carriers Standards Association (ECSA)
- The Subcommittee on Information Processing Systems of the American National Standards Institute (ANSI)
- The Electronic Industries Association (EIA)
- The Institute of Electrical and Electronics Engineers (IEEE)
- The Corporation for Open Systems (COS)

The FCC is a government agency charged with regulating U.S. telecommunications. Part 66 of FCC *Rules and Regulations* provides standards for interconnection to the public network. Part 15 of the *Rules and Regulations* governs permissible levels of electromagnetic emissions.

The USCCITT is a State Department group which serves as a public advisory committee. It recommends U.S. policies and contributions to the CCITT. The USCCITT has a national committee, a joint working party on ISDN, and four study groups dealing with telecommunication policies and services, the World Administrative Telegraph and Telephone Conference, the worldwide telephone network, and data and ISDN.

The USCCIR is another State Department group which functions as a public advisory committee which recommends U.S. policies and contributions to the CCIR. The USCCIR has four study groups dealing with fixed service using communications satellites, mobile services, fixed service using radio relay systems, and broadcasting service.

ECSA is a carrier organization which sets standards for the U.S. public network. The focal point for network standards is the T1 Committee on Communications. This has subcommittees dealing with carrier–customer-premises interfaces; the integrated services digital network (ISDN), internetwork operations, administration, maintenance, and provisioning; performance; carrier-to-carrier interfaces; and specialized subjects.

ANSI is a national standards institute which coordinates voluntary standards activity. It deals with digital telecommunications standards by the X3 Committee on Information Processing Systems. This has subcommittees dealing with data communications, data encryption, data interchange, OSI, input/output (I/O) interface, and office and publishing systems.

The EIA is a trade association which develops industry standards. It has committees concerned with facsimile, data-transmission systems and equipment, telephone terminals, and optical communications.

The IEEE is a professional organization which develops industry standards. Its primary involvement is through the 802 Committee on Local-Area Networks (LANs).

COS is an industry joint venture that tests and promotes standards for open networking. It is composed of provisional committees set up to examine various aspects of OSI such as testing and network management.*

---

*Datapro Research Corp., "Communications Standards and Standards Organizations," Delran, N.J., McGraw-Hill, MT 70-700, June 1990, pp. 101–106.

1.6 Regulatory Implications*

Since 1866, the telecommunication industry in the United States has been subject to government regulation. The nature of the telecommunications industry is such that strict guidelines have been deemed necessary to ensure fair business practices as well as to guarantee uninterrupted service to the public over a broad geographic area.

Although advances in technology over the past several years have considerably altered the telecommunications market, the basic telephone service network is still considered a public utility. An industry is classified as a public utility if it is a "natural monopoly" and if the public is reliant on the services it provides. In a natural monopoly, only a single enterprise can operate efficiently in a particular market. An industry that supplies services essential to public welfare, services for which there is no ready substitute, is said to be *in the public interest*. The efficient operation of the telephone industry certainly continues to be in the public interest, although the monopolistic aspect of the industry has come to be viewed quite differently in the past 10 years.

When the telephone industry was developing in the United States there would have been no advantage to encouraging several different telephone companies to compete in the same locality; the duplication of facilities and services would have been a waste of resources. Moreover, the public demand for telephone services is inelastic; a company having a monopoly in the field could raise its prices drastically without eliciting a proportionate decrease in demand. Therefore, the government imposes regulations not only to ensure that a monopoly does exist, but also to see to it that the company allowed to operate as a monopoly provides services at reasonable rates. To accomplish this, the telephone industry in the United States is regulated by the Federal Communications Commission and by utility commissions in each of the 50 states.

The basis for the government regulation of certain business activities is Art. 1, Sec. 8, of the United States Constitution, in which Congress is granted the power to "regulate Commerce with foreign Nations, and among the several States...." The regulation of telecommunications in the United States began in 1866 with the passage of the Post Roads Act. Under this act, the Postmaster General was authorized to fix rates for government telegrams. This act also grant-

---

*Datapro Research Corp., "A History of the Telecommunications Regulatory Environment," Delran, N.J., McGraw-Hill, MT 10-620, June 1987, pp. 101–118.

ed rights-of-way over public lands. In 1887, a provision of the Act to Regulate Commerce empowered the Interstate Commerce Commission (ICC) to order the interconnection of the lines of telegraph companies in the interest of better service to the public. In 1907, the states of Wisconsin and New York passed legislation creating commissions with powers to regulate public utilities, including telephone companies. By 1920, more than two-thirds of the states had established regulatory commissions and today all 50 states plus the District of Columbia have commissions.

The Radio Act of 1912, administered by the U.S. Department of Commerce, reserved certain frequencies for government use and set rules for the transmission of distress signals from ships at sea. The act also provided for the licensing of the first radio stations. The following year, the ICC issued an order requiring telephone companies to keep accounts according to certain guidelines laid down by the commission. In 1918, the ultimate act of government regulation was imposed on the communications industry—a complete takeover by the U.S. Post Office Department of all the telephone and telegraph systems in the country. The federal government, finding itself involved in a world war, had decided that such an extreme measure was necessary in the interest of the national security. On Aug. 1, 1919, all telephone and telegraph systems were returned to private ownership. The government would not repeat this action during World War II, apparently concluding from the experience of the previous war that the communication systems of the country were safe in the hands of civilians.

By the 1930s, it had become apparent that a single regulatory body was needed to deal with the changing conditions in the communications field. The regulation of all interstate and international communications was finally consolidated by the Communications Act of 1934 which established the FCC as an independent government agency responsible directly to Congress and charged with regulating interstate and international communications by radio, television, wire, and, in recent years, satellite and cable.

In 1949, the Justice Department filed a suit charging that effective regulation of telephone rates was hampered because higher prices charged by Western Electric, a subsidiary of AT&T, for equipment automatically increased the investment on which Bell System companies were allowed to earn a reasonable return. The suit requested that Western Electric be required to separate itself from AT&T, and furthermore, be divided into three separate companies that would compete with other manufacturers for Bell System business. In 1956, it was decided that, while Western Electric would not be separated from AT&T, it would limit its manufacturing operation to the type of equipment purchased by the Bell System and would refrain from entering other markets.

In 1962, legislation called the Communications Satellite Act brought space age communication technology under FCC jurisdiction. This law empowered the FCC to (1) ensure that effective competition prevailed in all procurements of equipment and services required for satellite systems by common carriers, (2) certify the technical characteristics of satellite systems to be used by domestic satellite common carriers, (3) authorize the construction and operation of each earth station, and (4) determine the vendor on the basis of the "public interest, convenience, and necessity."

Following the settlement of the 1949–1956 antitrust suit by the consent decree, the FCC began to reexamine its policy of allowing AT&T to prohibit subscribers from connecting to the telephone lines any equipment not supplied by the telephone company. The FCC's change in attitude was reflected in the case of the Carterfone, an inductive-acoustical device manufactured by Carter Electronics Corp. of Dallas and designed to interconnect private two-way radios with the telephone system by means of a base station. The FCC decided in June 1968 that the Carterfone and other telephone attachments could be connected to the public telephone system but conceded that the telephone companies would be allowed to install protective equipment between the line and the foreign device. The Carterfone decision opened the door to increased competition in the telephone industry and made possible a new interconnect industry that would become a viable competitive force in the communication equipment market.

Following a 5-year inquiry, now known as Computer Inquiry I, into regulatory and policy questions raised by the increasing interdependency of computers and communications, the FCC issued its final decision on March 18, 1971. It held that, with minor exceptions, common carriers desiring to provide data processing services to persons other than the carrier itself, or its affiliates, could do so only through complete separate affiliates. Additionally, the FCC concluded that no public interest would be served by the regulation of data process services, but rather that the market would best develop and flourish in a competitive environment. In 1971, in its specialized common carrier proceedings, the FCC adopted a policy of increased competition among existing and new common carriers in the sale of data transmission and other specialized communications services to the public.

In 1975 the FCC made a move to open up the mobile telephony market by reallocating the 806- to 946-MHz portion of the radio spectrum to land mobile communication uses. In 1976, the FCC took steps to ease the regulatory burden on microwave expansion applications by simplifying rules and forms and by automating the application processing procedure. Further, orders from the FCC in 1975 and 1976 dictated that "foreign" (non-AT&T) equipment could be installed without the intervention of protective devices, provided that foreign equipment was certified and registered by the FCC to ensure that connection of the devices would not harm the network.

Between 1971 and 1977, the FCC conducted a two-part investigation into AT&T interstate rates and the impact of interrelationships among the Bell System's organizational entities [Bell Operating Companies (BOCs), Western Electric, Bell Labs, AT&T Long Lines, AT&T General Departments, etc.]. The FCC established 8.5 percent as an appropriate rate of return in 1972 and decided that the operating companies should have more autonomy than they previously had to procure equipment from manufacturers other than the AT&T internal manufacturing affiliate, Western Electric.

Between 1976 and 1980, the FCC undertook Computer Inquiry II and decided to no longer attempt to define data processing separately from data communications. Instead, it established two classes of service: basic and enhanced. The decision also defined two classes of customer premises equipment (CPE): embedded and nonembedded. Basic service and embedded equipment would continue to be subjected to state tariffing while the enhanced services and nonembedded equipment would be detariffed effective Jan. 1, 1982. The third major point stated that all regulated carriers (including AT&T) would be permitted to provide enhanced services and nonembedded CPE only through a fully separated subsidiary. This ruling has subsequently been modified to allow regulated carriers to sell, but not manufacture, CPE, effective Jan. 1, 1984. AT&T's response to this point was the creation of a new, fully separated subsidiary, AT&T Information Systems, which began operations on Jan. 1, 1984.

Having opened various interstate communication markets to competition, the FCC was faced with a situation in which traditional wireline carriers were competing in some markets while still maintaining a monopoly in others. This environment offered some carriers the opportunity to undercut service prices in competitive areas by subsidizing those operations with revenues from markets in which they held a monopoly. As such practices are forbidden by the Communications Act, it was up to the FCC to see to it that rates for services were based on the cost of providing those services. Since the Bell System and other interstate carriers were using facilities in common to furnish services, the cost of the facilities had to be apportioned to the various services.

In 1976, the FCC adopted a set of guidelines for cost allocation which was known as Fully Distributed Cost Method 7, and AT&T was ordered to file tariffs according to these standards. After modifying its previous standards, the FCC adopted in 1980 an interim manual for cost allocation, which AT&T was required to use in computing its rates. Accordingly, AT&T in December 1981 filed new tariffs for mobile telephone service (MTS), wide-area telephone service (WATS), and private-line services that were designed to earn the designated interstate rate of return.

1969–1982: Jurisdictional separations. Jurisdictional separation is the name of the process by which the cost of facilities for providing local and long-distance telephone service is divided between the state or federal regulatory jurisdiction and the local or interstate companies providing the service. Since telephone facilities are used to supply both intrastate and interstate service, methods have been devised for separating plant investment, operating expenses, taxes, and reserves between the intrastate and interstate operations of telephone companies. Since the separation process is the basis for rate regulation by the FCC and the state commissions, changes in the procedures would have an effect on the rates paid by the public for local and long-distance telephone service.

In 1969, the FCC adopted a report and order which prescribed separation procedures to be used by the Bell System and independent telephone companies. Previously, separation procedures developed by cooperative efforts of the FCC and state regulatory commissions through the National Association of Regulatory Utility Commissioners (NARUC) had been used by the regulatory agencies for rate making purposes, but without the official sanction of the FCC. In 1970, the FCC approved revisions of the prescribed separation procedures. The revision had been proposed by the FCC-NARUC Joint Board on Jurisdictional Separations, consisting of three FCC commissioners and four state utility commissioners, convened by the FCC for that purpose.

The FCC created another Federal-State Joint Board in 1980, instructing it to revise the separation procedures pertaining to exchange plant investment and expenses. The board began by reviewing all local telephone exchange allocation procedures and investigating the effects of the deregulation of terminal equipment. The FCC was particularly concerned about the increase in the assignment of cost to the interstate jurisdiction because of the use of the subscriber plant factor (SPF) to weight interstate minutes of use. SPF assigns costs attending the use of the local telephone exchange plant by message services (MTS and WATS) at more than 3 times the level that would result from a straight, unweighted relative use assignment.

In 1982, acting on the recommendations of the Federal-State Joint Board, the FCC issued amendments revising the separation procedures to allow a more realistic assignment of local plant costs to the interstate jurisdiction for use of the local plant. One of the proposals from the Joint Board adopted by the FCC was a plan to reduce the allocation of CPE costs to the interstate jurisdiction over a 5-year period. The extended phase-out period would be for the removal from the rate base of embedded CPE—equipment installed and already subject to regulation prior to Jan. 1, 1983. But CPE manufactured or acquired after January 1 would not be subject to regulation under Title II of the Communications Act. The FCC requested this plan to provide a means for putting into effect an earlier decision to deregulate CPE.

1981–1982: Legislation to revise the Communications Act. In 1981, bills were introduced in both houses of Congress to propose major revisions to the Communications Act of 1934. As the last session of the 97th Congress drew to a close in 1982, neither of these bills (S898 and HR5158) had become law. Because of ensuing events involving the AT&T antitrust settlement, no new legislation calling for comprehensive revisions in the Communications Act of 1934 was introduced after these bills expired. However, improving telecommunication policy remains an active project in both the Senate and the House.

1974–1983: AT&T antitrust suit and divestiture agreement. In 1974, the Justice Department made a move that was to have tremendous impact on the telecommunications industry. On November 20, the Antitrust Division of the U.S. Department of Justice filed a complaint that accused AT&T, Western Electric, and Bell Laboratories of conspiring to "prevent, restrict and eliminate competition from other telecommunications common carriers, private telecommunications systems, and manufacturers and suppliers of telecommunications equipment," in violation of the Sherman Antitrust Act. The suit asked for divestiture of the entire stock interest that AT&T held in its manufacturing arm, Western Electric Co., and further requested that AT&T either give up its long-distance telephone business or retain the business but surrender its interests in its 22 local telephone companies (the Bell System).

AT&T maintained that most of the actions with which it was charged were within the jurisdiction of the FCC and state regulatory authorities. At the request of the court, the commission rendered its opinion on this question. The FCC concluded that the

Communications Act did not deprive the court of jurisdiction to hear the Justice Department suit. AT&T Chairman John B. deButts warned at a press conference that economies of scale arising from the size and integrated structure of the Bell System would be lost by its dismemberment, inevitably resulting in higher costs and decreased service.

Seven years after filing its antitrust suit against AT&T, the U.S. Department of Justice made an agreement with the company whereby the suit would be dropped if AT&T gave up its 22 local operating companies. The tentative settlement agreement was announced Jan. 9, 1982, and was filed as a modification to the 1956 consent decree.

The terms of the original tentative settlement included the following points:

- AT&T would have to divest itself of all segments of its business that provided exchange access and local exchange services. It could keep the segments of its business that provided CPE and interexchange services. The divestiture could be done in "any appropriate manner." When tallied, the segments to be divested amounted to over $100 billion in assets.

- No relationships between the divested companies and AT&T or one another could exist, except for a central services organization (CSO) that might provide such services as planning, engineering, and administration to the divested companies more efficiently than each could provide for itself. The original intent of this exception was to keep customer rates from skyrocketing as a result of the divestiture. In July 1983, the courts confirmed the provisions for the CSO in spite of heavy protest that the organization could gain such leverage that it would in effect represent a restoration of the AT&T corporate power among the operating companies and in the telecommunication marketplace. This decision was upheld, though, and what would later become known as Bell Communications Research was born.

- The divested companies would have to provide equal access to all interexchange carriers. For example, users would have to have equally convenient access to any long-distance carrier, instead of simply dialing a three-digit area code to access AT&T Long Lines and inconveniently dialing multiple strings of numbers for MCI and other independent carriers.

- The divested companies would not be permitted to discriminate in favor of AT&T for procurement of products or services, dissemination of technical information, facilities planning, and other services.

- The divested companies would be able to provide basic services only and would be prohibited from providing interexchange services, information services, or any type of nontariffed services. Services would be confined to specified local access and transport areas (LATAs).

Although AT&T and the Justice Department had agreed to this settlement, it was up to U.S. District Court Judge Harold H. Greene to review its terms and dismiss the antitrust suit. Over 100 interested parties petitioned the court and were granted the privilege of participating in the proceedings, since the outcome of the case would be so far-reaching. Arguments concerning many aspects of the settlement were heard, including such issues as when and how AT&T's assets and liabilities would be accounted for, allocated, and transferred and who would have what powers to set exchange access charges.

A number of modifications resulted from the review, including the decision that the divested local operating companies would be allowed to provide, but not manufacture, CPE. Local operating companies would be allowed the right and the wherewithal to produce and distribute telephone directories. Local operating companies would be permitted to provide interexchange service and manufacture equipment, if they could show evidence that their entry into the market would not stifle competition. AT&T would be prohibited from offering electronic publishing services for 7 years, except in areas where it had been providing electronic directory information as of Jan. 8, 1982. AT&T would have to relinquish virtually all rights to the Bell name and logo to the operating companies. Among other things, this ruling required the renaming of the new American Bell subsidiary.

AT&T and the Justice Department agreed to Judge Greene's modification of terms, and the antitrust suit was subsequently dismissed on Aug. 24, 1982.

Following the dismissal of the antitrust suit, AT&T was directed to submit within 6 months a plan outlining how the divestiture of its local operating companies would be carried out. Once submitted and approved by the Justice Department and the court, the plan was to be acted on within 1 year. Six months prior to the dismissal, a tentative plan had been presented by AT&T, which proposed grouping the local operating companies into seven independent, regional corporations, each with its own stock, chief executive officer, and board of directors. Judge Greene decided to hold hearings and solicited comments on this plan. In September 1982, NARUC announced that seven state commissioners had been appointed to a committee to serve as a liaison between state regulatory agencies and the Justice Department,

FCC, and AT&T while the reorganization plan was being ironed out. The committee was formed to provide a means for state regulators to participate in the plan's development.

The divestiture plan was filed in December 1982 and approved on Aug. 5, 1983. It called for the Bell operating companies (BOCs) to be divested effective Jan. 1, 1984.

1986: The reorganized AT&T. The antitrust suit filed by the U.S. Justice Department was an attempt to enhance competition for terminal equipment and long-distance service by separating the 22 BOCs from the more competitive segments of AT&T's business. Before this separation, AT&T subsidized competitively threatened products and services with regulated (and profitable) products and services. AT&T thus charged artificially high rates for some services and artificially low rates for others (e.g., it subsidized local telephone service by charging higher than necessary long-distance rates). This is what the divestiture was all about—to force AT&T to charge its costs where they occurred.

The divested AT&T is approximately one-fifth its original size. The company, under the terms of the divestiture agreement, originally was required to offer network services and CPE through separate subsidiaries, but the FCC has since removed this restriction. As a result, AT&T in mid-1986 began consolidating its long-distance and network operations unit with its equipment unit.

Because of the divestiture, AT&T is now free to enter nearly any competitive market it chooses.

The 22 BOCs that existed prior to divestiture are now grouped into seven regional Bell holding companies (RBHCs). The RBHCs are completely disassociated from AT&T and are free to deal with any equipment and facilities vendor they choose, either AT&T or any of its competitors. The RBHCs can sell but not manufacture CPE. They have retained their Yellow Pages service, along with their local exchange and exchange access service, and have diversified into many new areas including real estate service, computer retail operations, software development, and local-area network design and installation. It is in these unregulated markets that the RBHCs see significant opportunities for growth.

The BOCs that together form the RBHC user market areas are called *local access and transport areas* (LATAs). A LATA is defined as a geographic area within which a local exchange company (which may be a BOC) may provide service.

Every area served by a BOC within a state must be included in a LATA, but LATAs are not allowed to cross state boundaries or include two standard metropolitan statistical areas (SMSAs) unless approved by the courts. LATAs encompass one or more continuous local exchange areas serving common social and economic purposes, even if these LATAs overrun municipal or other local government boundaries. The RBHCs may not do business or carry traffic beyond their LATAs except into independent local operating company territory and where corridor exceptions occur (a corridor is allowed to avoid excessive network rearrangement costs for the RBHCs).

There are six different types of calls that can be made under the LATA concept:

- Interstate–inter-LATA is a long-distance call between two different LATAs in two different states. Service is provided by the interexchange companies (IXCs) and regulated by the FCC.

- Interstate–intra-LATA is a call that crosses state boundaries but remains within one LATA. This service is provided by the local exchange carrier (LEC) and the IXCs and regulated by the FCC.

- Intrastate–inter-LATA is also a long-distance call, but between two different LATAs within the same state. It is provided by the interexchange companies and regulated by each state's public utilities commission (PUC).

- Intrastate–intra-LATA is a call between two local exchange areas within the same state and the same LATA. This circuit is provided by the local exchange carrier and regulated by the PUC. It is a service that AT&T has filed tariffs for but is open to IXCs for filing as well.

- Non-LATA to LATA is a call between a non-BOC exchange carrier and a BOC exchange carrier and can be either interstate or intrastate.

- Non-LATA to non-LATA is a call between two non-BOCs and can also be either interstate or intrastate.

Once divestiture occurred, it was left up to the Justice Department to review the progress being made toward introducing competition into the long-distance network while maintaining the protections offered to the BOC regulated monopoly. Several issues continue to be in flux and will probably continue to evolve into the twenty-first century.*

As it pertains to connectivity issues, divestiture meant an end to the end-to-end maintenance of service for the customer. When CPE

---

*Datapro Research Corp., "A History of the Telecommunications Regulatory Environment," MT 10-620, June 1987, pp. 101–118.

provision became optional, maintenance of the system became clouded. To correct this proposition, it was held that only those facilities provided to the customer by the BOCs were to be maintained by them. Of course the transport of service to the equipment was carried out over a mix of customer premises wiring and BOC wiring.

Since CPE jurisdiction was determined by the customer premises boundary, it was logically extended to the wiring for the CPE. As a result, customer premises wiring cost (installation and maintenance) has fallen to the owner of the premises and/or system.

1.7 Connectivity Alternatives

Generally, premises connectivity is confined to the physical links of twisted pairs, coaxial cable (coax), and optical fiber, which are the subjects of this book. There are wireless alternatives for the premises environment that include AM phantom circuits over AC wiring, infrared, and short-haul microwave. In campus environments and other specialized environments, even satellite communication is employed. However, discussions of these communication techniques are common, and adequate knowledge of their implementation can be gained from a wealth of other publications. This work will therefore concern itself with the information transport mediums that are physical in nature and commonly applied in today's environments.

1.8 Book Overview

The remaining chapters of this book will discuss in detail the following topics:

Chapter 2 Wire, Cable, and Coax Applications.   This chapter will look into industry and vendor standards, the configurations of public and private networks, connectivity environments and medium application, medium specifications and capacities, the design of connectivity systems, forecasting of connectivity requirements, system security, exception design, testing and diagnostic techniques, and cost estimating.

Chapter 3 Optical-Fiber Applications.   This chapter presents the background of optical connectivity, details the evolution of the discipline, describes the types and specifications for waveguide applications, and describes testing, cost estimating, and hardware.

Chapter 4 Support Structure and Connectivity Hardware.   This chapter defines the supporting structure configurations found in premises and campus distribution systems. Also included are the commonly accepted definitions for division of plant locations, environmental hazard potentials, separation requirements, and electrical protection.

Chapter 5   Noise Suppression and AC Power and Lighting Considerations. This chapter will deal with sources and types of voice and data signal interference caused by AC and other power and equipment, recommend minimum lighting and power provisioning of the various connectivity housings and station locations, and demonstrate the effects of AC appliances on power circuits.

Chapter 6   Telecommunication Systems and Services. This chapter discusses the types of telecommunication systems available, the vendors of services, and the transmission techniques involved in the delivery of the services. System integration techniques are also presented.

Chapter 7   Data Processing Applications. This chapter discusses the various data system applications over the various low-voltage mediums, medium capacities, networking architecture, protocol integration, and power backup, and presents selected LAN topographies.

Chapter 8   Management and Control Systems. In this chapter, system administration requirements are discussed, including facility inventory systems, facilities numbering plans, system performance documentation, and trouble analysis and repair diagnostic procedures.

Chapter 9   Planning Guidelines. This chapter discusses the various techniques and variables associated with economic evaluation by alternative study comparisons. It also gives strategies for contract bidding, bid specification development, and an example of a connectivity bid proposal.

Chapter 10   Job Scheduling. This chapter explains the coordination dynamics of low-voltage wire project implementation. The focus is on the job scheduling process and trade task coordination. System turn-up and acceptance procedures are included to enhance the job scheduler's knowledge.

2

Wire, Cable, and Coax Applications

2.1 Public and Private Networks*

The United States telephone network has been described as "the world's most complicated machine." When a telephone call is made, it may travel over many different types of channels, and complex switching facilities are needed to set up its path.

The telecommunications system can be divided into four main parts:

Instruments. The term *instrument* is used for the device the subscriber employs to originate and receive signals. The vast majority of instruments are telephone handsets. Today, however, an endless array of other devices are being attached to telephone lines, including computer terminals used for data transmission.

Local loops. These are the cables that enter the subscriber's premises. On a telephone network they connect the telephone handsets or other devices to the local switching office (central office). Telephone loops today consist of wire-pair cables and optical fiber interfaced to optical converters. Every subscriber has a private circuit to the local switching office, and nobody else uses this unless the subscribers are on a party line. The United States has several hundred million miles of telephone subscriber loops. Coaxial cables are also laid into homes by cable television organiza-

---

*James Martin, *Telecommunications and the Computer,* 3d ed., Prentice Hall, Englewood Cliffs, N.J., 1990, pp. 148–153.

tions, and these cables have many potential uses other than television. In the future, very high capacity fiber-optic cables will be used extensively in the local loop.

Switching facilities.   An elaborate network of switching offices enables any telephone to be connected to almost every other telephone. Most of the switching and control functions are carried out entirely by computers. Some electromechanical switching facilities will continue to exist in less populated rural areas for a few more years.

Trunk circuits.   Trunk circuits are transmission links which interconnect the switching offices. Such links normally carry more than one telephone call. On high-traffic routes, they carry many thousands of calls simultaneously. A variety of different transmission mediums are employed, including wire pairs, coaxial cables, microwave radio, satellites, and fiber optics.

The technology of transmission has traditionally been managed separately from the technology of switching. As both switching and transmission are increasingly controlled by computers and with the use of satellite and fiber optics, the technologies of switching and transmission are merging.*

2.2 Local Distribution†

Readers should distinguish between local distribution facilities and trunk networks. Trunk networks carry signals between the common carrier offices. Local distribution facilities carry signals from the subscriber's instrument or terminal to the local common carrier office and back.

Because the trunks interconnect the common carrier offices, sophisticated equipment can be used for organizing them, such as switching equipment and equipment for combining many signals. Other value-added common carriers, for example, use computers to interleave the data packets that are sent and select the route over which to transmit them. The transmission links between common carrier offices can have a high capacity because they carry large traffic volumes. High-capacity facilities such as fiber-optic cable, microwave radio, and satellites link the offices.

---

*James Martin, *Telecommunications and the Computer,* 3d ed., Prentice Hall, Englewood Cliffs, N.J., 1990, pp. 148–153.
†*Ibid.,* pp. 153–158.

The telephone system has a pair of wires (local loop) going from its local telephone offices to each subscriber. In most cases only one subscriber is connected to this pair of wires. The utilization of the local loop is very low. It stands idle most of the day, and when it does transmit it is used at a fraction of its capacity. However, the wire pairs exist, going into almost every home and office building in North America and having a total book value of between $20 and $40 billion.

The telephone companies claim that local loop operation is a "natural monopoly." It would not make sense for competing organizations to run separate pairs of wires into subscribers' homes. Other ways of transmitting speech and data to and from the end users are possible, however. These include using community antenna television (CATV) cables or various types of radio. A single CATV cable interconnects many hundreds of subscribers: hence, it is installed at a lower cost per subscriber than telephone loops, each serving one subscriber. Many commercial buildings have their own satellite, microwave, or high-speed point-to-point links, bypassing local loops.

Local distribution can be a problem for the independent carriers. They can build trunking systems economically but often cannot build the last leg to and from the subscriber economically. To overcome this problem, many of the larger independent common carriers have built their networks so that the main access points are as close as possible to the majority of corporations that may wish to use their facilities. To gain access to the independent carrier's network, a corporation must lease the appropriate circuit from the local telephone company to connect with the other carrier. If the independent carrier is offering a data transmission service operating at 1.544 Mb/s (T1), the user must lease an equivalent circuit from the local telephone company to interconnect with the other carrier. In most large urban areas, this is not much of a problem, and as networks become increasingly digitized, the problem is less acute. However, in the local domestic subscriber loop, the sheer volume of users makes the introduction of faster transmission facilities very expensive. Local distribution is the Achilles' heel of many new telecommunication services and will remain so until all subscribers are connected to a high-speed broadband fiber-optic network.

Local loops to a company building normally terminate at a switching facility in the building which has extension lines to telephones. The switching is done automatically by private branch exchange (PBX). A tie line, or tie trunk, is a private, leased communication line between two or more private branch exchanges. Many companies have a leased system of telecommunications with switching facilities.

To telephone a person in a distant company location, an employee must first obtain the appropriate tie line to that person's private branch exchange.

All lines can transmit data as well as voice. Where one voice line is used, the transmission speed is limited to a few thousand bits per second, depending on the equipment used at each end of the line. This is the data capacity of an analog telephone line.

In many large corporations today, the tie lines interconnecting the various major cities are digital. The capacity of the digital circuit depends on the quantity of traffic. The most common digital circuit interconnecting remote PBX systems in use today in North America operates at a speed of 1.544 Mb/s. This can provide up to 24 separate voice or data channels of 64,000 b/s. The European or international equivalent is 2.048 Mb/s, divided into 32 channels of 64,000 b/s. Between PBXs having very high traffic flow, a number of 1.544-Mb/s links (in multiples of 1.544 Mb/s to 45 Mb/s) can be used, some of which will be used only for voice traffic and others for high-speed data to and from remote computers. Some corporations that build and operate their own networks instead of leasing circuits from a carrier can install much higher capacity links.*

2.3   Environment and Applications

Wiring is the least noticeable, yet most critical, component of any communication architecture. It is the common thread of both voice and data communication. With the divestiture of the Bell System, confusion has arisen concerning ownership of existing wire plant, with several concerns bidding for control over building wiring systems. Throughout the short history of divestiture, independent telephone companies and vendors of equipment have assembled unique products, both hardware and operating systems, that discourage user defection because the cost of conversion to a competitor's product is almost prohibitive. Given the increasing complexity of the various systems available, conversion has, to all intents and purposes, become unthinkable. Consequently, while competition for new applications and customers has been intense, the profits have been made on upgrading the systems to support expansion.

Accordingly, most building owners have elected to establish their own wiring standards to complement the many equipment applications available without eliminating any vendor from consideration.

---

*James Martin, *Telecommunications and the Computer,* 3d ed., Prentice Hall, Englewood Cliffs, N.J., 1990, pp. 153–158.

2.3.1 Horizontal building systems (Fig. 2.1)

The typical horizontal layout is fed from the building network demarcation point, or minimum point of presence (MPOP), by house cable which is terminated in the apparatus closet. It is then cross-connected, by means of jumper wires or patch cords, to either a secondary house cable going to a satellite closet or directly to an end user through key telephone common equipment.

From the satellite closet, the feed is cross-connected to either another house cable going to another satellite closet or directly to the end user via station wiring. These apparatus and satellite closets may be used to terminate and cross-connect other services such as data, video, security systems, and environmental control systems.

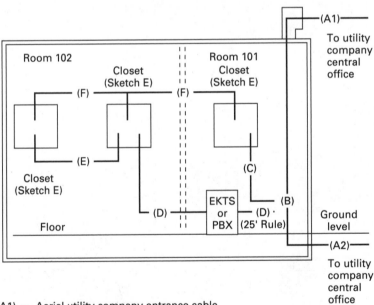

(A1) = Aerial utility company entrance cable
(A2) = Underground utility company entrance cable
 (B) = Minimum point of presence (MPOP)
         – Demarcation point for EKTS or PBX
         – Main building interface for CENTREX
 (C) = CENTREX house cable (utility maintained)
 (D) = EKTS/PBX house cable (customer maintained)
 (E) = Closet tie cable
         – CENTREX (utility maintained)
         – PBX/EKTS (customer maintained)
 (F) = Interagency tie cable (customer maintained)

Figure 2.1 Typical horizontal building system.

2.3.2 Vertical building systems (Fig. 2.2)

The layout of a multistory building consists of the vertical distribution of various types of services over riser cables between the demarcation or MPOP and the distribution points on each floor (apparatus/

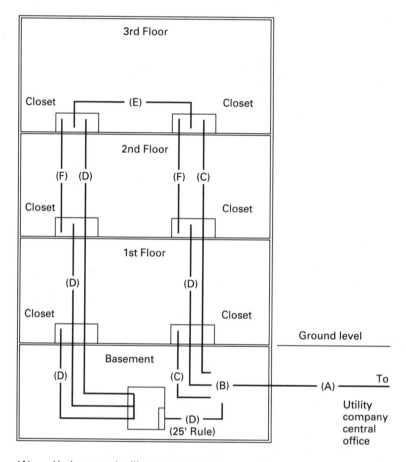

(A) = Underground utility company entrance cable
(B) = Minimum point of presence (MPOP)
     – Demarcation point for PBX
     – Main building interface for CENTREX
(C) = CENTREX riser cable (utility maintained)
(D) = PBX riser/house cable (customer maintained)
(E) = Closet tie cable
     – CENTREX (utility maintained)
     – PBX-EKTS (customer maintained)
(F) = Interagency tie cable (customer maintained)

Figure 2.2 Cable detail for typical vertical building.

satellite closets). This network of riser primary distribution consists of riser cables connecting the apparatus closets of each floor with the demarcation or MPOP. From the apparatus closets, house distribution cables will feed tandem satellite closets.

Intraagency network access cables provide a direct link to remote locations of the same building and can appear at the building side of the demarcation, in apparatus closets, or at a combination of locations.

2.3.3 Campus distribution systems (Fig. 2.3)

In campus distribution, vertical and horizontal buildings are tied together for the purpose of connecting a common communication system and/or control center. This system may be voice only or a combination of voice, data, environmental control, video, and security systems.

The four common methods of installing interbuilding cables are:

- Underground conduit
- Buried cables
- Tunnel systems
- Aerial facilities (rare)

Each method has its own advantages and disadvantages in terms of price, performance, and appearance. The location of the interbuilding cable entrance depends on the proximity of other buildings on the premises. Factors influencing the selection of a distribution system are cable distribution patterns between buildings, other utility service entrance points, and the location of the equipment room housing common equipment. Each campus environment is unique and will require the services of a qualified engineer for design.

2.3.4 Demarcation points (MPOP)

The demarcation point, also known as the *minimum point of presence* (MPOP) and *network access point,* is the beginning of the building distribution system (Fig. 2.4). This is the point at which the telephone company (telco) cable first enters the premises and is terminated on telco cross-connect blocks. In California, the utility companies have also determined that the demarcation point may be located at the first termination of the entrance cable, which may be at any location within the building. This principle applies only to existing buildings

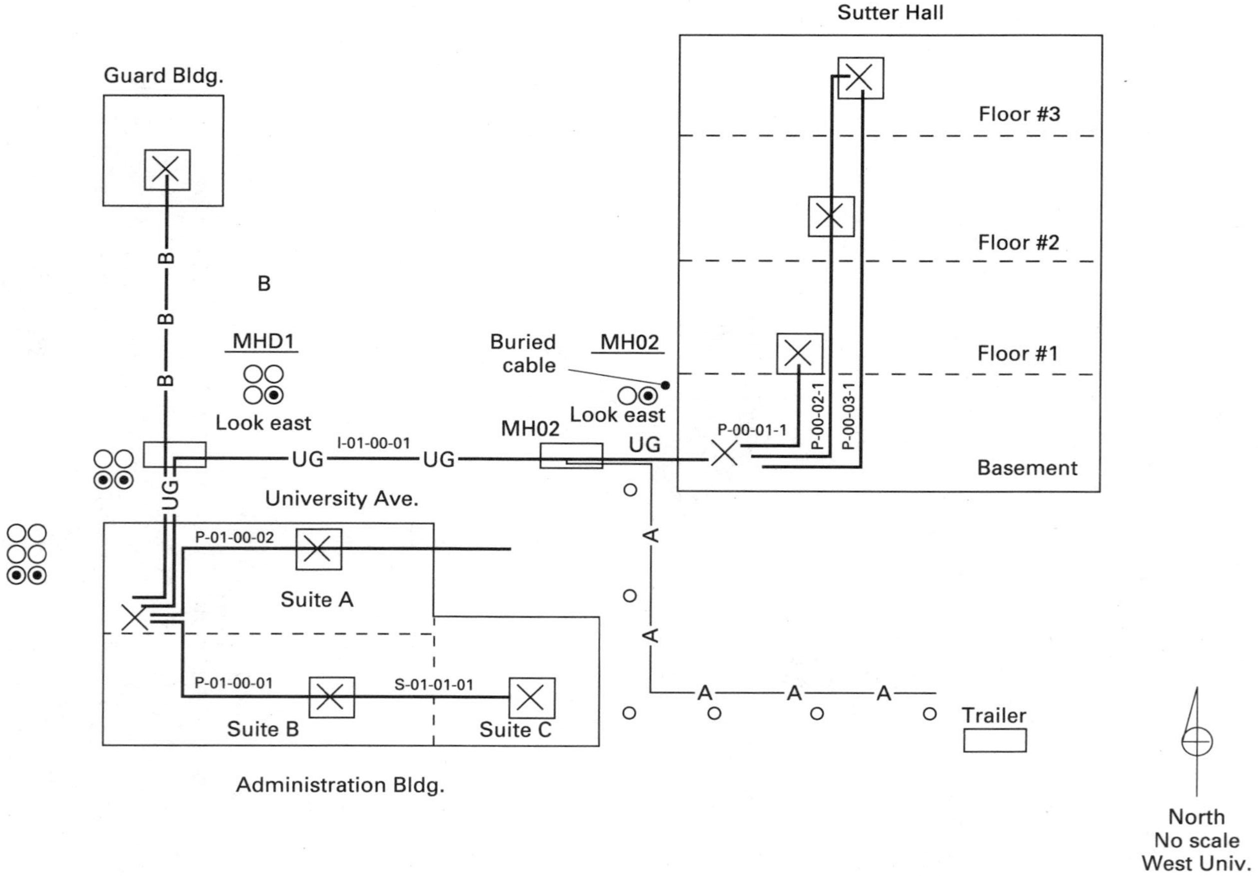

Figure 2.3 Campus distribution system.

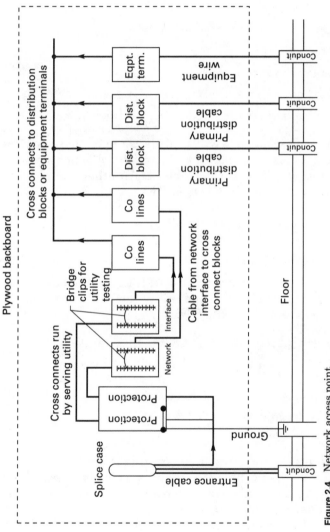

Figure 2.4 Network access point.

2.9

and is negotiable, according to California Public Utility Commission opinion. In new buildings, the demarcation point is considered to be in the basement or first appearance of the utility company cable in the building.

The components that make up the demarcation are:

- Environmental factors
- Service entrance
- Electrical protection
- Registered network interface block
- Distribution and cross-connect terminals
- Cable and wire support structure

The local telephone company will provide three of these components: the service entrance cable, the electrical protection devices, and the registered network interface block. All other components must be provided by the building owner. The exception to this is in entrances and protection devices required in the campus distribution system for buildings other than the MPOP.

2.3.5 Environmental specifications

The location of the building entrance area (MPOP) will be agreed to between the building architect and the utility company. The formula for determining the area of the MPOP is:

- Minimum space required: 8 ft wide $\times$ 8 ft high $\times$ 4 ft deep
- 10,000 to 800,000 ft^2 building area: Add 1 ft^2 floor space (8 ft high) for every 2000 ft of building space (e.g., MPOP floor area for a 10,000-ft^2 building is $4 \times 8 + (10,000/20,000) = 37$ ft^2)
- Over 800,000 ft^2 building area: Add 1 ft^2 floor area (8 ft high) per 3000 ft^2 of building space

In small buildings having a usable floor area of up to 20,000 ft^2, wall-type terminals should be used. If the usable floor area is in excess of 20,000 ft^2, terminals can be either wall-mounted or floor-mounted on terminating distribution frames.

When wall-mounted terminations are used, one or more $\frac{3}{4}$ in, 4-ft $\times$ 8-ft plywood backboards should be securely fastened to the wall for the purpose of securing the termination blocks. Allow 2 ft $\times$ 3 ft of additional backboard space for such terminations as security and environmental control systems. Assure that local fire regulations are met when specifying backboard type.

2.3.6 Additional MPOP specifications

- Two of the walls should have a minimum of one 120-V AC duplex outlet for 2-pole, 3-wire grounded plugs. A separate dedicated-service, 20-A circuit breaker for each duplex is required.

- Ceiling height in the MPOP shall provide a minimum clear space of 8 ft.

- The area shall be free of corrosive, explosive, or combustible gases.

- The area shall be dry and not susceptible to flooding.

- Lighting to be a minimum of 30 candle feet per square foot of area.

- Access should be secure to prevent unauthorized entry.

- Fire protection equipment should be provided as required by local code. The room must have a 2-hr fire rating with sprinklers located outside.

- Voice, data, and electronic key telephone system (EKTS) equipment require 24-hr air conditioning. Humidity and static electricity control are advised.

- Floor area covering shall be of material that will not produce static electricity or dust.

- Doors shall not be less than 36 inches wide and 80 inches high.

2.3.7 Service entrance

The service entrance point is the point at which the telephone company cable enters the building. If the building is the MPOP in a campus distribution system, cable will have to be run between the first point of presence to all buildings included in the distribution plan.

There are three types of entrance systems:

- *Underground conduit entrance:* The most common, this type of entrance (Fig. 2.5) consists of underground conduit that protects the cable and minimizes the need for subsequent trenching on the property. The conduit is provided by the building owner on the building property to the point of intersection with the utility company conduit at the property line.

- *Buried entrance:* Direct buried cables (Fig. 2.6) are placed in an unprotected trench from the utility intersect point through the

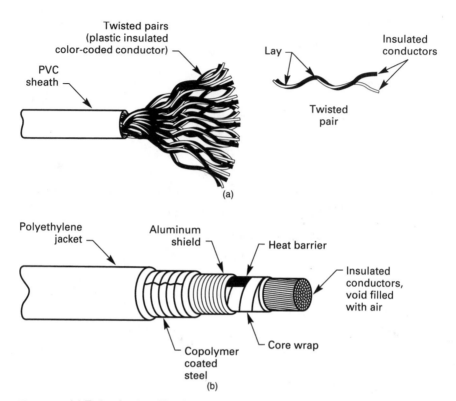

Figure 2.5 (*a*) Twisted-pair cable. (*b*) Air-core ductpic cable, stalpeth sheath.

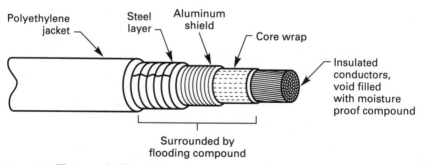

Figure 2.6 Waterproof ASP cable.

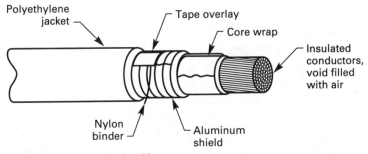

Figure 2.7 Air-core alpeth cable.

foundation wall. The trench is provided by the building owner unless the cable is to be plowed in. A protective slab or cover plate should be placed over the cable to prevent service interruptions caused by subsequent digging or excavation in the area of the buried cable.

- *Aerial entrance:* The least common, an aerial entrance consists of a cable that provides service, above ground, from a pole to the building (Fig. 2.7). Aerial installation is the least desirable because of its exposure to weather, electrical contacts, and vandalism.

The general specifications for underground entrance conduits are as follows:

- The conduit will be corrosion resistant.
- The conduit run will not have more than two 90° bends (or a total of 180°) between the utility company intersect point and the building contact.
- If metallic, the conduit will be reamed, bushed, and capped.
- Ducts through the foundation wall will be long enough to prevent shearing. The ducts will extend 10 ft beyond the foundation wall outside the building and be flush on the interior wall or floor.
- Conduit will be placed at least 18 inches below the surface on private property and to code on public rights of way (normally, at least 36 inches).
- Separate telephone conduits from power conduits or facilities by at least 3 inches of concrete or 12 inches of well-packed (97 percent compaction or better) earth.

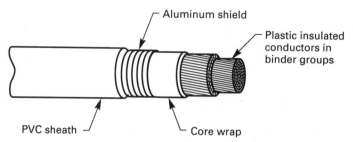

Figure 2.8 Twisted-pair building cable.

The following types of conduit are acceptable:

- Galvanized steel with asphaltic coating applied
- Plastic (PVC schedule 40 or better)
- Any equivalent, code-approved material with a 65-year service life

2.3.8   House and/or riser cable definitions

Riser cable is that cable which connects apparatus (floor distribution) closets to the demarcation (MPOP) interface (Fig. 2.8). In vertical buildings, riser cable normally refers to the cable that progresses vertically to individual floor apparatus closets, but it can also be found in horizontal situations where the cable connects apparatus closets with the demarcation (MPOP) interface. It is always configured to a home run topology. Riser cable is sometimes referred to as *black cable* by telephone utility representatives.

House, or *distribution,* cable is that cable which runs between closets on the same or different floors in a multistory or horizontal building. House cable does not connect directly to the demarcation (MPOP) interface, but serves to provide reinforcement facilities to congested closets or to provide interconnect potential for devices originating at different locations within the same building. House cable is occasionally referred to as *gray cable* by utility company representatives.

Both riser and house cables comprise the backbone of the building cable distribution system.

2.3.9   Cable and wire specifications

For voice and low-speed data, all riser and/or house multipair cables will utilize twisted-pair construction. These cables consist of twisted individual pairs, grouped to form a core or unit for a larger cable. Twisting the conductors also provides greater separation between pairs. This minimizes cross talk between adjacent pairs and decreases the capacitance unbalance to ground, both of which result in signal distortion. Twisted pairs are also less susceptible to electromagnetic interference. This cable will accommodate all voice requirements and data transmission to T1 levels (1.544 Mb/s). The following specifications list the characteristics of the cable.

Voice and Low-Speed Data

Cable category	Multipair cable, unshielded
Cable type	24 AWG solid conductors, twisted pairs

Individual conductors	
Gage	24 preferred (26 gage when space requires)
Conductors	Bare copper
Insulation	Semirigid plastic
Jacket	Plastic (UL 910 for plenum)
Color code	Standard telco

Nominal parameters (24 gage)	
Mutual capacitance	20.4 pF/ft
DC resistance	25.7 Ω/1000 ft
Characteristic impedance	600 Ω at 1 kHz
	92 Ω at 1 MHz
Attenuation	0.75 dB/1000 ft at 1 kHz
	8.0 dB/1000 ft at 1 MHz

Cable sheath	
AR series riser cable	Composed of dual expanded plastic-insulated conductors with plastic over expanded polyethylene, and sheathed with a bonded aluminum plastic (alvyn).

Sizes available
12 to 1800 pairs

Approvals
UL listed—type CMR (CMP for plenum)
UL classified 1666—NEC Art. 800-3(b)
IEEE 383, Vertical Flame Test

High-Speed Data—1.544 to 32 Mb/s

Follow specifications for "Data Specifications."

High-Speed Data—32 Mb/s or Greater

Follow specifications in "Optical-Fiber Applications," Chap. 3.

Cable and Wire Specifications

Station wire is that wire which connects the station equipment (voice and data) to the cross-connect block in the apparatus or satellite closet (see *Wire Plan Manual,* vol. 3, *Distribution System Specifications*).

Voice Specifications (Fig. 2.9)

The specifications below will accommodate Centrex, PBX, electronic key telephone system (EKTS), and ISDN voice wire serving configurations. The voice wire will also allow for data transmission to T1 speed (1.544 Mb/s) and port speeds to 19.2 kb/s. Key sets and attendant equipment requiring more than four pairs of wire will have to be designed separately. Recommended longest loop = 250 ft. Absolute maximum = 100 m (330 ft).

Wire category	Hookup wire (station wire)
Wire type	24 AWG solid conductors, twisted pairs

Individual conductors	
Gage	24
Conductors	Bare copper
Insulation	Semirigid plastic insulation
Jacket	N/A (UL 910 equivalent for plenum)
Color code	Standard telco

Nominal parameters	
Mutual capacitance	16.0 pF/ft
DC resistance	25.7 Ω/1000 ft
Characteristic impedance	600 Ω at 1 kHz
	100 Ω at 1 MHz
Attenuation	0.63 dB/1000 ft at 1 kHz
	6.8 dB/1000 ft at 1 MHz
Pairs	4
Approvals	Utility recognized—1935; 90°C, 300 V
	[plenum—UL classified 910 NEC Art. 800-3(d)]

4 pair - station wire

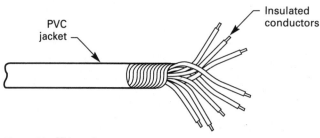

Figure 2.9 Voice wire.

Data Specifications

The specifications below will accommodate all voice requirements (see voice specifications) and data transmission up to 32 Mb/s or up to 100 Mb/s with line conditioning. The data station wire (Fig. 2.10) is intended for high-speed data local-area networks (LANs) but may be used for modem, fax, and even voice requirements. Refer to Wire Plan Manual, Vol. 4 for specific LAN applications. Recommended longest loop = 250 ft. Absolute maximum = 100 m (330 ft).

Cable category	Multipair cable, shielded
Cable type	22 AWG solid conductors, twisted pairs

Individual conductors

Gage	22
Conductors	Bare copper
Insulation	a. Foamed cross-linked polyethylene with flame retardant skin
	b. Foamed fluoropolymer for plenum
Shield	a. Two pairs individually shielded with aluminum polyethylene terephthalate
	b. Overall tinned copper braided shield (90% coverage)
Jacket	PVC (UL 910 equivalent for plenum)
Color code	Standard telco color code

Nominal parameters

Mutual capacitance	<16 pF/ft
Capacitance unbalance	1500 pF/km at 1 kHz
DC resistance	16.2 Ω/1000 ft
Characteristic impedance	150 Ω, 3- to 20-MHz sweep
Attenuation	45 dB/km max. at 16 MHz
	22 dB/km max. at 4 MHz
Cross talk (near end)	40 dB min., 12- to 20-MHz sweep
	58 dB min., 3- to 5-MHz sweep
Pairs	2 to 4
Approvals	Utility recognized—1935, 90°C, 300 V
	[plenum—UL classified, NEC Art. 800-3(d)]

2 pair - multiconductor, shielded wire

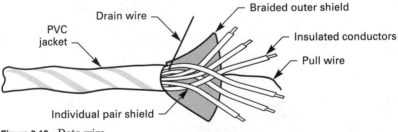

Figure 2.10 Data wire.

2.3.10   Wire and cable terminations

Wire and cable terminations will be accomplished by using standard telecommunications industry hardware.

Riser cable will be terminated on high-density (110 type) connectors at the main building terminal and apparatus closets.

Voice station wire will be terminated on high-density (110 type) connectors at the apparatus (serving) closets and on 8-pin modular jacks (RJ45 type) at the station location (Fig. 2.11).

Data wires will be terminated on 66M150-type blocks at the apparatus (serving) closets equipped with 2- to 50-pin modular connectors, and terminated on 4-pin modular jacks (RJ11 type) at the station location (Fig. 2.12).

2.4   Entrance/Interbuilding Cable Sizing

Entrance cable is the cable that is installed by the serving telephone company for access to the telephone central office. This cable is installed during the construction of a new building. The number of conductor pairs required to serve a new building should be determined as early in the building design as possible. This will be accomplished by the utility company on the basis of information provided from the builder or client organization. The following formula will also be applied to sizing interbuilding distribution cables in the campus environment.

The criteria generally used for determining the size of entrance cable are as follows:

- 1.5 pairs per telephone

- 3.0 pairs per 4-wire private-line voice or data circuit

- 1.5 pairs per dial access modem

The number of universal information outlets (end-user station voice and data jacks) to be located in a building can be estimated at 1 outlet per 110 ft^2 of usable office space. If the building is not initially set up as all offices (e.g., part warehouse) but could be converted to all offices at a later date, that space should be counted as office space for cable sizing purposes.

Additional requirements for private line voice and data requirements above those mentioned earlier must be based on immediate and probable future needs.

2.5   Station Wire Sizing

Station wire is the wire that runs between the universal information outlet and the distribution cable cross-connect (apparatus or satellite)

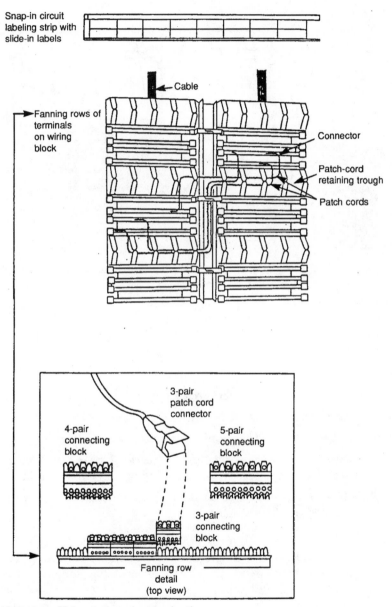

Figure 2.11 Voice station wire termination.

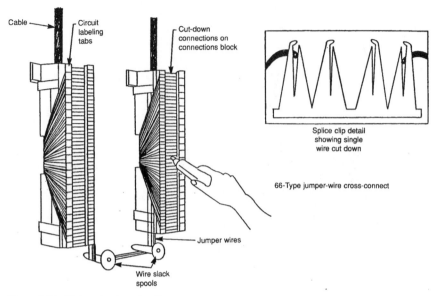

Splice clip detail
showing single
wire cut down

66-Type jumper-wire cross-connect

Figure 2.12   Data wire termination.

closet. The following are the minimum requirements for each station wire system:

- The telephone modular jack requires 4 pairs.
- The data terminal modular jack requires 2 pairs.
- Station wire to 4-wire private-line equipment requires 2 pairs.
- Station wire to dial access modems requires 1 pair but 2 pairs should be the minimum size wire used to allow for reapplication.
- RS232 connections require from 2 to 25 conductors. Check with vendor for exact pair requirements.

With these requirements, and the probability for a reduction in pair requirements with future technologies, the 4-pair voice and 2-pair data configuration recommended earlier will meet 95 percent of the flexibility, data speed, and maintenance needs of the workstation environment in office buildings.

2.6   Riser/Distribution Cable Sizing

Distribution cable is the cable that runs between the cable closets on a horizontal plane. Riser cable is the distribution cable that runs vertically between cable closets on different floors in a multistory building.

The criteria for sizing this cable are as follows: Provide 5 pairs for each universal information outlet which can be either directly cross-connected or cross-connected through secondary distribution cables. This will provide sufficient pairs for normal voice and data service plus additional services such as security, environmental controls, special services, and maintenance or spare pairs. If there is a large amount of special services, as in a data center, then provide 3 pairs for every 4-wire circuit, 1.5 pairs for every 2-wire circuit, etc.

2.7 Intraagency Tie Cable

Intraagency tie cables are cables placed strictly for communications within the confines of that agency. They will not connect to the entrance cable, either directly or indirectly, through a switching device, such as a PBX, EKTS, or multiplexer (MUX), which has entrance cable access. It is important to understand the distinction between a cable which does have connections to the entrance cable and a cable that does not. First, tie cables may be of the unshielded variety, like equipment wire. Second, FCC rules governing registration of telephone equipment and signal levels pertain only to connections with access to the public telephone network. Third, existing utility company record-keeping systems do not cover tie cables; these cables should be sized the same as distribution and riser cables.

2.8 Cable Performance Tables

See Tables 2.1 through 2.6 for more cable performance data and Tables 2.7 through 2.9 for manufacturers and products.

TABLE 2.1   Diameter, Weight, Resistance of Solid Bare Copper Wire, American Wire Gage

Gage, AWG or B&S	Nominal diameter in	mm	Circular mil area	Weight lb/m	Resistance at 68°F, Ω/m
10	0.1019	2.60	10,380	31.43	0.9989
11	0.0907	2.30	8,234	24.92	1.260
12	0.0808	2.05	8,530	19.77	1.588
13	0.0720	1.83	5,178	15.68	2.003
14	0.0641	1.63	4,107	12.43	2.525
15	0.0571	1.45	3,260	9.858	3.184
16	0.0508	1.29	2,583	7.818	4.016
17	0.0453	1.15	2,050	6.200	5.064
18	0.0403	1.02	1,620	4.917	6.385
19	0.0359	0.912	1,200	3.899	8.051
20	0.0320	0.813	1,020	3.092	10.15
21	0.0285	0.724	812.1	2.452	12.80
22	0.0253	0.643	840.4	1.945	16.14
23	0.0226	0.574	511.5	1.542	20.36
24	0.0201	0.511	404.0	1.223	25.67
25	0.0179	0.455	320.4	0.9699	32.37
26	0.0159	0.404	253.0	0.7692	40.81
27	0.0142	0.361	201.5	0.6100	51.47
28	0.0126	0.320	159.8	0.4837	64.90
29	0.0113	0.287	126.7	0.3836	81.83
30	0.0100	0.254	100.5	0.3042	103.2
31	0.0089	0.226	79.7	0.2413	130.1
32	0.0080	0.203	63.21	0.1913	164.1
33	0.0071	0.180	50.13	0.1517	206.9
34	0.0063	0.160	39.75	0.1203	260.9
35	0.0056	0.142	31.52	0.09542	331.0
36	0.0050	0.127	25.00	0.07588	414.8
37	0.0045	0.114	19.83	0.0613	512.1
38	0.0040	0.102	15.72	0.04759	648.6
39	0.0035	0.069	12.20	0.03774	847.8
40	0.0031	0.079	9.61	0.02993	1060.0

Information from National Bureau of Standards Copper Wire Tables—Handbook 100.

TABLE 2.2 Comparative Properties of Plastic Insulations

	PVC	Low-density polyethylene	Cellular polyethylene	High-density polyethylene	Polypropylene	Cellular polypropylene	Polyurethane	Nylon	Teflon
Oxidation resistance	E	E	E	E	E	E	E	E	O
Heat resistance	G–E	G	G	E	E	E	G	E	O
Oil resistance	F	G	G	G–E	F	F	E	E	O
Low-temperature flexibility	P–G	G–E	E	E	P	P	G	G	O
Weather, sun resistance	G–E	E	E	E	E	E	E	E	O
Ozone resistance	E	E	E	E	E	E	E	E	E
Abrasion resistance	F–G	F–G	F	E	F–G	F–G	O	E	E
Electrical properties	F–G	E	E	E	E	E	P	P	E
Flame resistance	E	P	P	P	P	P	P	P	O
Nuclear radiation resistance	G	G	G	G	F	F	G	F–G	P
Water resistance	E	E	E	E	E	E	P–G	P–F	P
Acid resistance	G–E	G–E	G–E	G–E	E	E	F	P–F	E
Alkali resistance	G–E	G–E	G–E	G–E	E	E	F	E	E
Gasoline, kerosene, etc. (aliphatic hydrocarbons) resistance	P	P–F	P–F	P–F	P–F	P	P–G	G	E
Benzol, toluol, etc. (aromatic hydrocarbons) resistance	P–F	P	P	P	P–F	P	P–G	G	E
Degreaser solvents (halogenated hydrocarbons) resistance	P–F	P	P	P	P	P	P–G	G	E
Alcohol resistance	G–E	E	E	E	E	E	P–G	P	E

P = poor, F = fair, G = good, E = excellent, O = outstanding.
These ratings are based on average performance of general-purpose compounds. Any given property can usually be improved by the use of selective compounding.

TABLE 2.3   Nominal Temperature Range of Insulating and Jacketing Compounds, °C

Compound	Normal low	Normal high	Special low	Special high
Chlorosulfonated polyethylene	−20	90	−40	105
EPDM (ethylene-propylene-diene monomer)	−25	106	—	—
Neoprene	−20	80	−65	90
Polyethylene (solid and cellular)	−80	80	—	—
Polypropylene (solid and cellular)	−40	103	—	—
Rubber	−30	80	−65	75
FEP Teflon*	−70	200	—	—
PVC	−20	80	−88	100
Silicone	−80	160	—	800
Halar†	−70	180	—	—
Tetgal†	−65	160	—	180
TFE Teflon*	−70	260	—	—

*DuPont trademark.
†Allied Chemical trademark.

TABLE 2.4   Loop Resistance

Conductor Gage	Material	Loop resistance, (Ω/kft) at 68°F*
17	Aluminum	16.5
19	Copper	16.1
20	Aluminum	33
22	Copper	32
24	Copper	52
26	Copper	83
28	Copper	130

*Use the following formula for resistance at a temperature T other than 68°F:
$R_T = R_{68}[1 + 0.0022(T - 68)]$

TABLE 2.5   Nominal Attenuation of Standard Exchange Cables

(Attenuation in dB/mi at 55°F)

Gage:	19	22	24	26	19	22	24	26
Type:	ADB	ADA	ADM	ADT	BHB	BHA	BKM	BKT
Frequency, kHz	Pulp				Air-core PIC			
1	1.3	1.8	2.3	3.0	1.2	1.8	2.3	3.0
48	4.8	7.6	10.8	13.6	4.2	7.2	10.8	15.4
96	6.4	9.2	13.2	16.4	5.5	8.5	12.2	17.2
136	7.4	10.5	14.9	18.4	6.6	9.6	13.0	18.2
168	8.2	11.4	16.0	19.6	7.4	10.5	13.8	19.2
208	9.1	12.7	17.4	21.1	8.1	11.4	15.0	20.6
256	10.3	13.7	19.1	22.8	9.2	12.6	16.4	22.0
772	20.1	26.9	35.9	43.3	16.8	23.2	29.6	39.6

Gage:	19	22	24	26	17		20	
Type:	AJB	AJA	AJM	AJT	ALCW		ALDW	
Frequency, kHz	Waterproof solid PIC				Waterproof DEPIC			
1	1.2	1.8	2.3	2.9	1.2		1.8	
48	4.1	6.6	9.9	14.5	4.2		7.1	
96	5.2	7.7	11.0	15.8	5.5		8.4	
136	6.1	8.6	12.0	17.0	6.6		9.6	
168	6.8	9.4	12.6	17.4	7.3		10.5	
208	7.5	10.3	13.5	18.3	8.0		11.4	
256	8.3	11.4	14.8	19.5	9.1		12.6	
772	13.3	19.4	24.6	31.2	16.0		22.3	

TABLE 2.6   Cable Sheath and Shield Resistance

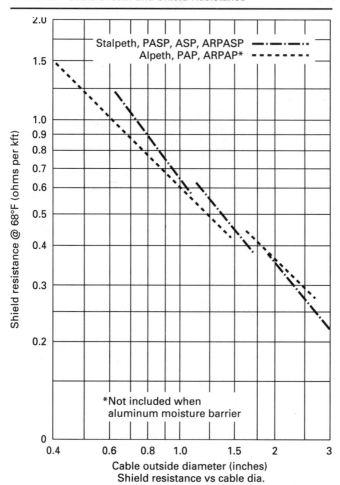

Shield resistance vs cable dia.

TABLE 2.7   Selected Makers of Independently Approved Cables

Manufacturer	ETL's EIA/TIA-568 Certification Program						UL's Performance-Level Certification Program					
	Category 3		Category 4		Category 5		Level 3		Level 4		Level 5	
	UTP	STP	UTP	STP	UTP	STP	UTP	STP	UTP	STP	UTP	STP
Alcatel Canada Wire Inc. North York, Ontario 416-424-5000	No	No	Yes	No	Yes	No	No	No	No	No	No	No
Belden Wire & Cable Richmond, Ind. 317-983-5200	No	No	Yes	Yes	Yes	Yes	No	No	No	No	No	No
Berk-Tek Inc. New Holland, Pa. 717-354-6200	Yes	No	Yes	Yes	Yes	Yes	Yes	Yes	Yes	Yes	Yes	Yes
Brand-Rex Co. Willimantic, Conn. 203-456-8000	No	No	No	No	No	No	Yes	No	Yes	No	Yes	No
Comm/Scope Inc. Hickory, N.C. 704-324-2200	No	No	No	No	Yes	No	No	No	No	No	No	No
General Cable Corp. South Plainfield, N.J. 908-769-3200	No	No	Yes	No	Yes	No	Yes	Yes	Yes	Yes	Yes	Yes
Hitachi Cable Manchester, N.H. 603-669-4347	No	No	Yes	No	Yes	No	No	No	No	No	No	No
Madison Cable Corp. Worcester, Mass. 508-752-7320	Yes	No	Yes	No	Yes	No	No	No	No	No	No	No
Mohawk Wire and Cable Corp. Leominster, Mass. 508-537-9961	Yes	No	Yes	Yes	Yes	Yes	No	No	Yes	Yes	Yes	Yes
NEK Cable Inc. Bohemia, N.Y. 516-567-5000	No	No	No	No	No	No	No	No	Yes	Yes	Yes	No
Northern Telecom Ltd. Mississauga, Ontario 416-897-9000	No	No	No	No	No	No	No	No	Yes	No	Yes	No
Philadelphia Insulated Wire Moorestown, N.J. 609-235-6700	No	No	Yes	No	Yes	No	No	No	No	No	No	No
Prestolite Wire Corp. Farmington Hills, Mich. 313-626-1336	No	No	No	No	No	No	Yes	No	Yes	No	Yes	No
Remee Products Corp. Florida, N.Y. 914-651-4431	No	No	Yes	No	No	No	No	No	No	No	No	No

TABLE 2.7   Selected Makers of Independently Approved Cables (*Continued*)

| | ETL's EIA/TIA-568 Certification Program | | | | | | UL's Performance-Level Certification Program | | | | | |
| | Category 3 | | Category 4 | | Category 5 | | Level 3 | | Level 4 | | Level 5 | |
Manufacturer	UTP	STP	UTP	STP	UTP	STP	UTP	STP	UTP	STP	UTP	STP
Servicios Condumex S.A. de C.V. Granda, Mexico 52-467-22185	No	No	No	No	No	No	No	No	No	No	No	No
Teldor Wires & Cables Ltd. Jezreel, Israel 972-6-768-444	No	No	Yes	Yes	Yes	No	No	No	No	No	No	No
Teledyne Thermatics Elm City, N.C. 919-236-4311	No	No	No	No	Yes	No	Yes	No	Yes	No	No	No

SOURCES: ETL Testing Laboratories Inc. and Underwriters Laboratories Inc.
Reprinted from *Data Communications,* November 1992.

TABLE 2.8   Wiring Options from the Big Four

Vendor	Product name	Cable types	Average cost per node*
AT&T Phoenix 602-233-5855	Systimax Premises Distribution System (PDS)	UTP and fiber	$170
Digital Equipment Corp. Maynard, Mass. 508-467-5111	Open DECconnect	UTP, STP, and fiber	$100 to $150 for Category 5 UTP; $120 to $170 for STP
IBM Armonk, N.Y. Contact local sales representative	IBM Cabling System	UTP, STP, and fiber	$275
Northern Telecom Ltd. Mississauga, Ontario 416-897-9000	Integrated Building Distribution Network (IBDN)	UTP and fiber	$100

*IBM price supplied by Commtran Consulting; others supplied by vendors.
Reprinted from *Data Communications,* November 1992.

TABLE 2.9 Selected 100 Mb/s-over-Copper Products

Vendor	Product name	Description	Cable type	Price
Crescendo Communications Inc. Sunnyvale, Calif. 408-732-5955	C1000	10-port hub	UTP and STP	$8995 to $10,995
	C300	Card for Sun, Micro Channel, or EISA PCs and workstations	UTP and STP	$1495
Digital Equipment Corp. Maynard, Mass. 508-467-5111	DEC FDDIcontroller 700-C	Card for DECstation/DECsystem 5000 series, DECsystem 5900, and VAXstation 4000 Model 60	STP	$1350
	DECconcentrator 500	18-port hub	STP	$6000 to $12,000
IBM Armonk, N.Y. Contact local sales representative	IBM FDDI Copper Base Adapter/A	Card for PS/2	STP	$3995
	IBM 8240 FDDI Concentrator	24-port hub	STP	$17,540 to $61,980
Interphase Corp. Dallas, Texas 214-919-9120	V/FDDI 5211 Peregrine	Card for VMEbus workstations	UTP and STP	$4195
	Fiberhub 1600	16-port hub	UTP and STP	$6400 to $13,900
Network Peripherals Inc. Milpitas, Calif. 408-321-7300	NP-AT	Card for PC ATs	STP	$2995
	NP-EISA	Card for EISA PCs	STP	$3495
	NP-MCA	Card for Micro Channel PS/2s and RS/6000 workstations	STP	$2695
	NP-SB	Card for Sun workstations	STP	$1995
	NP-VME	Card for VMEbus workstations	STP	$2795
Plus Tec Corp. Bountiful, Utah 801-292-3162	Plusnet-100 Adapter	Card for EISA, ISA, and Micro Channel PCs	UTP and STP	$595
	Plusnet-100 Multi-Port Repeater	16-port hub	UTP and STP	$1500 to $2700
Schneider & Koch & Co. Datensysteme GmbH Karlsruhe, Germany 49-721-7920	SK-FDDI Concentrator	12-port hub	STP	22,510 to 30,290 deutsche marks ($15,960 to $21,475)
	SK-Net FDDI-TI	Card for ISA PCs	STP	3950 deutsche marks ($2800)
	SK-Net FDDI-TE	Card for EISA PCs	STP	4450 deutsche marks ($3155)
Synoptics Communications Inc. Santa Clara, Calif. 408-988-2400	Lattisnet 3000	40-port hub	STP	$8990 to $54,445
	Model 2912	14-port hub	STP	$15,500

Reprinted from *Data Communications*, November 1992.

3

Optical-Fiber Applications

3.1 Origins of Optical Transmission

In the late 1970s, fiber optics emerged from the laboratory and rapidly gained acceptance as a transmission medium for metropolitan and long-haul communications circuits. As costs declined, fiber optics has become an important medium for local-area networks as well.

Fiber-optic conductors are tiny pipes made by depositing layers of silicon dioxide on the inner surface of a glass tube about 3 ft long and 1½ in in diameter. When the deposition is completed, the tube is collapsed under heat and drawn into a thin strand of highly transparent glass. The optical characteristics of glass fiber are so precise that a light signal entering one end of the fiber can travel for a mile and exit with nearly the same intensity.

3.2 Fiber-Optic Modes

When glass fibers first emerged from the laboratory, their loss, or the amount of attenuation of the light from one end to the other, was high. This meant that regenerators were required at frequent intervals, adding to the cost, and thus limiting the feasibility of fiber optics as a transmission medium.

The loss is not only the result of lack of transparency of the glass, it also comes about because of reflection of light waves as they travel down the fiber. Light waves glancing from the walls of a fiber take a longer path than waves that are emitted in a straight line. With some light waves arriving slightly out of phase with other waves in the

TYPICAL 8/125 SINGLE MODE FIBER

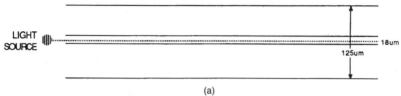

(a)

TYPICAL 62.5/125 MULTIMODE FIBER (STEP INDEX)

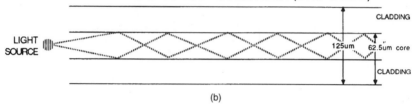

(b)

TYPICAL 62.5/125 MULTIMODE FIBER (GRADED INDEX)

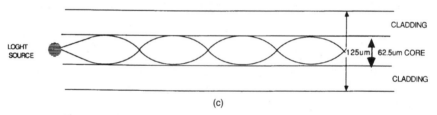

(c)

FIGURE 3.1 Types of fiber: (a) single-mode; (b) multimode step index; (c) multimode graded index.

same pulse, rounding of the pulse occurs. Before the point is reached where zeros and ones can no longer be distinguished in the pulse burst, the pulse must be intercepted, the signal rebuilt, and the pulse retransmitted in its original form. A new generation of fiber, known as *single-mode,* greatly reduces these multipath reflections and allows much greater repeater spacing than with the older multimode technology (Fig. 3.1). While the multimode fiber requires greater planning, its light source and repeater costs have become quite attractive for use in building and campus environments, while single-mode applications have dominated the public network architecture.

3.3   Optical Communication Links

Optical communication links may be of several different types, but the major categorization is guided propagation versus unguided propagation. The first category encompasses mainly the various types of fiber-optic links while the latter category may refer to simple links intended for only a few hundred meters, or to long-distance links using high-power solid-state or gaseous lasers and complex lens systems.

The concept of using visible-light-transparent fibers to guide light over fairly long distances dates back to the late 1960s. Work was begun at the time on the development of very transparent optical materials. Most of the common "transparent" materials were so lossy or scattered the light so much that light could not pass through a length of these experimental materials of less than one to several meters without being almost totally absorbed. This high absorption is due primarily to the presence of impurities within the glass which serve to absorb or scatter the light. Early attempts at producing pure glasses resulted in materials with losses still on the order of 1000 dB/km, but progress was made rapidly, and by 1970 glasses were produced with losses on the order of 20 dB/km. By 1975 this was down to about 2 dB/km, and now optic materials with losses considerably less than 1 dB/km are available. At the present time the most desirable range of wavelengths for propagation along optical fibers is the near-infrared region from about 800 nm to about 1500 nm for laser applications.

Recently, the well-advertised fiber "superhighway" has been presented as the medium for future intelligent communications, and it will be. Today, the superhighway uses what used to be referred to as *metropolitan fiber rings* (Fig. 3.2). These local fiber networks were placed in the early 1980s and were never fully utilized because of the high cost of connecting existing services to fiber converters. The change was invisible to the subscriber, and technology, along with regulation and Consent Decree interpretations, precluded the regional holding companies (RHCs), or Bell operating companies (BOCs), from introducing intelligence into the network, thus negating the potential benefits of this transport medium.

Today, services attributed to the superhighway are available at reasonable cost to high-volume subscribers. However, the use of optical fiber, in most instances, requires the end user to supply the interface conversion equipment to change the photon signal to an electronic pulse compatible with existing equipment. Before getting too excited about conversion, the reader is urged to realistically evaluate the need for light information services over and above what is currently available and in use within the organization today, and to quantify the justifications for a change. Remember that your organization probably is using electronic equipment that has evolved into the most efficient for existing embedded digital electronic public and local networks (Fig. 3.3).

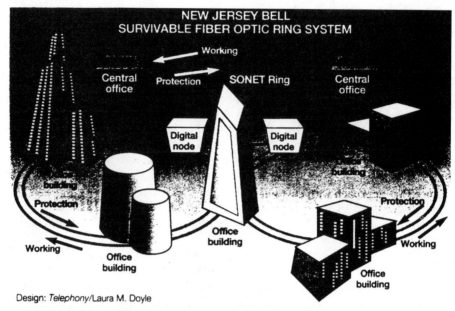

Design: *Telephony*/Laura M. Doyle

FIGURE 3.2 Metropolitan fiber ring—survivable fiber-optic ring system. (*Reprinted from* Telephony, *May 11, 1992. Copyright 1994 by Intertec Publishing Corp., Chicago, IL. All rights reserved.*)

To fully utilize the capacities of fiber optics, photonic end-user processing equipment will have to be developed, tested, and demonstrated in real-life conditions before it is considered as reliable as its electronic equivalent. Large companies will also be concerned with the current strategy of centralized information storage in the utility company central office equipment and of potential liability, or loss, should the integrity of these files be breached. If you decide to convert, make sure you *see* the features you seek working in other than a laboratory environment.

3.4   Fiber-Optic Cable Systems

Optical-fiber placement options have properties that make them particularly appealing. As the bandwidth of transmission increases, core size generally decreases. Fibers are usually immune to signal-distorting interference and considerably more tolerant of environmental attack. The coaxial and twisted-pair mediums, in stand-alone comparisons, are more costly to place, require considerably more engineering, and consume many times more space. Currently, fiber-optic

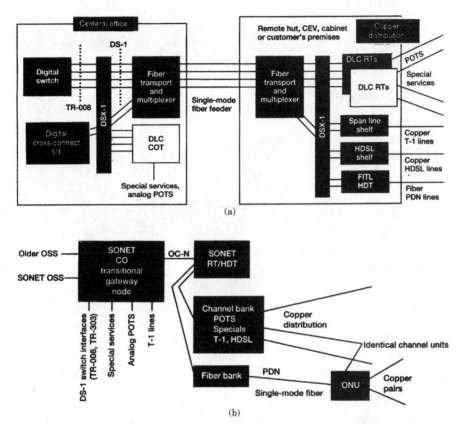

FIGURE 3.3 (*a*) Present method of operation with fiber transport loop (FITL) and high density subscriber line (HDSL) capabilities added. (*b*) Transitional approach to the target architecture. (*Reprinted from* Telephony, *September 21, 1992. Copyright 1994 by Intertec Publishing Corp., Chicago, IL. All rights reserved.*)

cable represents the leading edge technology in regard to local-area network (LAN) architecture. Fiber-optic applications to voice communications were introduced in the early 1980s, but were restricted to point-to-point, single-mode fiber links. In 1985, technology advanced to the point that optical-fiber links were introduced as a cost-competitive short-haul alternative to coax and twisted pairs in the telecommunication utility company local loops and office-to-office trunk routes. Today, costs for distributed optical links have been reduced to the point that applications are practical, if not least-cost, in local-area data networks. A physical comparison of fiber, coax, and twisted pairs follows.

Comparative item	Fiber	Coax	Twisted pair
Dialectric isolation	Yes	No	No
Interference immunity:			
Radio-frequency interference (RFI)	Immune	Limited	Limited
Electromagnetic interference (EMI)	Immune	Limited	Limited
Ground currents	None	Poor	Poor
Cross talk immunity	Excellent	Limited	Poor
Nuclear hardness	Moderate	Yes	Yes
Bandwidth	200 mHz/kf	40 mHz/kf	0.2 mHz/kf
Attenuation	0.2–10 dB/km	Frequency dependent	Frequency dependent
Withstanding temperature	1000°C	300°C	300°C
Operating temperature	−55 to +155°C	−55 to +80°C	−65 to +200°C
Bending radius	Good	Poor	Good
Strength:			
Conductor	300,000 lb/in^2	70 lb/in^2	70 lb/in^2
Cable	100 lb	29 lb	25 lb
Sparks and fire hazard	No	Yes	Yes
Lightweight materials	Yes	No	No
Cable cost	Low	High	Low
Interface cost, $/circuit	Low to high, $10–$300	Low to high, $1–$300	Low, $1
Complexity	Low	Moderate	Low
Signaling formats:			
Digital	Excellent	Good	Moderate
Analog	Good	Poor	Good
Network topology:			
Distributed tree	Poor	Excellent	Excellent
Distributed star	Excellent	Good	Good
Point-to-point	Excellent	Excellent	Excellent
Ring	Excellent	Excellent	Poor

3.5   Types, Specifications, and Capabilities

3.5.1   Single-mode fiber

Single-mode optical fiber is 6000 to 12,000 nm in diameter (0.000236 to 0.000472 in). This is very small diameter glass fiber when one considers that an average human hair is about 0.004 in in diameter. The single-mode fiber's small diameter intensifies the mechanical problems of splicing, connecting, and so on and makes it more difficult to couple radiation into it. In addition, the single mode of propagation requires that the light source feeding the fiber be highly collimated, which usually means that a laser source must be used. However, single-mode optical fiber accepts and propagates light more efficiently, thereby reducing the number of the angles the light travels along and increasing the capacity of the optical link.

3.5.2 Multimode and graded-index optical fibers

Multimode stepped-index fibers are 50,000 to 100,000 nm in diameter, leading to about 10 possible modes of propagation. The multimode fiber has the advantage of being able to accept light coming into the end of the fiber within a much wider cone of acceptance than can the single-mode fiber. Many modes of this type of fiber allow the half angle of the acceptance cone to be about 14°, while with single-mode fiber, the only radiation that will propagate is that which enters at approximately 0°.

Graded-index-type fiber has a core which is as large as multimode fiber, but it has an index of refraction which decreases with the radial distance from the center of the core. This leads to a multiplicity of curved paths. Light entering at different angles will propagate in curved paths due to the gradual bending of the rays by the radically changing index of refraction. Although the higher-number rays travel longer paths than do the lower-numbered ones, their velocity of propagation increases as they pass through regions of the core. If the change within the radial distance in the core is carefully engineered, it is possible to make all the rays have nearly the same average axially directed velocity component. This greatly reduces the pulse dispersion that is present due to different rays traveling different distances.

Technology in the local-area network environment has slowly evolved to the uniform selection of the 62.5-μm core/125-μm cladding multimode fiber. This is the recommendation of IBM and is normally compatible with and meets the transmission requirements of other equipment vendors providing interface equipment for their products.

3.6 Light Sources and System Components

3.6.1 Light sources for fiber-optic communication links

The two most commonly used sources for optical-fiber communication links are the light-emitting diode (LED) and the injection laser. In the near infrared regions of current interest, the LED is commonly of aluminum/gallium arsenide (AlGaAs) or indium/gallium arsenide/phosphide (InGaAsP) construction and produces radiation from a "well," in the more conventional design, etched perpendicularly into the diode junction area. The efficiency of the LED is rather poor. For a given temperature, the optical output is approximately a linear function of the diode current. This makes the LED easy to operate. The LED can be switched on and off [pulse code modulation (PCM)] at rates up to 100 Mb/s and is reasonably low in cost. These last two advantages are balanced by the disadvantages that the optical power is not well-collimated and its spectrum is spread out over a relatively wide range of wavelengths, as compared to the spectrum of the solid-state injection laser. These characteristics cause problems of coupling the power into the fiber and cause multimodal and chromatic pulse dispersion. These limitations make the LED most useful with multimode or graded-index fibers for applications over relatively short distances (less than 2 km) at moderate data rates.

The injection laser is essentially another form of the AlGaAsP diode with the radiation taken out of the edge of the junction region. When highly polished and made closely parallel to each other, the sides of the semiconductor chip form an optical cavity within the junction region, permitting the diode to operate in the stimulated-emission-of-radiation mode. This has several net effects: a higher conversion efficiency is obtained, the radiation is more monochromatic in nature, and the output is highly collimated. All these make the injection laser suitable for use with single-mode fibers. Since it can put out on the order of 10 mW of highly collimated near-monochromatic radiation and can be pulsed at more than 10 Gb/s, it is possible to construct long-distance high-data-rate single-fiber optical data links up to many kilometers in length without intermediate repeater stations. The main problems with using the injection laser are its higher cost, its lower level of reliability, and shorter life span. It is likely that in the near future, these criticisms will be greatly mitigated.

From the preceding it is apparent that two basic types of systems are emerging at the present time: (1) a low-cost, short-haul, moderate data rate system based on LEDs and a multimode or graded-index type of fiber and (2) a long-distance, high data rate, more costly system based on injection lasers and single-mode fiber. For cost and application reasons, all building and campus systems will employ LED-driven multimode fiber-optic links until an intermediate type of system is developed based on the edge-emitting diode, a device somewhat similar in construction to the laser, but less monochromatic and efficient.

3.6.2 Optical detectors

Two basic types of optical detector are presently being used to convert the optical signal coming out of the optical fiber at the receiving end into an equivalent electrical signal. These are the PIN diode and the avalanche diode. Without going into a highly technical discussion of the physics involved, both detectors are diodes which receive light input in much the same manner and transfer the photon energy received into electrical impulses within an electrically charged field. Basically, these detectors sense electrons which are created by the impinging photons of the optical radiation in a region where there is a large electrical field. The electrons are thus accelerated quite rapidly to a high enough velocity that, when they collide with the structure of the crystal lattice, they produce other "secondary" carrier pairs which also contribute to the reverse (output) current of the diode. The net multiplication is about 10 times the charge transfer caused directly by the photon itself. The difference between the two types of diodes is the amount of electronic amplification required and the dB loss.

In the local-area network (LAN) environment, sensor circuit design and dB budget loss requirements are stated by the manufacturer. Since clients are unlikely to fabricate their own optical equipment, the detailed engineering and design of optical detectors will not be included in this work.

3.6.3 Other components of the optical-fiber link

In addition to the optical source, which converts an electrical signal into a photon signal at the transmitting end, the photodetector which converts it back into an electrical signal at the receiving end, and the optical fiber which conveys the optical signal from transmitter to receiver, the optical communication link may also involve a number of other types of components depending on the use to which it is put. Two of these will usually appear: fiber-optic cable connectors and splices. Both these perform essentially the same function: to connect one piece of optical fiber to another. The difference between them is that connectors are usually used in situations where it may be desirable to disconnect the connection easily, whereas splices are used where the connection is to be more or less permanent.

Some links may also involve fiber-optic switches, direction couplers, duplexers, and multiplexers as well as other special-purpose optical devices. Optical amplifiers will also be part of the long-distance, laser-driven optical link.

3.6.4 Fiber connectors and splices

Because of optical fibers' small size, joining them either temporarily or permanently requires extreme mechanical precision. This is especially true in the case of the single-mode fiber. Theoretically, if the ends of two optical fibers are well-polished perpendicular to the axis of the fiber and if the two polished ends are exactly aligned in both displacement angles, the optical radiation will be transmitted across the junction without loss.

Three devices which are designed to do the job of aligning the fibers are biconic, SMA, and ST connectors. They rely on accurately machined plugs, one of which fits over each end of the fiber cable section to be joined, and are securely connected by epoxy adhesive in the case of splicing or by patch panel configurations in the case of connections.

- The *biconic connector* is a high-performance field-mountable fiber-optic plug based on the well-known, time-proven biconic design. It is used for cable-to-cable or cable-to-equipment connections. It can be installed on interconnect cable or buffered or unbuffered fiber. Operating temperatures are -40 to $+165°F$ and their tinsel load is approximately 50 lb.

- The *ST connector,* or bayonet connector, is a plastic- or ceramic-tip fiber-optic connector plug. It uses a spring-loaded bayonet-type twist-and-lock mounting arrangement and will accept 125-μm outer diameter fiber unless otherwise stated. The plug may also be field-mounted. This connector was developed by AT&T and is the most popular connector in service today.

- The *SMA connector* is the same as the ST-type except that it employs a knurled nut mounting arrangement. The major problem with this type of mounting is that technicians tend to overtighten the connection, thus destroying the optical alignment.

All three connectors are field-rated at 0.5 dB per connection. The ST connector appears to be the connector of choice for interface connections to most optical interface cards.

The fiber itself passes through a very small opening which is (to within matching or molding tolerances) along the axis of the cone. When the two plugs are then inserted into an accurately machined or molded receptacle, their axes are closely aligned, and thus the fibers which lie along these axes are also closely aligned. If the ends of the fibers are polished after the cables are installed in the plugs, there will be very little loss in transmitting light from one of the fibers to the other.

For connecting more than one fiber at a time, some variation of the groove-block type of connector or splicer may be used. The technique of mechanically aligning a group of fibers is quite often applied to cables containing multimode or graded-index fibers. One common standard that is used has 12 fibers per plane and up to 12 planes (a maximum of 144 fibers in one groove block). If used for a splice, the fiber ends from the two cables are polished, then butted together within a single block, then clamped into place. For a connector, the individual fibers from one cable are inserted into a block with their ends protruding. The protruding fiber ends are trimmed and polished flush with the block surface, which will mate with a similar surface of the block attached to the end of the other cable. An accurately machined connector assembly and indexing pins keep the two groove blocks in accurate alignment.

Another type of connecting or splicing technique is to use some type of focusing system between the two fibers. The simplest of these techniques is to round the ends of the two fibers to form in effect two semispherical lenses. This is done by melting the ends of the two fibers in an electric arc. The lenses then help to focus and collect the light that might otherwise be lost by displacement or angular misalignment of the fibers.

Another technique for aligning fibers accurately involves the use of three smooth straight rods which, when placed together parallel to each other and enclosed within a piece of heat-shrink tubing, clamp the two fiber ends into close alignment. This low-cost, quick-join technique is usable for the larger-diameter fibers. Still another approach to permanently connecting two fibers is to melt (weld or fuse) the two ends together with an electric arc while the two ends are held in precise mechanical alignment by a welding jig. This method is quite often used on long-distance single-mode optical-fiber links. Although the operation is time-consuming and expensive, the splice thus produced is very low in loss and is worth the effort for the high-speed transmission integrity it ensures.

3.6.5 Wavelength multiplexers

Wavelength multiplexing employs an optical grating to separate a beam containing two wavelengths of radiation into two spatially separated beams. Since the process is reversible, the grating can also be used to combine two separate beams of different wavelengths into one common beam. Or the same structure can be used as one type of optical duplexer to enable one fiber to carry information in both directions simultaneously. However, some type of lens system to first collimate and then focus the radiation going to and from the grating is necessary to perform this function.

Another type of duplexing structure employs a flat-faced fiber coming from the transmitter, on which is formed an optical interference (wavelength-selective) filter that passes the transmitter-wavelength radiation, but which reflects radiation at other wavelengths. The transmission fiber has a rounded end to help it collect radiation coming out of the end of the fiber from the transmitter, and the fiber going to the optical receiver also has a rounded end to help it collect the radiation from the other end of the link which comes out of the transmission fiber and either hits the receiver fiber directly or is reflected off the interference filter on the face of the fiber coming from the transmitter. Since some of the power from the local transmitter will invariably get into the fiber going to the local receiver, it is necessary either that the local optical detector be insensitive to the local source's wavelength or that an optical rejection filter of some type be incorporated in the local receiver.

Duplexing and two-wavelength multiplexing schemes using multimode or graded-index fibers commonly use radiation at the two near-infrared wavelengths of approximately 800 and 900 nm. For multiplexing more wavelengths of optical radiation using the grating method, it has been found possible to combine as many as 10 different signals via wavelength multiplexing, using the infrared region from about 1140 to 1500 nm, with the spacing between the channels being about 36 nm and the bandwidth being about 20 nm.

3.6.6 Optical directional couplers

Directional couplers are made by fusing the claddings of two multimode fibers which have first been heated and stretched so as to reduce the diameter of the fiber cores, thus causing some of the optical radiation power contained in the higher modes in the core of a fiber to leave the core and enter the cladding, from which it is then partially transferred to the cladding of the other fiber through the fused region and thence into the core of the other fiber at the point where the core expands back to its original diameter. Under precise conditions, the amount of the coupling can be made wavelength-dependent, and incoming radiation at two different wavelengths on one of the fibers can be made to separate, with the radiation at the one wavelength being almost completely coupled into the other fiber and radiation at the other wavelength continuing to propagate along the original fiber. Such an arrangement is utilized as a wavelength multiplexer or duplexer.

3.7 Optical Loop Calculations

Loop calculations for optical lines are very similar to those for ordinary wire or coax loops. Assuming that enough optical power will be available to the photodetector at the receiving end of the link, the maximum data rate which may be carried on a link will be dependent on the maximum rate at which the optical source can be switched on and off, the response time of the photodetector, and/or the pulse dispersion which occurs on the optical-fiber transmission medium. In some cases two or even all three of these factors may be effective, and in that case their effects must be combined by adding the rise or fall time of the pulse at the source, the response time of the detector, and the dispersion time of the pulse. The maximum data rate is then the inverse of this sum if the assumption is made that there is enough power at the detector.

Once this target figure is obtained, the power budget of the link may be calculated to determine if this amount of power will be available to the receiving-end photodetector. This is done in the conventional manner by stating the power output of the source in dBm and then subtracting the various decibel losses [the optical-fiber loss, the loss at each anticipated connector or splice and the loss margin figure (safety figure) to account for inversion losses or degradation such as a decrease in the source output, a decrease in the sensitivity of the detector, extra splices needed to repair cable breaks, etc.]. If the resultant expected power at the receiving end is not enough to satisfy the minimum power requirement equation, then the required minimum power level must be lowered by decreasing the data rate or using a more sensitive detector, or else the received power level must be increased by increasing the source power or decreasing the link losses.

In most short-distance links, it will usually be found that there is enough received power that the data rate is limited by the other three factors discussed above. On long-distance links, the received power level and the available detector sensitivity, plus possible pulse dispersion on the optical fiber line, will be the factors which limit the available data rate. On short-distance links, where a very high rate is desired, multimode fiber is often used since it has good dispersion characteristics and is still very easy to work with insofar as splicing and connecting are concerned. The single-mode fiber used with the injection laser and the avalanche-diode photodetector is usually reserved for very long distances where the high cost of the laser and problems of working with the small-diameter single-mode fiber are worthwhile because they eliminate the need for repeaters on links up to several tens of kilometers in length. The larger fibers are usually used where fairly short distances are to be covered, where connecting and disconnecting of fibers may be required, and where large bandwidths (high data rates) are needed.

3.8 Practical Integration*

The use of a fiber-optic link for long-distance repeaterless trunking of digital data, digitized voice, and so on has already been discussed as one of the main areas of application for the laser/single-mode fiber systems. The information superhighway will bring this mode to the user location. The use of optical fibers for shorter distances where very wide bandwidths are required has also been mentioned. One such area is in local-area (data) networks. In this case, several practical fiber-optic multiplexing and/or channel units have been developed that allow for the economic insertion of optical-fiber links into Ethernet and token ring data LANs. Great strides have been made in point-to-point multiplexing of RS232 signals along with several vendor-specific protocol radio-frequency and digital optical transmission capacities. Additional economies have been realized in the transmission of voice-grade service to the point where optical links have an economic advantage in rehabilitation, reinforcement, and new construction expense over backbone and feeder metallic circuits in many building and campus environments.

Other places where fiber-optic links excel in performance are in those environments where there is considerable electromagnetic interference, such as in a power plant, substation, factory, or large aircraft. The fiber link can pass data through this type of environment with little chance that it will be damaged by an electrical switching spike or other electrical impulse noise. Also, since no detectable optical energy escapes from a fiber cable, and since it is very easy to detect at the receiving end that any usable amount of optical energy has been lost or removed from the link, the security of the data passing over a fiber-optic link is much greater than it is for a metallic alternative.

Optical integration of existing service is technically possible in all communications environments. Cost is the only impediment and should be calculated, using all unique and common costs, so that the embedded investment total-to-whole comparison is not overlooked.†

---

*James Martin, *Telecommunications and the Computer,* 3d ed., Prentice Hall, Englewood Cliffs, N.J., 1990, pp. 639–657.

†James Martin, *Telecommunications and the Computer,* 3d ed., Prentice Hall, Englewood Cliffs, N.J., 1990, pp. 639–657.

4

Support Structure and Connectivity Hardware

4.1 Component Definition

The low voltage distribution system is the heart of any communication system. The components that make up this system tie together all information from the public network to the end user, interconnecting all devices and networks in between.

Chapter 2, "Wire, Cable, and Coax Applications," discussed the demarcation point [minimum point of presence (MPOP)] specifications. This chapter will address the specifications of the support facilities and their design as it pertains to the nonpublic components of the private network.

4.2 Apparatus and Satellite Closets

A closet is defined either by its function or by its size. Apparatus closets house cross-connect or interconnect hardware that provides circuit connecting facilities between the MPOP, primary distribution and satellite closets, and station outlets. An apparatus closet is designed to contain cables, cross-connecting terminals, and voice and data electronic equipment (Fig. 4.1). A satellite closet (or terminal) contains the cross-connections necessary to terminate station wire and tie cables from an apparatus closet. It is normally fed from an apparatus closet or equipment room. A closet is a walk-in closet when it is 4 ft or more deep and at least 8 ft high. A closet is a shallow closet when its depth varies from 18 to 30 in.

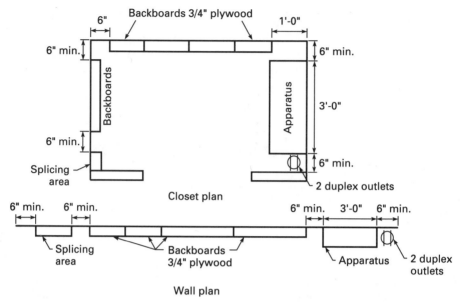

Figure 4.1 Typical apparatus closet layout.

4.2.1 Apparatus closet specifications

Apparatus closets are designed according to the size and makeup of the distribution system, the space required for electronic equipment, special data patch panel arrangements, and any other voice or data terminating equipment or facility. Although the square footage of the floor area to be serviced determines the size of the apparatus closet, a minimum area of 18 ft^2 with a continuous wall of at least 5 ft should be provided.

These closets are usually aligned vertically in a multistory building with one on each floor to create what is called a *riser shaft*. Cables can then be run vertically through the floor between closets. Access to an approved ground source is required for the grounding of metallic cable shields and equipment.

Environmental requirements for these closets are similar to those for the MPOP. The general requirements are as follows:

- Provide adequate access for authorized personnel entry.

- Ensure that closets are dry and free from danger of flooding.

- Walls should be lined with ³⁄₄-in plywood for mounting of cross-connect blocks and equipment frames.

- Doorways must be a minimum of 3 ft wide and 6 ft 8 in high, excluding the measurement of doorsill or center posts.

- Provide appropriate 120 Vac, 20 A, isolated ground, duplex outlets.
- Provide appropriate lighting with switch near door.
- Provide sufficient space to serve up to 10,000 ft^2 of usable floor space.
- Walls, floor, and ceiling of the closet shall be treated to eliminate dust.
- False ceilings should be avoided.
- Preferably, the closet should not be located adjacent to duplicating-type equipment because of potential electromagnetic interference (EMI).
- When ceiling distribution systems are to be used, the closet shall have adequate conduit or openings through the beams or other obstructions to the accessible ceiling space to permit the placing of wires and cables.

4.2.2 Satellite closet specifications

Satellite closets usually contain feedthrough cross-connect cable and wire and auxiliary power sources for terminal equipment. Satellite closets are used when there is a need for more than one serving closet on a floor; they supplement apparatus closet capacity by providing additional cable termination space, shorten loop length between local equipment and the end user, and accommodate force expansion by supplementing exhausted space in existing closets. While not recommended, satellite closets can house local-area network (LAN) and electronic key telephone equipment, if space and power are available.

Specifications for designing the satellite closet can be tailored from the apparatus closet specification to meet the particular needs of the location. The number and size of satellite closets depend on the floor space to be served, not to exceed 5000 ft^2. The closet should provide access to universal information outlet termination. Adequate lighting facilities, with the switch inside the door, are usually required for walk-in satellite closets. Shallow closets should be no less than 18 in deep to facilitate the placement of equipment. Lighting is desirable, but not required, inside shallow closets. AC power and ground source access are mandatory, regardless of the satellite closet size.

4.2.3 Satellite cabinets

Satellite cabinets serve the same purpose as satellite closets and house cable terminations in surface-mounted or flush-by-wall cabinets or terminal boxes. Since satellite cabinets are limited in size, the floor area they are capable of serving is also limited. For satellite cabinets and closets, mechanical protection must be ensured by running conduit with no more than 180° bends (no reverse curves) between the satellite locations and the apparatus closets.

4.3 Testing and Patch Panels

Testing and patch panels and wire termination blocks provide a break between facilities. The blocks are constructed so as to allow for plug-in jumper wires with modular ends, to be used for the completion of a circuit. They may be placed anywhere in a circuit but are usually placed between distribution cable and/or equipment access blocks and the universal information outlet (Fig. 4.2).

Advantages of the patch panels are:

- Simplicity of facility modifications
- No tools required
- Fast access for testing and trouble isolation

The disadvantages of patch panels are:

- Patch blocks are more expensive than punch-down blocks (exception: 110-type blocks). Modular patch cords are much more expensive than jumper wire.
- A high degree of management control is required to maintain accurate records.
- Better housekeeping is required than for punch-down jumpers.
- The wide variety of connectivity topology available may lead to the premature commitment of hardware for system applications with a potential for pin-out reversals for noncompatible patch panel terminations.

While it is possible to manipulate the end terminations of voice and data circuits to accommodate patch panel uniformity, it is usually best to defer the placement of panels until system selection studies have been completed.

4.4 Demarcation Points

4.4.1 Hardware

Backboard terminations, cross-connects, and equipment mountings have an industry-adopted universal color identity so that there is

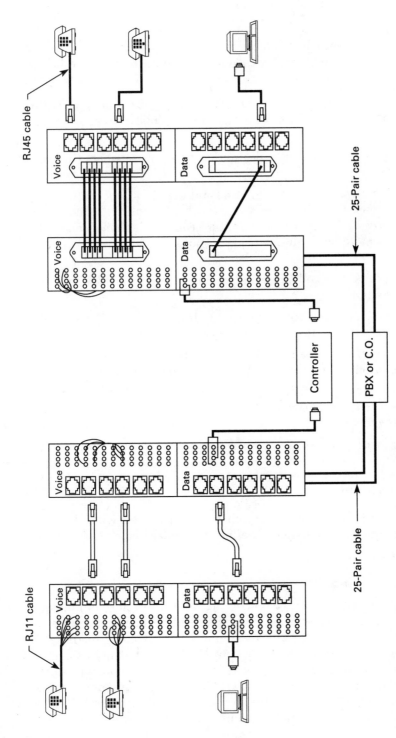

Figure 4.2 AMPIX applications—simple station cross-connect.

4.5

standardization of space allocation and use. At any telecommunications closet or backboard location, the colors have the following meaning:

Green: Public network central office connections (Fig. 4.3). Utility-company-owned cable, termination blocks, splice cases, and protection devices.

Orange: RJ21X (MPOP) termination and cross-connect (interface) blocks.

Blue: Connections to and from the end-user location. Station outlet connections: the wiring from universal information outlets to provide access to riser and distribution cable, telephone and/or data equipment in the equipment room, apparatus closet, or satellite closet.

Red: Trunk and toll line connections to and from key telephone equipment.

Purple: Trunk and line connections to and from PBX equipment and/or telecommunication system common equipment.

Yellow: Termination and mounting hardware for data equipment and/or special data services equipment and connections.

White: Primary distribution and campus cable. Cable which runs between equipment rooms and apparatus closets and the inter-building facilities.

Gray: Tie cable connections. That cable and wire which runs between cross-connects, apparatus closets, and satellite closets or between satellite closets and other departments within the same building or site.

When high-density termination blocks are used with their own front-mounted, color-coded labeling, backboard color coding is not required.

4.4.2 Physical security

Physical security refers to restricting access to telecommunication and security wiring systems. Access should be limited to those persons authorized to perform maintenance and facility modification activity on the wiring and equipment housed in the various cabinets and rooms. Depending on the sensitivity of the equipment, increasingly sophisticated systems and procedures may be employed. At a minimum, all closets and terminal rooms should be controlled by key access. All locations should be equipped with a lock and key system

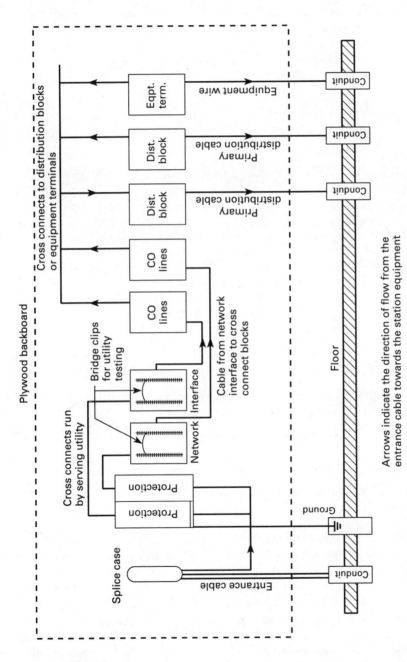

Figure 4.3 Network access point.

Plywood backboard

Cross connects to distribution blocks or equipment terminals

Bridge clips for utility testing

Cross connects run by serving utility

Eqpt. term.

Dist. block

Dist. block

CO lines

CO lines

Network Interface

Cable from network interface to cross connect blocks

Equipment wire

Primary distribution cable

Primary distribution cable

Protection

Protection

Splice case

Entrance cable

Ground

Conduit

Conduit

Conduit

Conduit

Floor

Arrows indicate the direction of flow from the entrance cable towards the station equipment

4.7

which is controlled by a responsible agency or employee. Access to these closets should be logged and controlled via manual record keeping procedures.

Should further access security be desired, existing methods used by sensitive public and private sector agencies, along with increased surveillance methods, may be instituted.

4.5   Building Environment

4.5.1   Horizontal overhead distribution methods

In-ceiling distribution systems, in the space between the suspended ceiling of the structural floor of the story or roof above, are used for housing cable and wire. From ceiling distribution, cables and wires may be run directly up to the floor above or down through poles or walls to the area below. The following methods are used:

- Zone
- Home run
- Raceway
- Poke-through

The ceiling system offers the advantage of common access to many areas of the floor space above or below. The disadvantages are:

- Inadequate clearance under air-conditioning ducts, electrical boxes, and lighting fixtures.
- Noise interference on wiring from electrical fixtures or other wiring.
- Insufficient ceiling support to hold the weight of cables and wires.
- In plenum, cable low-flame insulation is required at a considerable increase in cost.

Zone.   With the zone method (Fig. 4.4), usable ceiling space is divided into areas or zones for cables. Cables may be pulled through a rigidly supported conduit from a nearby serving closet to the center of each zone. When plenum cable is used, conduit is not required, although it may be used to provide mechanical separation and protection. From the center of the zone, cables are run to nearby walls or utility columns and then to the floor below or above. At the center of the zone, the cable terminates in an adapter that converts a large cable to a small one that terminates at each universal information

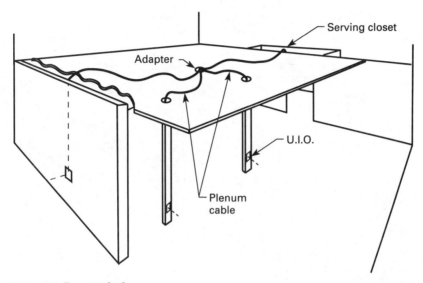

Figure 4.4 Zone method.

outlet. This is a flexible and economical method for distributing cables in the ceiling.

Home-run. In this method (Fig. 4.5), plenum cable runs directly from the equipment room or serving closet to the desired outlet. This method is relatively economical and offers the most flexibility for distributing cables and wires in a ceiling. It also eliminates the possibility of signal interference caused by mixing analog and digital signals in the same cable sheath.

Raceway. Raceways are open or closed metal trays that are suspended in the ceiling area from the floor or roof structure above (Fig. 4.6). They are generally used in larger buildings or where the distribution system is complex enough to demand the extra support.

Raceways, sometimes referred to as header raceways, are used to bring cables into a desired area. Later raceways branch off from the main header raceways to provide access to the zoned floor space below. Cables are then run to utility columns, or poles, or partition walls through short lengths of flexible conduit, or left exposed and terminated in information outlets at terminal equipment locations. Raceways provide mechanical protection and support and are effective for large installations. However, they require extensive planning,

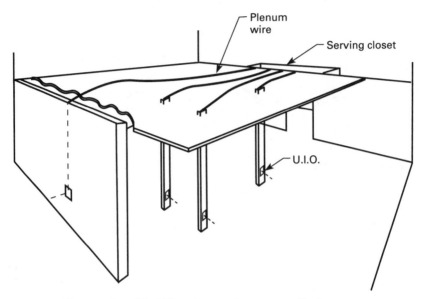

Figure 4.5   Plenum wire with utility columns—home-run method.

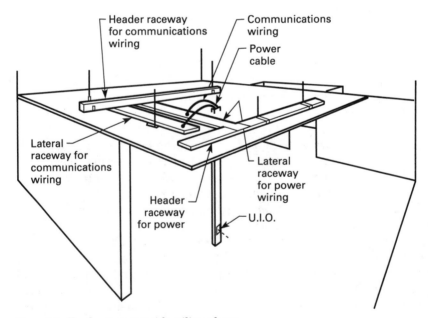

Figure 4.6   Header raceway with utility column.

are expensive to install, may ultimately limit flexibility, and can add excessive weight to the overhead support structure.

Poke-through.   Poke-through cable placement (Fig. 4.7) involves drilling or breaking a hole through the floor and poking cables and wires up from the ceiling space below. Because constructing fire barriers at the opening or maintaining the effectiveness of an existing barrier is very difficult, this method reduces the fire resistance between floors. It may also weaken the floor, and, during installation and maintenance activity, may interrupt work on the floor below. Poke-through is not recommended and is used only as a last resort in situations where other conventional distribution methods cannot be used.

4.5.2   Horizontal floor distribution

The alternative to placing horizontal distribution cables in the ceiling area is to install them under the floor. The four methods used to distribute cables in floor systems are:

- Shared-use underfloor channel
- Cellular floor
- Raised floor
- Underfloor conduit

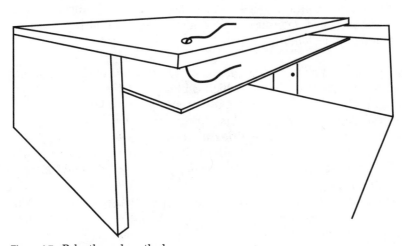

Figure 4.7   Poke-through method.

An underfloor duct system consists of a series of metal distribution channels, often enclosed in concrete, and open or closed metallic feeder channels or troughs. Depending on communication and power cabling requirements, the thickness of the floor, and the floor space to be occupied, underfloor duct methods can be used on one or two levels.

A one-level underfloor duct system consists of distribution and feeder ducting on a single plane. A two-level system consists of a double layer, with the feeder duct running underneath the distribution level.

Underfloor channel. Rows of distribution ducts typically house cables from apparatus closets to satellite locations or other desired building areas where cross-connects, interconnects, or universal information outlets can be placed (Fig. 4.8). Feeder ducts intersect junction boxes at right angles and extend from the junction box into the serving closet.

The junction boxes, used to access cable for pulling operations, have a metal access plate that can sometimes be seen in the middle of a carpeted or tiled floor. Partitions inside the boxes keep power and communication cable separate and ensure that they are not run together in the same channel.

The underfloor channel method is one of the most secure methods for horizontal distribution. It must, however, be considered in the initial stages of building space planning and installed before the building is completed. Other advantages of the underfloor duct method are mechanical protection for cables, reduced electrical interference, increased security from wire tapping, concealment, preservation of appearance, and reduction of safety hazards. Disadvantages of underfloor ducts include initial expense, construction scheduling conflicts, special treatment of openings for service access, limited fixed capacity, and added floor loading.

Cellular floor. The cellular floor consists of a series of channels, like corrugated cardboard, through which cables pass (Fig. 4.9).

Cellular floors form a ready-made raceway for the distribution of power and communication cables. Alternating the power and communication cells can provide a flexible layout. In some cases, it may be desirable to use a third parallel cell through which low voltage signaling circuits are fed. Reasonable space between cells is normally 5 to 6 ft.

Depending on the floor structure, the distribution cells may be constructed of either steel or concrete. In both cases, as with underfloor ducts, header ducts may be used as feeders to pull cables from the distribution cells to satellite locations.

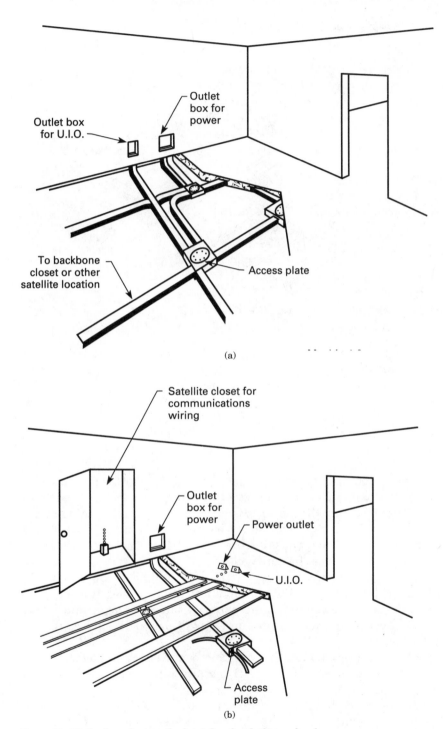

Figure 4.8 Underfloor duct method. (*a*) One-level; (*b*) two-level.

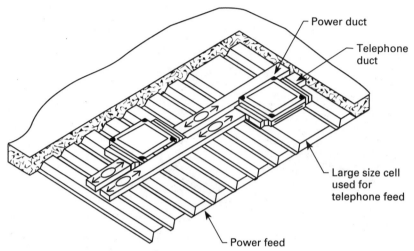

Figure 4.9 Cellular steel floor with underfloor raceway.

Cellular floors offer all the advantages of underfloor ducts, such as security and mechanical protection for cables. They also provide high capacity for large cables and add little extra floor weight. They have the disadvantage of being expensive to design, install, and administer. They are fixed, and reinforcement of congested cells is extremely expensive.

Raised floor.   In this distribution method, the entire floor consists of square plates that rest on steel or aluminum locking pedestals attached to the building floor (Fig. 4.10). The plates typically consist of a bottom steel layer adhesively bonded to a laminated wood core covered by cork, carpet tiles, or vinyl tiles. Any square can be removed for access to the wire beneath.

This approach provides for virtually complete flexibility in distributing telephone and other communication cables. Raised floors accommodate easy installation, capacity for all present and future needs, access to wiring over the entire floor area, and easy fireproofing.

Disadvantages include the sounding-board effect created by walking on raised floors, high initial installation expense, added floor weight, poor control over cable runs, and decreased room height. When the space beneath the raised floor is used as an air plenum, plenum cable must be installed at a substantially increased cost.

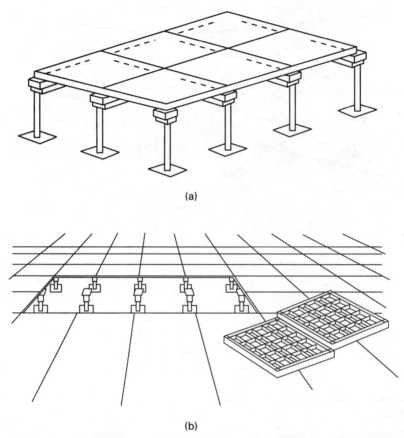

(a)

(b)

Figure 4.10 (a) Raised flooring. (b) Raised flooring with panels removed to gain access.

Underfloor conduit. An underfloor conduit system is made up of a number of metal or plastic pipes, home-run from an apparatus or satellite closet to potential terminal locations in the floor, wall, or columns of an office space (Fig. 4.11).

If a sufficient number of outlets are installed, the conduit system is suitable for buildings which have relatively stable terminal equipment locations.

The major advantage of an underfloor conduit system is its low initial installation cost for areas with few outlets, particularly where the outlet locations are definitely established. It is also a concealed system that preserves appearance. The major disadvantage of the sys-

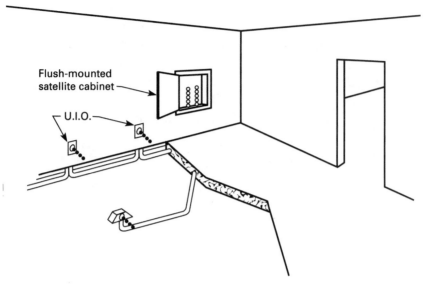

Figure 4.11 Underfloor conduit method.

tem is the limited flexibility; if the desks or tables are not located over the floor outlets or adjacent to wall outlets, additional exposed wiring is required from the outlet to the workstation location.

4.5.3 Horizontal distribution in older buildings

The horizontal distribution methods discussed so far are intended primarily for use in new buildings or buildings under construction. There may, however, be a need to change the current distribution system or install a new one in an existing building. Many of the methods of distribution, such as cellular floors or underfloor duct, would be costly or difficult to install in an existing building because of the need to tear up the existing floor, reinforce the support members, and install the new distribution system and floor. The following distribution methods are less expensive and easier to use in old or renovated buildings.

- Baseboard raceways
- Overfloor ducts
- Molding raceways
- Flat, undercarpet cable

Baseboard raceways. Just as the name implies, baseboard raceways are metal, wood, or plastic channels that run along the baseboards of a room (Fig. 4.12). They provide easy access to cables and are used for small floor areas where most information outlets will be located along the walls. The front panel of the baseboard raceway is removable, and outlets may be placed at any point along the channel. Power and communication cables are in separate channels, divided by a continuous, grounded metallic partition.

Overfloor ducts. In this method, overfloor rubber or metal ducts enclose and protect house wiring across floor surfaces. Cables are laid inside the metal ducts, which are secured to the floor. Covers are then fastened to the base of the ducts. Rubber ducts consist of lengths of specially molded rubber, cemented to the floor, through which cables are pulled. Overfloor ducts provide easy, quick installation, and are suited for low-traffic areas such as individual offices or work locations near walls. They should not be placed in halls or high-foot-traffic areas for obvious safety reasons.

Molding raceways. Molding raceways are metallic, wood, or plastic moldings placed on the walls of hallways and rooms close to where

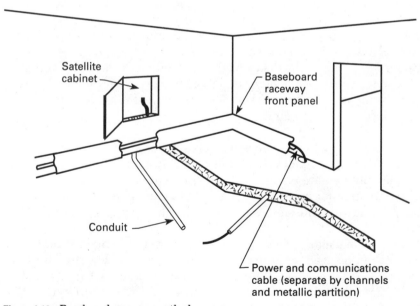

Figure 4.12 Baseboard raceway method.

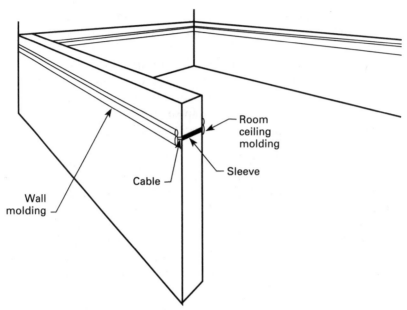

Figure 4.13 Molding raceway method.

the ceiling joins the wall (Fig. 4.13). Conduit may connect the hallway molding to a concealed satellite location. Behind the molding, small sleeves go through the wall to allow cable to pass through into the room where additional molding conceals the cable going to workstation information outlets. Although this method is generally outdated, it is acceptable in some older buildings where preservation of appearance is important. This method has limited flexibility due to the limited capacity of the molding.

Flat cable. Flat or undercarpet cable is designed for installation in open office spaces and other locations where standard, round inside wiring might create an obstruction. This type cable is often run under carpet or in other areas that have limited run space. It eliminates the need for utility columns and other unsightly supports in the office environment.

This cable usually comes in a 26-gage, 4-pair, unshielded configuration. It can be used for either voice or data service. However, since the cable pairs are run parallel to each other, the data runs should be short (less than 60 ft) in order to reduce inductive interference. This type of cable is not recommended for use with any service which uses a digital signal.

4.6 Campus Distribution Systems

Interbuilding distribution facilities may be supported by aerial, buried, or underground conduit structures (Fig. 4.14). As each outside plant environment will vary, it is recommended that the planner or system manager contract the design and installation of these facilities to a firm qualified to provide such design. The purpose of this section is to familiarize the reader with the basic components of each type of structure, provide construction specifications sufficient for bid preparation by engineering vendors, and give the relative benefits and disadvantages of each system.

4.6.1 Outside plant alternatives

In general, the least-cost structure alternative will be aerial plant. However, undergrounding of interbuilding cable facilities is mandatory in most new business parks, apartment complexes, and commercial subdivisions. While possible, it would be highly unlikely that any owners would choose to reverse this policy when their own property is involved, if for no other reason than lack of legal precedent. Therefore, the only aerial applications that should be seriously considered for private installations would be those for temporary service for construction activity.

Direct buried cable and underground conduit are the normal methods of serving buildings from the public network access point (NAP; more recently called minimum point of presence, MPOP). Cables will be sized for each building by the entrance cable sizing method. The method of delivering these cables to each building site will be determined by calculating the ultimate number of pairs required for each building, determining how many cables are required for these pairs, and then determining the least-cost method of placing the cables.

Normally, several cables will be required at numerous sites. In order to maintain an orderly inventory of these cables, separate cable sheaths should be home-run to each building from the MPOP. Where more than two cables are contemplated, it is generally more cost-effective to install conduit and manholes to facilitate their placement rather than to plan for subsequent trenching. In addition, conduit is placed in an independent trench which separates it from power, gas, and water, thus providing superior separation and protection from potential ground faults, hazardous gas contamination, or water intrusion. The telecommunication and information facilities housed in conduit are also better protected from dig-ups when maintenance is being performed on other underground facilities.

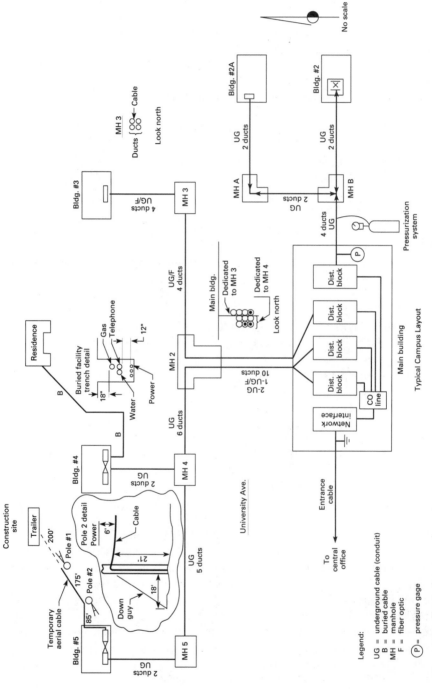

Figure 4.14 Typical campus layout.

Legend:

UG = underground cable (conduit)
B = buried cable
MH = manhole
F = fiber optic
(P) = pressure gage

4.6.2   Aerial construction

Aerial facilities (Fig. 4.15) include poles, wires, messenger guys, down guys, anchors, and cables. In high-wind areas, dampers are attached to the cables to inhibit cable oscillations, known as *cable dancing.*

As discussed earlier, aerial facilities should only be placed to provide temporary service to construction sites. When dealing with temporary service situations, the reader should contact the electrical power utility and arrange for them to design the pole line construction detail. As telephone service to the site will normally be from a telephone company central office, the utility company will coordinate the aerial cable or wire joint placement, removal, and billing details.

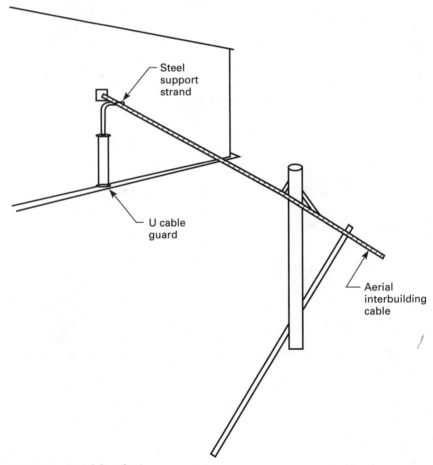

Figure 4.15 Aerial distribution.

4.6.3 Buried cable

Buried cable is placed in trenches connecting the various buildings in a campus environment. If these trenches are longer than convenient reel lengths of cable, provisions will have to be made for splices or special-order cable lengths. Splices can be directly buried or placed in manholes, splice boxes, or pedestals. For economic reasons, it is best to include all buried utilities in a single trench. This will require a concerted coordination effort between the utilities and the lead contractor to assure timely performance by the various placement entities.

Additionally, in design of buried facilities, consideration must be given to:

- Trench location (right-of-way, natural obstacles, other underground utilities, future underground placements, etc.)
- Depth of trench (minimum 18 in cover)
- Separation from other utilities
- Cable and sheath selection (mechanical protection, pressurization, insulation, etc.)
- Local placement and protection ordinances
- Utility company joint trench clearance requirements

Because there are no general rules to design a buried cable distribution system, it is recommended that, when this method is under consideration, the design be contracted to a firm that is not only qualified to perform the work in accordance with state, federal, and local government ordinances but also familiar with utility company requirements and has experience in the layout, construction bidding, and the utility company coordination requirements of buried cable design.

4.6.4 Underground conduit

A typical underground conduit system consists of conduits (pipes) and underground cable access boxes, or manholes, which provide space for splice cases and facilities for cable pulling operations. The inside diameter of the conduit can be from 2 to 4 inches. The conduits are placed from the network access point to the various buildings on the site. Manholes are placed along the route to facilitate cable pulling and splicing. The manhole can also intersect an existing conduit structure to provide access to new buildings along an established route. The following considerations are involved in the design of a conduit system (Fig. 4.16):

- Number of ducts (primary and secondary)
- Location of structure
- Section length (distance between pulling locations)

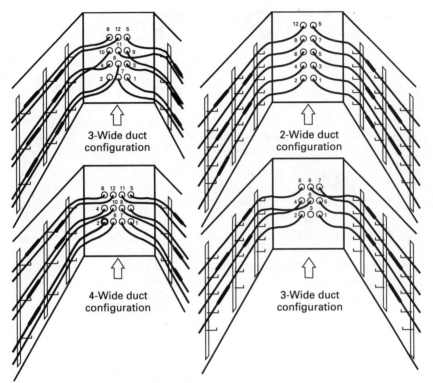

Figure 4.16 Racking arrangement and order of duct selection for double-racked man-holes (*left*) and single-racked manholes (*right*).

- Pulling tension of weakest sheath
- Curve and radius stress
- Duct arrangement and assignment order
- Separation from other structures
- Placement, backfill, and compaction requirements
- Manhole size and location
- Manhole frame and cover load requirements
- Duct termination and cable racking requirements
- Cable pressurization manifold placement and splice isolation requirements

Because of the number of variables associated with any conduit system, it is recommended that a qualified firm, specifically experienced in the design, permit, and construction demands of utility company conduit design, be engaged for any proposed campus conduit distribution system design.

4.7 Electrical Separation Requirements

The installation of power and communication cables must be in accordance with the National Electrical Code (NEC),* Arts. 725-5 and 800-3; the National Electrical Safety Code (ANSI C2); the Uniform Building Code; utility company tariff letters, and applicable government orders issued by the regional public utilities commission. These codes and standards stipulate the separation of system components and wiring between different classes of voltages. Adherence to these practices is important to reduce the risk of accidental shock, fire, and system integrity compromise due to inadvertent contact with dangerous voltage levels. The basic separation requirements are as follows:

4.7.1 National Electrical Code position

According to the NEC†:

> Conductors from the protector to the equipment or, where no protector is required, conductors attached to the outside or inside of the building shall comply with the specifications listed below.
>
> (a) Separation from Other Conductors
>
> (1) *Open Conductors.* Conductors shall be separated at least 2 inches (50.8 mm) from conductors of any electric light or power circuits or Class 1 circuits (NEC 725-38).
>
> *Exception No. 1*—where the electric light or power or Class 1 or communication circuit conductors are in a raceway or in metal-sheathed, metal clad, nonmetallic-sheathed, type AC or type UF cables.
>
> *Exception No. 2*—where the conductors are permanently separated from the conductors of the other circuit by a continuous and firmly fixed non-conductor, such as porcelain tubes or flexible tubing, in addition to the insulation on the wire.
>
> (2) *In Raceways and Boxes.* Communication conductors shall not be placed in any raceway, compartment, outlet box, junction box, or similar fitting with conductors of electric light or power circuits or Class 1 circuits.
>
> *Exception No. 1*—where the conductors of the different systems are separated by a partition.

---

*National Electrical Code® and *NEC®* are registered trademarks of the National Fire Protection Association, Inc., Quincy, MA 02269,

†Reprinted with permission from NFPA 70-1989, the *National Electrical Code®*, Copyright ©1989, National Fire Protection Association, Quincy, MA 02269. This reprinted material is not the complete and official position of the National Fire Protectin Association, on the referenced subject which is represented only by the standard in its entirety. Readers should assure that correct NEC standards are applied when subsequent issues are available.

Exception No. 2—conductors in outlet boxes, junction boxes, or similar fittings compartments where such conductors are introduced solely for power supply to communication equipment or for connection to remote-control equipment.

(3) *In Shafts.* Conductors in the same shaft with conductors of electric light or power shall be separated from light or power conductors by not less than 2 inches (50.8 mm).

Exception No. 1—where the conductors of either system are encased in noncombustible tubing.

Exception No. 2—where the electric light or power conductors are in a metal raceway, or in metal-sheathed, metal-clad, nonmetallic-sheathed, or type UF cables.

4.7.2 General utility company separation requirements

Separation protection is needed for cable located between protector and telephone equipment or station wire, double protection for data communication equipment and circuits. This applies to cable from fuseless or fused protector to telephone equipment and telephone wiring requiring no protector. Separations apply to crossings and to parallel runs. Minimum separations between cable, outside or inside buildings, and type of plant involved are as shown:

Separation from	Type of plant involved	= Minimum separation
Electric supply	Bare light or power wire of any voltage	5 ft (Note 1)
	Open wiring not over 300 V	2 in (Note 2)
	Wires in conduit, or in armored or nonmetallic sheath cable or power ground wires	None
Radio and television	Antenna lead-in and ground wires	4 in (Note 2)
Signal or control wires	Open wiring or wires in conduit or cable	None
Communication wires	Community TV system coaxial cables with shields at ground potential	None
Telephone or block wire	Fused protectors	2 in (Note 2)
	Fuseless protector or no protector required	None
Telephone ground wires		None
Sign	Neon signs and associated wiring from transformer	6 in (Note 3)

Separation from	Type of plant involved	= Minimum separation
Lighting	Fluorescent, neon, incandescent, or high-intensity discharge fixtures	5 in
Electrical equipment	Power panels, transformers, circuit panels	3 ft to 480 V, 4 ft all other
Lightning	Lightning rods and wires	6 ft
Pipe	Steam, hot water, or heating	Note 4
Stationary grating, metal shutter grillwork, etc.		Note 2

Note 1: Power is to be turned off when work is done above bare wire. Ladders shall be placed to maintain a 5-ft minimum clearance.

Note 2: Use B cable guard or two layers of vinyl tape (10 mm) extending at least 2 in beyond each side of object to be crossed.

Note 3: Avoid neon sign location if alternative run is possible.

Note 4: Excessive heat may damage plastic insulation; avoid heating ducts and other heat sources.

4.8 Electrical Protection, Grounding, and Cable Sheath Bonds

4.8.1 Electrical protection

Building and campus environments may require electrical protection to guard against the effects of voltage surges and sneak currents. In general, communication cables outside a building (including utility company cables and interbuilding campus cables) may be exposed to voltage surges caused by lightning. Exposure to lightning also depends on the electrical characteristics of the earth in the vicinity of the cable. Exposure to surges from power line contact and sneak currents induced by nearby power lines depends on the proximity of communication cables to power wires. For complete electrical protection against voltage surges and sneak currents, electrical protection devices, protective grounds, and bonding must be considered.

Complete surge protection includes carbon-block or gas-tube–type protectors connected to individual wires of exposed cable, connections to building grounds, and a bonding arrangement that connects all building grounds together. Electrical surge protectors are installed inside the building at each end of an interbuilding cable. These are typically located in the building entrance area or the equipment room. Sneak current protection is required for circuits in terminal equipment that operates with a low-resistance path to ground.

Appropriate protective grounding arrangements must be considered in the architectural design phase of a building. Commonly, requirements will specify at least a $\frac{3}{4}$-in conduit to an approved ground connection point with at least a no. 6 ground wire enclosed. Protective devices must be connected to these grounds. Many types of equipment, such as a PBX system or data controller, require a ground connection in addition to the normal power ground.

Bonding must be considered since all grounds must be bonded together to reduce electrical ground potential differences that contribute to electrical hazards in building wiring. Bonding involves tying protective grounds, power grounds, and equipment grounds together with an appropriate size wire.

The telephone company provides electrical surge protectors where their outside cables are exposed to lightning and power line crosses. The protectors reduce the hazards to equipment but may not be adequate for full electrical protection. As a result, additional surge protection may be required for complete protection. Sneak current protection, when required by the equipment vendor, must also be provided as part of the distribution system.

Local building codes may vary, and the system manager should assure that the installation contractor is in conformance to all applicable codes.

4.8.2 Electrical protectors

Voltage surge limiters (Fig. 4.17) control the magnitude of high voltage surges on building wiring. A voltage surge limiter uses air gap or gas tube protectors and consists of a resin-filled block with screw-in gap protectors, a 26-gauge input stub cable that serves as a fusible link, and two ground lugs. The air gap protector consists of carbon blocks inside a metal housing with a threaded cap. When the voltage on the conductor being protected exceeds a predetermined level, it is limited by arcing in the air gap between the carbon blocks. If the voltage is short-lived, such as that caused by lightning, the protector returns to an open-circuit condition. When the voltage persists, however, arcing continues long enough to melt a lead alloy spacer inside the unit on the spring-loaded carbon block, causing it to move up against the other carbon block and establish a direct path to ground. The gas tube protector consists of a surge protector and a fusible disk, mounted inside a housing with a threaded brass cap. It functions like the air gap protector, except that it uses two metallic electrodes sealed in a glass envelope filled with an inert gas.

Current limiters interrupt currents of abnormally long duration that can overheat and damage equipment connected to a system. The

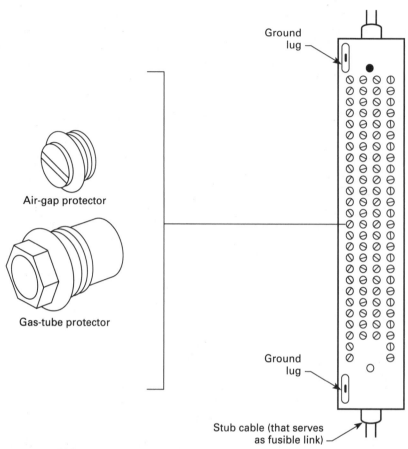

Figure 4.17   Voltage surge limiter (accepts either air gap or gas tube protectors).

voltages associated with these currents are not strong enough to trigger voltage surge limiters, and usually result from accidental contact between communication cable and power lines or induction from nearby power conductors. A current limiting device, or sneak current protector consists of a 66-type connecting block and 50 fuses (Fig. 4.19). Sneak current protectors also include heat coils and fusible links. Heat coils, similar to fuses, shunt sneak currents to ground instead of simply interrupting the circuit; fusible links (short lengths of fine-gauge wire) open the circuit and are normally part of a voltage surge limiter.

Protector panels provide either voltage limiting or voltage limiting and sneak current protection. Designed for compatibility with either

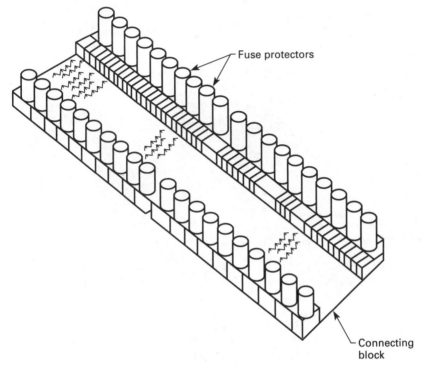

Figure 4.18 Current limiting unit with fuse protectors.

high-density or 66-type distribution hardware (Fig. 4.18), these panels use interchangeable plug-in protector units that typically contain two air gap or gas tube protectors, each in combination with a heat coil. A swivel stub cable (fusible link cable about 25 ft long with fine-gauge wire pairs) serves as a fusible link and is prewired through the panel to either a high-density or 66-type termination field. A similar version of these devices has the stub cable prewired through the protector panel to pin connectors to which a 25-pair cable can be connected.

4.8.3 Electrical grounds

The building grounding electrode system is described in the National Electrical Code, Art. 250. There must be an identifiable common grounding electrode which is used as the source of ground for all enclosures and equipment in or on the building. This is the same electrode that is used to ground the power service entrance.

Each separately derived power system within the building should have its own identifiable grounding electrode for the specific purpose

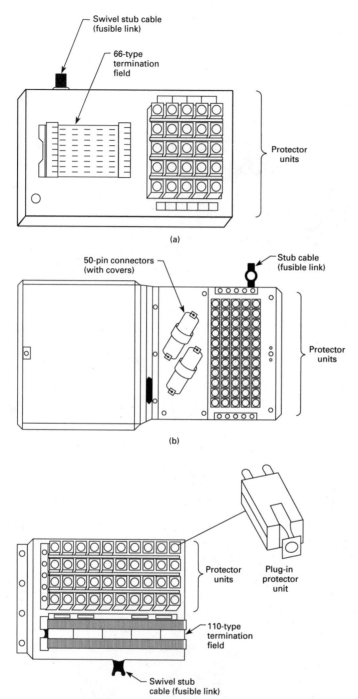

Figure 4.19 (*a*) 66-type compatible protector panel. (*b*) Protector panel with pin connectors. (*c*) 110-type compatible protector panel.

of grounding that system. This grounding electrode should be bonded to the grounding electrode for the service entrance equipment.

Grounding electrode system. The following elements can provide suitable grounds for cabling networks (Fig. 4.20). They must be able to properly dissipate electrical currents from the communication system, as well as from other electrical devices attached to the grounds.

- *Building steel:* Building steel is the structural steel that makes up the frame of the building. It may be used for grounding only if the steel frame is buried in the earth at a suitable depth. Although possible, using building steel can be difficult if building construction has been completed before the grounding system is installed.

- *Concrete-encased electrodes:* Concrete-encased electrodes are made up of copper wire embedded in the building concrete footing or near the bottom of the building's concrete foundation. The footing should contain at least 20 ft of the bare copper wire encased in a minimum of 2 in of concrete. This method of grounding, which is most effective when the building foundation is in rock, must be planned before the foundation is constructed.

- *Ground ring:* At least 20 ft of bare copper wire is buried at least 2 ft 6 in in the earth and next to or encircling the building.

- *Metallic power service conduit:* The power service enclosure, or power grounding electrode, installed by the power company.

- *Cold-water pipes:* Metallic pipes (typically copper or galvanized steel) with no plastic couplings, grounded at meters, and continuing in length for at least 10 ft underground. The metropolitan or rural public water system usually meets these requirements and is a ready-made ground electrode system. Cold-water pipes must provide short paths for all ground connections to dissipate lightning surges and redirect them to the earth. This is probably the least expensive method available but may be problematic due to the increasing use of plastic pipe.

When none of the grounds listed above is available or accessible, one or more of the following types of ground devices may be used:

- *Gas pipe system:* Metallic underground gas pipes that cannot be interrupted by insulating sections or joints, or that have outer conductive coatings. Use of this system must be acceptable to and permitted by the gas supplier and local jurisdictional authorities.

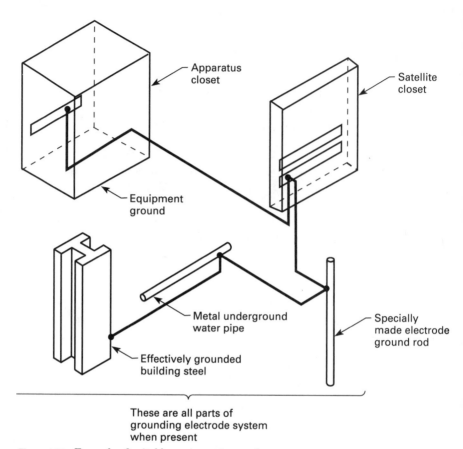

Figure 4.20 Example of suitable equipment grounds.

- *Ground rods, ground pipes, and plate electrodes:* Called *node elec-trodes,* these devices can be installed after building construction is completed and are designed for that situation. Ground rods are metallic, 8-ft rods driven into the earth that disperse about 90 per-cent of the electrical surge within 6 to 10 ft of the rod; ground pipes are steel pipes that act electrically the same as ground rods; plate electrodes are squares of steel buried in the earth. Such electrodes are inexpensive and easy to install, but should be considered only as alternatives to the primary building grounds described previously.

Telephone switching ground. Ensure that the basic switching ground is specified by the supplier of the switching or data equip-ment. The equipment room ground, which is the reference ground used to ground the telephone and data cables, should be installed in strict accordance with the equipment manufacturer's specifications.

For high-rise buildings (more than three stories), the preferred cable is telephone house cables. These cables are constructed with a corrugated aluminum or copper shield that serves as the grounding conductor to the upper floors. If corrugated shielded cables are not used, a 4-mm- (0.162-in-) diameter (6-gauge) or larger copper wire must be run parallel to the house cable to provide riser ground points.

Wiring closet grounds for voice and data systems. Each wiring closet must have a ground system using an electrical ground bus whenever one or more of the following conditions exist:

- Cable pairs enter or leave the shield of a riser cable
- Electrical communication hardware is installed in the closet to support horizontal distribution
- A shielded distribution cable ends in the closet
- Electronic equipment is used that requires a power return path

Wiring closets should be placed toward the center area of the building to limit the exposure to voltages caused by lightning and to reduce the length of the ground wire. A building exposed either directly or indirectly to lightning has a greater voltage rise on the perimeter of the building than in the center. Therefore, communication systems are better protected in the center of the building.

Wiring closet grounds for data communication systems. A suitable ground must be provided to each wiring closet that may house data equipment, distribution panels, or patch panel racks. The connection between the wiring closet and the ground point must be such that the resistance measurement from the wiring closet ground to the building grounding electrode system is 2 ohms or less. Any wire used in this ground circuit should not be less than 4-mm- (0.162-in-) diameter (6-gauge) copper wire. All metal panel enclosures, boxes, and raceways in the equipment closet must be grounded.

Grounding cable shields at splice cases. All telephone cable sheaths, whether aboveground or underground, must be bonded to maintain electrical continuity at each splice case. Use an approved bonding method with shield connectors and bonding straps suitable for handling the voltages and currents that may potentially be induced in the cable sheath. In addition, all metal splice enclosures must be bonded to the cable sheaths and grounded, using factory-approved bonding clips. They must be connected to a 1/0 bus bar system with annealed copper wire not less than 4 mm (0.162 in) in diameter (6 gauge). Annealed copper wire is wire that has been heated and then cooled to make it less brittle.

When telephone cable pairs in a riser system are removed from a metallic sheath cable for the purpose of splicing, the sheath must be bonded to an approved building or electrical ground in the closet in which the splice occurs. In addition, in each apparatus closet, all equipment cable sheaths and metallic splice cases must be appropriately bonded together, using approved hardware and bonding connectors and cable suitable for the purpose.

Grounding building entrances. The sheaths of metallic cables installed outside the building must be grounded at both ends regardless of whether the cables are installed aboveground or underground.

The metallic cable sheath at building entrances should be bonded, using factory-approved bonding methods and copper wire not less than 4 mm (0.162 in) in diameter (6 gauge), to the ground bus system immediately after entrance.

Grounding aboveground cable sheaths and cable messengers. Aboveground (aerial) cables are, by their nature, exposed to electrical hazards. These hazards may be in the form of lightning strikes, causing either direct or induced voltages. Also, hazardous voltages may come from electrical power sources. Cables are also subject to induced noise, such as radio-frequency interference, that may disrupt communication on the local network.

If cables are placed on a joint-use pole line (shared with other utility services such as telephone, power, and CATV), the grounding practices of the joint-use utilities must be observed. The practices normally require that the cable sheath and the messenger cable be bonded to the multigrounded neutral at distances not to exceed 2000 ft. Surge protectors may be required at the ends of all conductors in the cables at the point where they enter and terminate in a building.

Grounding underground cable. Underground cables also need to be protected from lightning. Underground cable is always exposed to some degree unless there is no lightning. The more conductivity (less resistivity) the soil has, the less severe the lightning hazard. Resistivity of less than 100 m $\cdot$ Ω (or 1 Ω resistance per 100 m^3 of earth) indicates a low exposure to lightning hazards.

Grounding at manholes. When manholes are used, they should be designed so that splice cases and shields can be bonded to a suitable manhole ground. These grounds may consist of either grids or rods to obtain a resistance to ground which is compatible with the cable or equipment but not greater than 25 Ω.

AC power branch circuits and grounding. Several types of equipment grounding conductors can be used in an electrical installation. Some of these types present a low impedance at the power frequency and a high impedance to electrical noise from radio-frequency interference, electrical transients, and electrostatic discharge (ESD) sources. This can affect the performance and reliability of the electronic devices and equipment.

Ground impedance values vary with the current and voltage rating of the ac circuit. For example, the ground impedance value for a 120-Vac, 15-A circuit is 1.0 Ω or less and the ground impedance value for a 240-Vac, 30-A circuit is 0.4 Ω or less. Exact grounding specifications and formulas may be found in NEC Art. 250.

5

Noise Suppression and AC Power and Lighting Considerations

5.1 Sources and Types of Signal Interference*

On any telecommunication link, there will be noise and distortion of some degree. A variety of techniques can be used in engineering the link to reduce this to an acceptable minimum.

Unfortunately, the acceptable criteria for data transmission differ from those for other uses of the link, such as voice. There can be a considerable degree of impulse noise and distortion on a voice line without the speech becoming unintelligible, annoying, or too unnatural. Similarly, telex lines can be noisy; if telex arrives with a few incorrect letters, it is still basically readable and understandable to a human being.

Voice transmission differs from data transmission in two fundamental ways. First, with voice we have an intelligent agency at each end of the line. If a burst of noise or other failure prevents the listener from hearing a word, they can either guess what the word should have been or else ask the speaker to repeat it. If they cannot hear, they will ask the speaker to speak more loudly or clearly. This highly flexible intelligence does not exist when machine talks to machine, so rigid procedures have been devised that are as fail-safe as possible.

---

*James Martin, *Telecommunications and the Computer,* 3d ed., Prentice Hall, Englewood Cliffs, N.J., 1990, pp. 639–657.

Second, the information conveyed by the human voice is at a very much slower rate than that at which we want the machines to "talk." The normal rate of speaking is equivalent to something like 40 b/s of written words, and this speech is "coded" with a very high degree of redundancy. We can usually follow the meaning of what is being said if we hear only about half of the words. On the other hand, data transmission over analog voice channels can take place at 9600 b/s and higher. It is because of the low rate of information in speech and because of the adaptability of the human ear that distortions and noise levels that would be damaging to data transmission have been quite acceptable in the engineering of the world's telephone lines.

In this section we outline the types of noise and distortion that are common and the effects they have. Some types can be overcome or minimized by suitable design of the terminal equipment or by slower rates of transmission. Others cannot be prevented but must be faced up to in the design of systems, by means of error detection and correction and retransmission techniques. Tight controls on the accuracy of data transmitted must be built into the computer programs.

5.1.1 Systematic and fortuitous distortion

Systematic distortion is that which occurs every time we transmit a given signal over a given channel. Knowing the channel, we can predict what is going to occur. The pulses may always be distorted in a certain way. Given frequencies will always have a certain minimum phase delay. Fortuitous distortion is something which occurs at random; so it is not predictable, except in terms of probability. Examples of fortuitous distortion are white noise, impulse noise, chatter from switching, cross talk, atmospheric noise, sudden changes in signal phase, and brief losses of signal amplitude. Fortuitous distortion refers to transient impairments rather than continuing conditions on the line.

Systematic distortion is, then, something which might possibly be compensated for electronically so that its effects are eliminated. Fortuitous distortion is more difficult to compensate for, though steps can be taken to minimize its effects and repair the damage it does. It may be possible to correct systematic distortion so that it never actually damages data. Fortuitous distortion, on the other hand, is occasionally going to produce an extra large noise burst or impulse which destroys or creates one or more bits at random.

White noise. This is the random hiss that forms a background to all electronic signaling. It cannot be removed, so it sets a theoretical maximum on the performance of any communication link and on the various modulation methods. The amplitude of the signal after attenuation must be kept sufficiently far above the white noise background to prevent an excess of hiss on radio or telephone circuits or an excess of errors on a data transmission circuit. On most analog lines, the signal-to-white-noise ratio is better than 30 dB, although occasionally a dialed call will encounter a much worse ratio than this.

Occasionally, there will be spikes of white noise higher than the majority of peaks of other noise types, such as cross talk, which may add to the white noise. The error rates in the various types of equipment due to white noise alone can be calculated theoretically.

If white noise were all we had to worry about, the design of data transmission systems would be more straightforward. Unfortunately, there are other types of noise and distortion that are far less predictable and more disastrous in their effects.

Impulse noise. Unlike white noise and the various types of systematic distortion described later, impulse noise can have peaks of great amplitude that saturate the channel and blot out data. Impulse noise is the main source of errors in data. The duration of the impulses can be quite long relative to the speed of data transmission—sometimes as long as 0.01 s, for example. This would be heard merely as a sharp click or crack to a human listener and would not destroy any verbal intelligence, but if data were being transmitted at 75 b/s, one bit might be lost. For speeds of 4800 b/s, a group of 50 or so bits would be lost.

Often a noise impulse removes or adds two or more adjacent data bits, and this means that odd/even parity checking may not detect the error. A more sophisticated form of error-detection code is needed.

There are many causes of impulse noise, some of which can be controlled but most of which cannot be, without a complete reengineering of the telecommunication facilities. Some impulse noise is audible during telephone conversations, and some goes unnoticed. Stray clicks and crackles are all too familiar. Impulse noise comes from a variety of different sources. It may come from within the communication channel itself or from a source external to the channel.

External noise is picked up by induction or capacitance effects. Sharp voltage changes in adjacent wires or equipment induce noise spikes in the communication channel. Many of the audible clicks which are of high amplitude and that damage one or several adjacent bits come from switchgear and telephone exchanges. Sometimes one can hear the rapid sequence of clicks generated by another person dialing. All relay operation is a potential source of noise if the shielding or suppression is not adequate. Any switches or relays that make or break circuits carrying current cause a sharp voltage change and can induce an equivalent voltage change in nearby sensitive circuits. Sometimes the power supply may induce hum or higher-frequency components into the communication channel.

The inductive or capacitive coupling through which noise is induced may be in the exchange. It may be coupling of adjacent cable pairs that are physically close. The noise generated by relays and switches may travel down wires a long way before reaching the low-level transmission signals and may come from plants in separate buildings. With a computerized digital exchange, most noise associated with switching is eliminated.

External noise can also come from atmospheric sources. Open-wire pairs (almost completely nonexistent now) hanging between telegraph poles can pick up atmospheric static. They are affected by distant lightning flashes and sometimes by contacts with trees or other foreign objects. Sometimes power lines or radar transmitters can cause interference. Radio interference has caused trouble with computer systems installed at airports, sometimes with the transmission of data and sometimes with the computer. Electric trains or electrical machinery sometimes cause periods of severe noise referred to as *electromechanical interference* (EMI).

Impulse noise also originates from within the communication channel itself. This may be caused by circuit faults such as poor-quality soldering and dirty relay contacts and jacks. Nonsoldered twisted joints may cause noise due to changes in temperature and slight movements of the joint. In all these cases, a variation in the contact resistance causes a fluctuation in voltage.

Many circuits carry more than one channel. In such systems there is usually a small amount of cross talk between one channel and another. The parameters of the multiplexing scheme and voice transmission equipment are chosen so that the effect of this is very small, and it is unlikely to affect data. However, an extra strong signal or impulse on one channel will exceed an overload point in the amplifiers and other devices, and various effects of this will be felt in the other channels. The signal that causes the overload may be a noise impulse, or it may result merely from the fact that all the signals being transmitted by the same multiplexing device happen to be at a peak at that instant. The sum of the peaks exceeds the capacity of the channel for distortion-free transmission. This fortuitous adding together of peaks occurs in most multiplex systems, especially during busy periods. Harmonics and modulation products spill into other channels.

Large impulses in any communication system tend to overload the amplifiers, repeaters, and other electronic equipment that they pass through on their journey. If an impulse is very large, it will momentarily render each amplifier or repeater inoperative, and this will tend to prolong the disturbance. It will tend to generate resonant frequencies characteristic of each amplifier and filter in its path. These frequencies are usually close to the maximum frequency the system is designed to handle and so again will tend to add to the disturbance. Amplifiers tend to convert severe amplitude disturbances into frequency disturbances so that they interfere with frequency modulation as well as amplitude modulation.

There are many different means of reducing the effects of impulse noise. First, good screening can be used and careful planning of the circuit paths can be exercised to minimize induction, especially from switching equipment and relays. Second, multiplex systems are designed so that cross talk and peak overloads are minimized. Third, the amplifiers, filters, repeaters, equalizers, and other equipment on the line are designed to lessen the noise effects sufficiently for voice. For data transmission, further equalization can be employed in the form of conditioning. Fourth, the choice of modulation method has an effect on the accuracy of transmission.

Computer system designers cannot change the properties of the line they are given. Choice of the most suitable modem, however, is sometimes in their hands. They can lease a line with "conditioning" to minimize distortion, and they can ensure that the cabling and equipment on the user's premises are located and designed so that noise is not picked up there (see Chap. 4, Sec. 4.7).

Cross talk. Cross talk refers to one channel picking up some of the signal that is traveling on another channel. Occasionally you hear faint fragments of someone else's conversation on the telephone. It occurs between cable pairs carrying separate signals. It occurs in multiplex links in which several channels are transmitted over the same facility. It occurs in microwave links where another antenna picks up a minute reflected portion of the signal for another antenna on the same tower. In the last two cases, the level of cross talk noise is very small because the system is designed with strict criteria for the maximum allowable cross talk.

Often the strongest source of cross talk is induction between separate wire circuits. Any long telephone circuits running parallel to each other will have cross talk coupling unless they are perfectly balanced, which is not the case in practice. Cross talk between wire circuits will increase with increased length of circuit, increased proximity, increased signal strength, and increased signal frequency.

Cross talk may originate in exchanges or switching centers where large numbers of wires run parallel to each other around the exchange. It may originate in a subscriber's building. It can be caused by capacitive as well as inductive coupling.

However, on most systems it is barely greater, and often less, than the level of white noise; so, like white noise, it does not normally interfere with data and is not annoying during speech. Occasionally, due to faults in the exchange, cross talk becomes louder, and it is possible to hear another person's voice on the telephone. There will also be momentary peaks in cross talk which do interfere with data. With these exceptions, however, the level of cross talk is known or measurable and can be treated like white noise for the purposes of selecting transmission parameters so that no interference with data will normally occur.

Intermodulation noise. There are certain undesirable types of data signals which can cause bad cross talk. On a multiplexed channel, many different signals are amplified together, and very slight departures from linearity in the equipment cause *intermodulation* noise. The signals from two independent channels intermodulate each other to form a product that falls into a separate band of frequencies, just as two sound waves may "beat" to form a sound oscillation of a different frequency. The result of this may fall into a band of frequencies reserved for another signal. Such products arising from large numbers of pairs of channels combine to form low-amplitude babble, which adds to the background noise in other channels. However, if one signal were a single frequency, when it modulates a voice signal on another channel, the voice might become clearly audible in a third, independent channel. One telephone user in this case would overhear the conversation of another. Privacy is important on the telephone, so the telephone company attempts to restrict the power of any single-frequency signal to a suitably low level.

The guilty single frequency could arise from data transmission in one of two ways. First, a repetitive code in a data signal could cause it, unless the modem were specifically designed to prevent this. Many data processing machines send repetitive codes to each other as part of their line-control procedures, for example, to keep machines in synchronization while data are not being transmitted. Second, the modem itself, if not designed to avoid doing so, might transmit a single frequency when not transmitting data. Often one can hear this when dialing a computer. When the connection is established, the apparatus at the other end will send a single-frequency *data tone* down the line to tell you that the connection is established. On some modems, this frequency will always be present when there are no data. Such techniques could cause intermodulation cross talk unless the signal strength is restricted. If it is restricted, we are effectively reducing the capacity of the channel.

This problem can be overcome by good modem design. When there are no data, the modem must not transmit a fixed-frequency tone, unless it is of very low amplitude. Further, the data must be randomized so that repetitive data patterns are not converted into a signal in which a single frequency dominates. Modems with these properties are now available and place no code restriction on the user. When they are used, they can be permitted to send a high-strength signal over the carrier network and so not cause undesirable intermodulation cross talk.

This is especially important for transmission over bandwidths greater than a single telephone channel. The power of the signal and the power of the noise are both proportional to the bandwidth. The limitation imposed on a single-frequency transmission is, however, independent of bandwidth. Therefore, with this restriction, the higher the bandwidth, the lower the signal-to-noise ratio will be for such signals.

Echoes.   Echoes on transmission lines are similar to cross talk in their effects on data transmission. Where there is a change in impedance on the transmission line, a signal will be reflected so that it travels back down the line at reduced amplitude, thus forming an echo. The signal-to-echo power ratio can occasionally become less than 15 dB, though it rarely falls below 10 dB. It can, however, be greater than white noise or cross talk.

Echo suppressors are used on long lines. They are not generally of value, however, in the problem of echoes in data transmission. Their action is triggered by the detection of human voice signals, and they are normally disabled when the circuit is used for data. In voice telephony, echoes of the speaker's voice become annoying when the speaker hears them with a time delay measured in tens of milliseconds. In data transmission, delays of a fraction of a millisecond are significant, and it is the listener rather than the talker who is affected by them. Multiple reflections down a 2-wire path, or echoes formed at the junction of a 2-wire and a 4-wire circuit, reach the receiving machine or modem. If they are of sufficient volume they can cause errors in data.

Sudden changes in amplitude.   Sometimes also the amplitude of the signal changes suddenly. This may be caused by faults in amplifiers, unclean contacts having variable resistance, loud, new circuits switched in, maintenance work in progress, or switching to a different transmission path.

These sudden changes can have an effect on certain data transmission systems and could result in the loss or addition of a bit. The effect depends on the type of modem in use.

Line outages.   Occasionally, a communication circuit fails to be operational for a brief period of time. These outages may be caused by faulty exchange equipment, storms, temporary loss of carrier on a multiplex system, or other reasons, producing a brief period of open or short circuit. Often, maintenance work on the lines, repeaters, or exchanges is the cause of brief interruptions. This gives rise to two concerns. First, the data may be damaged by signal losses of a few milliseconds. Second, signal losses of 10 s or more may cause a serious break in system availability.

Radio fading.   Radio links are subject to fading. Many long-distance telephone circuits travel over microwave paths, and fading sometimes occurs. Small fades are compensated for by the radio automatic gain control. Large fades, however, may cause a serious degradation of the signal-to-noise ratio. Heavy rain and snow may cause fading. Violent gales may cause slight movement of the microwave dish. On rare occasions, objects such as helicopters or flocks of birds may come into the transmission path. Sometimes microwave paths have been interfered with by new buildings which may cause microwave reflections. High-frequency radio is subject to many variations and sudden changes that are not found in wire transmission or microwave, such as deep and variable fading.

On local walkie-talkie radio transmission, the signal passes directly from the receiver to the transmitter. Long-distance radio, including ordinary AM radio, relies on reflection by the ionosphere, and the path variations are very much greater. There are many propagation paths, and these suffer large daily and seasonal variations. The different propagation paths interfere with each other and so cause severe amplitude and phase variations. Different frequencies will suffer these effects in different amounts, and the effects constantly change so that they cannot easily be compensated for. Long-distance radio with ionospheric reflection is normally not regarded as good enough for data transmission, although, where it is the only facility available, it is used at a low transmission rate with special modems.

Changes in phase.   The phase of the transmitted signal sometimes changes. Sometimes impulse noise causes both attenuation and phase transients. Transients which affect only the phase are also common, especially on long lines. Sometimes the phase slips and returns and sometimes it slips without returning. Typically, the phase change occurs in less than 1 ms, but sometimes a gradual rotation of phase occurs.

Systematic distortion. The types of noise and distortion discussed above are all *fortuitous*. Those below are *systematic* and can be compensated for partially or completely in the design of the electronics.

In general, as the speed at which data are transmitted increases, the need for uniformity in the transmission characteristics becomes greater. On private leased lines, steps can be taken to ensure that this uniformity measures up to certain standards. When dialing for a connection, however, it is not certain which path the call will take, and there are likely to be certain connections in the network on which the distortion will be high. Loaded cables (exchange cables over 15.6 kf in length), for example, that are not properly terminated or that have discontinuities give very undesirable transmission characteristics. It is possible that a short section of a nonloaded cable can be switched into the link, again increasing its distortion level. Some of the loading coils may be faulty or missing. The proportion of faulty or poor-quality sections in a public network differs from one area to another. Some networks still have a high proportion of old telephone plant in them. Some of the early cables used very heavy loading. This gives bad delay distortion, which is discussed below. In most of the major industrial countries, this old plant is being replaced with equipment having more suitable characteristics for data transmission, but in some areas the replacement rate is slow.

When dialing for a connection on the public network, then, there is a small probability that a line with several nonuniform characteristics might be obtained. This would give a higher proportion of errors than normal, especially if a high transmission speed were being used. Redialing might establish a different route between the two points and avoid the bad segment. Often, however, the error rate is not high enough to indicate to the user at the time that the line is bad. The user may never know that the connection is experiencing 2 or 3 times, or perhaps even 10 times, the normal error rate.

Loss. The loss of signal strength on a circuit is typically about 16 dB. It may vary because of aging equipment, amplifier drift, temperature changes, and other causes. Adjustments are made for changes during routine maintenance.

Attenuation distortion. The attenuation of the transmitted signal is not equal for all frequencies as, ideally, it should be. The increases in attenuation at the edges of the band are indeed what demarcates the bandwidth.

The attenuation of a typical cable pair, within the voice band, is approximately proportional to the square root of the frequency. To compensate for this variation in amplitude and to reduce attenuation, the cable may be loaded by adding inductance at intervals (3 kf from the central office and every 6 kf thereafter for loops over 15.6 kf in standard exchange loops). Similarly, on multiplex systems carrying many voice channels, filters are designed to yield a flat amplitude-frequency curve. However, some variations in amplitude across the band remain, and this is referred to as *amplitude-frequency distortion*. The effect of this is to distort the receivable signal slightly.

Harmonic distortion. Harmonic distortion refers to distortion in which the signal attenuation varies with amplitude. For example, a 1-V signal may be attenuated by one-half while a 5-V signal is attenuated by two-thirds. If a sine wave is transmitted on such a channel, it is flattened at the peaks. Pulse shapes are therefore not reproduced faithfully at the receiver. If harmonic distortion is considerable, the demodulation operation is affected.

The flattening of the sine wave is equivalent to adding harmonics of low amplitude to the transmitted signal. If a sine wave of frequency f is transmitted, harmonic distortions would be equivalent to also transmitting a low-amplitude sine wave of harmonic frequencies $2f$ and $3f$. Thus, a signal with harmonic distortion might have a signal-to-second-harmonic ratio of 25 dB and a signal-to-third-harmonic ratio of 35 dB. Keeping these ratios low is important for high-speed data transmission. Telephone companies attempt to control the harmonic distortion of their circuits.

Delay distortion. The phase of a signal also is not transmitted linearly. The signal is delayed more at some frequencies than at others. This is referred to as *phase-frequency distortion* or *delay distortion*. Some frequencies reach the receiver ahead of others. Signals on wire pairs are propagated with somewhat different speeds at different frequencies.

Delay distortion is a serious form of distortion in data transmission. It has only been corrected to a limited extent on the voice channels that we may wish to send data over because human understanding of speech is not greatly affected by it. The ear is a relatively slow-acting organ. It is normally necessary for a sound to exist for 0.2 s in order to be recognized. If we have delay distortion of 0.05 s, the speech is still intelligible, and normally the distortion is not nearly that great.

If there were no delay distortion, the curve of phase of the received signal plotted against frequency would be a straight line. In reality, it is a curve. Envelope delay is defined as the slope of such curves and may be measured in microseconds. In effect, envelope delay at a given frequency is the delay that would be suffered by a very narrow bandwidth signal transmitted at that frequency. Because there is delay distortion, envelope delay varies with frequency. It is sometimes simply referred to as the delay at a given frequency.

Frequency offset. Signals transmitted over some channels suffer a frequency change. That is, if 1000 cycles per second are sent, 999 or 1001 might be received. This is sometimes caused by the use of multiplex systems for carrying each voice band at a different frequency. The oscillators used for generating the carrier for modulation and demodulation are not precisely at the same frequency. When demodulation occurs, the entire band suffers a frequency change because the carrier used for demodulating does not have exactly the same frequency as the one used for modulating. Frequency shift is sometimes overcome by transmitting the carrier with the signal. A change of about 20 Hz could be permitted without causing unpleasant distortion of the human voice. However, the CCITT (Consultative Committee on International Telegraph and Telephone) recommendation is that frequency offset should be limited to ± 2 Hz per link, and most circuits conform to that. A circuit with five links in tandem might, then, have a frequency shift up to ± 10 Hz. Under faulty conditions, this value may be exceeded for short periods, but is rarely exceeded enough to interfere with data transmission. On the former Bell System, frequency shift on the long-haul carrier system is generally held to less than ± 1 Hz per line section. The overall private line data channel objective is ± 10 Hz, because there are several line sections in tandem.

Bias and characteristic distortion. Repeaters are often used to reconstruct pulses, producing a new, clean, square-edged pulse out of what had become a distorted pulse. Similarly, the output of a modulation system is "sliced" to give square-edged pulses. A form of systematic distortion, referred to as *bias distortion,* can result in all the pulses being lengthened or shortened. If all the 1-bit pulses are lengthened, it is called *positive bias* or *marking bias,* and, if they are shortened, it is called *negative bias* or *spacing bias.* Bias distortion changes sign when the 1 and 0 bits are interchanged or, in other words, does to a 1 what it would do to a 0. Bias distortion may be caused by a decision threshold for the pulse regeneration being set at the wrong value and, thus, can usually be adjusted.

A similar type of systematic distortion is called *characteristic distortion.* Here, the effect is not reversed when the 1 and the 0 are interchanged but is similar for either of them. For example, a single 1 or 0 may be shortened in transmission, whereas a long mark or space representing adjacent 1s or 0s may lengthen. This may be caused by bandwidth restriction or intersymbol interference.

Conditioning. Many of the impairments which affect circuits can be controlled so that they do not exceed certain limits. Some are controlled by telephone company internal practices, which may limit the impairments to levels in the CCITT recommendations. An additional degree of control is sometimes applied to leased voice-grade circuits to improve their properties for data transmission. This is referred to as *conditioning.*

There are two types of conditioning used in North America, each controlling different impairments. C conditioning applies attenuation equalization and delay limits. D conditioning controls harmonic distortion and the signal-to-noise ratio. Many high-speed modems automatically equalize the circuit they operate on, effectively doing the same thing as C conditioning. With such modems, C conditioning does not help performance. Less expensive modems can benefit from C conditioning.

The effects of attenuation distortion and delay distortion are thus reversible. The effects of noise and harmonic distortion cannot be reversed by the modem. Consequently, D conditioning can be used, which specifies limits to the permissible noise-to-signal ratio and harmonic distortion. D conditioning is intended for use with modems that transmit 9600 b/s over voice-grade lines.*

Quantum noise (fiber optics). A different type of noise occurs in fiber-optic communications, referred to as *quantum noise.* This noise occurs in a digital system when the number of detected photons per second is equal to the number of hole-electron pairs generated by incident light. In laser transmission of analog signals, the modulation index must be kept to less than 60 percent or the signals will be destroyed. Generally, a 50 percent modulation will allow some margin of protection.†

---

*James Martin, *Telecommunications and the Computer,* 3d ed., Prentice Hall, Englewood Cliffs, N.J., 1990, pp. 639–657.

†Donald C. Baker, *Local-Area Networks with Fiber-Optic Applications,* Prentice Hall, Englewood Cliffs, N.J., 1986, pp. 56–68.

5.2   Power and Lighting Considerations*

Traditionally, telecommunications services originated at the utility company Central Office. In this environment, 48 V DC was provided to each telephone circuit and allowed sufficient power for efficient voice and data transmission. As technology evolved, many customers migrated to on-premises telephone systems which required other than utility-provided power, thus mandating the provision of ac power at the equipment location. As further innovations developed, more demand for on-premises AC power access has evolved, resulting in inefficient or inadequate branch circuit design.

The National Electric Code (NEC) has attempted to provide guidelines for designing these circuits, but falls short of giving a generic standard because of the variety of equipment available and the potential integration impacts of several diverse systems.

Since clean uninterrupted electrical power is critical to the smooth operation of a communication equipment room, extreme care should be taken in planning power requirements. It is generally the customer's responsibility to provide all electrical connections, and consideration for future growth should be a key part of the design. The power source should be isolated from other power systems in the equipment center. Power specifications can be obtained from the equipment vendor. Assure that the guidelines in the National Electric Code, and any other applicable local codes, are followed. If the nature of the operation requires uninterruptible power, specialized power systems such as motor-generators (MGs) and uninterruptible power systems (UPSs) should be considered. Provide a dedicated ground circuit for maximum protection. Install lightning protection equipment if (1) primary power comes from overhead lines, (2) the utility company installs lightning protectors on the power source, and (3) your area is subject to electrical storms or similar types of power surges. Provide sufficient convenience outlets for other equipment. Make sure you have an emergency power switch installed near the entrance to the equipment room to shut down all power systems in an emergency.

Electrical power, as supplied by the utility companies, is not constant. Greater demands have forced the utilities to modify the ways power is distributed, and these procedures may cause brownouts and blackouts. While these disturbances make up less than 5 percent of the total types of power disturbances, their effect can be disastrous.

---

*Datapro Research Corp., "An Overview of the Telephone Room Environment," MT50-110, Delran, N.J., September 1987, pp. 107–108.

Two other types of voltage problems you must deal with are line noise and voltage fluctuations. Line noise, which constitutes about 90 percent of power problems, typically includes voltage spikes and is usually caused by oscillation disturbances, utility network switching, voltage transients caused by faults or lightning hits, and switching on and off heavy equipment like air conditioners, freezers, elevators, and various types of office equipment. Voltage fluctuations, which occur regularly, are usually caused by transmission line drops due to impedance, intrabuilding drops, power company brownouts, sags/spikes caused by fault-clearing equipment, sudden heavy start-up loads, or delayed action by utility regulating equipment. These items account for 10 to 15 percent of power problems. If line voltage is too high, the equipment could be damaged, and if it is too low, you could lose data and get faulty call processing from your system.

Since these voltage problems occur quite frequently, telephone and data systems have been designed to absorb many of them, within specified limits, before serious damage occurs. Systems are designed to handle steady-state variations of ±6 to ±10 percent, depending on the manufacturer. Transients generally should not exceed +15 percent or −18 percent of the nominal voltage level and should return to a steady-state condition within a half second or less. Verify the specific limitations of your equipment with the vendor well in advance, so you can plan on an appropriate method of maintaining clean power. The following is a list of power conditioning equipment and a brief description of each:

- *Line filters* smooth out certain types of voltage fluctuations at frequencies above 50 kHz.

- *Ultraisolation transformers* attenuate both common mode noise (between the ground wire and current-carrying conductors) and transverse-mode noise (between the current-carrying conductors) for transients ranging from 400 Hz to 5 kHz and voltage spikes from 10 to 100 kHz.

- *Line voltage regulators* eliminate up to +15 percent and −25 percent voltage fluctuations, measured at the outlet.

- *Power line conditioners* eliminate 95 to 99 percent of all line fluctuations as long as the hit lasts only a few milliseconds.

- *Motor-generators* can sustain power for anywhere from 100 to 500 ms ($\frac{1}{2}$ s).

- *Uninterruptible power systems* can protect telephone systems from virtually 100 percent of all potential power problems. UPSs are designed to maintain adequate power for as much as 15 h, which is enough time to wait for power restoration or system shutdown. For users that anticipate prolonged outages, a diesel or gas turbine generator should be substituted for, or added to, the storage batteries.

5.2.1 Power planning considerations

Since the provision of an adequate supply of clean power to the equipment room is required, it is recommended that a thorough analysis of the site's power requirements be undertaken. There are many alternatives to chose from, and you can expect vendors of the different systems to present very convincing arguments in their favor. Before you look at auxiliary equipment, consider these two suggestions: (1) install a dedicated power line to the equipment room; isolating it from other lines could be sufficient to yield clean power, or (2) install a second power line from a different substation to serve as a hot standby in case the primary line fails. Of course, if the entire region has an outage or experiences hits, neither of these suggestions offer sufficient protection.

If you have determined that your area has a history of only occasional irregularities, you probably will not need anything as sophisticated as a line conditioner, MG, or UPS. Filters, voltage regulators, and isolation transformers are fairly inexpensive, require little maintenance, and are generally easy to install.

However, if your area is prone to brownouts and electrical storms and has a high concentration of power-hungry equipment, or if your business will not tolerate even the slightest disruption of service, it is recommended that you investigate line conditioners, MGs, or UPSs. Consider these items during your investigation:

- Voltage levels currently provided
- Maximum length of outage tolerated
- Minimum length of outage expected
- Location of equipment; UPS typically requires a separate room (NEC 645-10 and 11 excepted)
- Maintenance; fewer moving parts generally indicate better mean time between failure (MTBF)
- Availability of spare parts
- Amount of power required to operate system
- Fail-safe capabilities
- Installation cost; delivery times

- System cost, typically measured in dollars per kilovoltampere (kVA)
- The cost to your company of power outages, spikes, and transients
- Provision of static control facilities
- Installation of water/humidity detection equipment
- System expansion
- Fire detection/suppression equipment
- Using an outside contractor/consultant.*

To assist the systems planner in directing the efforts of those involved with the initial design stages of space required for the housing of communication and information equipment, should a detailed study not be available when the information is required, the following guidelines are offered.

5.2.2 Lighting

Adequate lighting fixtures shall be provided to give 50 to 70 fc of light 30 in above the floor in equipment rooms. Adequate emergency lighting shall be provided at all locations.

5.2.3 Grounding

The telephone equipment grounding must be bonded to building and power grounds and/or metallic cold-water pipe. The ground requires a measurement reading of 0 to 5 Ω.

5.2.4 AC power requirement

A minimum of one 4-plex, 120-V, 20-A, isolated ground per equipment location. Add one such outlet for each 16 ft^2 of floor or wall equipment space.

5.2.5 Equipment and power clearance

Electrical codes require adequate clearances between power and communication equipment when installed in joint locations.

A minimum of 4 ft of clear working space must be provided between all the access sides of communication cabinets and any power wiring

---

*Datapro Research Corp., "An Overview of the Telephone Room Environment," MT50-110, Delran, N.J., September 1987, pp. 107–108.

or equipment energized at 750 Vac or less. There is no space requirement on the sides of cabinets not used for access.

A minimum of 6 ft of clear working space must be provided between all the access sides of communication cabinets and any power wiring or equipment energized at between 750 Vand 8700 V.

A minimum of 3 ft of clear working space must be provided between all the accessible sides of communication equipment in areas where power is not involved.

5.3   Main Terminal or Equipment Rooms

The equipment room may be located at the demarcation point [minimum point of presence (MPOP)] or in a separate room. Generally, major on-premises telecommunication and data processing equipment is located in these rooms and is equipped with battery backup or standby power devices, which provide direct power to the primary unit. The general rule is to provide three 120-V duplex, three-pronged, grounded convenience outlets on two walls (six outlets) for the main terminal room and one, 20-A, 120-Vac branch circuit for each system and convenience outlet on the same branch circuit for each anticipated system expansion unit in the equipment room.

5.4   Apparatus and Satellite Closets

These closets shall be equipped with 120-Vac duplex receptacles separately fused at 15 to 20 A. The rule of thumb is to provide one receptacle for each 4000 ft^2 of usable floor area being served from the closet.

5.5   Station Locations

The following tables indicate the approximate power consumption of office automation workstation devices for voice and data station equipment. Consumption rates are shown for other devices to display the impacts of noninformation devices on the station ac power outlet.

Each workstation should be provided with at least 125 percent amperage for the devices expected; the average work location should perform reasonably well on a 10-A ac circuit, plus capacity for accessories. However, circuit breaker capacity will not ensure a continuous source of power, and optional convenience devices can overload the workstation outlet rather quickly. Accordingly, it is recommended that an 8-h standby power supply be provided at locations where data preservation is critical and that optional equipment not be allowed on that circuit. Standby power will not be necessary for single-line centrex service and Electronic Key Telephone System (EKTS) equipped with backup battery supply.

AC wiring for office furnishings is covered in Art. 605 of the National Electrical Code.

5.5.1 Voice telephone equipment

Description	Power consumption, W at 120 Vac	Current required, A
Electronic telephone	50	0.5
Key telephone	70	0.6
Speaker attachment	0	
Auto dialer	10	0.1

5.5.2 Data

Description	Power consumption, W at 120 Vac	Current required, A
Terminal or cathode-ray tube	50	0.4
DTE/DCE*	320	2.6
Printer	240	2.0

*DTE = Data Terminal Equipment; DCE = Data Control Equipment.

5.5.3 Accessories

Description	Power consumption, W at 120 Vac	Current required, A
Solid-state radio	10	0.1
Digital clock	5	0.05
VCR	30	0.2
Fluorescent desk lamp	66	0.6
Coffee maker	900	7.6
Microwave oven	1300	10.8
Hair dryer	1350	11.3

6

Telecommunication Systems and Services

6.1 Private Languages and the Sale of Telecommunication Services

As indicated in Chap. 1, the telecommunication industry has gone to great lengths to protect its products and services from competition. Through custom and habit, the industry continues to be a mystery to noninsiders, even after divestiture. Accordingly, product identification has become an art, whereby a vendor gives a product a different identity from similar competitive products and often deemphasizes certain attributes where they are proclaimed by the competition.

Today, with the Federal Communications Commission (FCC) and local public utility commissions (PUCs) relaxing their prohibition on allowing intelligence in the network by the Bell operating companies (BOCs), the embedded local operating utility is capable of providing practically every feature available on customer premises equipment (CPE). With the *fiber superhighway* (formerly known as *fiber rings*) becoming available in metropolitan areas and extending into the urban distribution facilities, the "information age technology" [formerly integrated services digital network (ISDN)] is practically on every subscriber's doorstep.

After reading the previous paragraph, it should be apparent that the industry is constantly renaming the same old product. Customers are led to believe that they are in the middle of an information "revolution," when, in fact, it is a technology evolution. Progress extends along a rather predictable arithmetical curve and rarely proceeds in quantum leaps.

In addition, there is a strong tendency to develop products for the pure joy of doing something that previously could not be accomplished. In the industry they call this "answering a question that hasn't been asked." When this happens, the developer attempts to create a demand. If a demand cannot be created, the product, if deemed marketable, will be "bundled" into a system and promoted as a "value-added" feature. Usually neither the buyer or seller of the product has the slightest idea of what they are going to do with it but are made to feel that someday it will be absolutely essential. This technique of delivering features is referred to as *pushing the envelope* and the inclusion of *embedded obsolete* features has contributed to the less than 15 percent feature usage of most CPE telecommunication systems. In most circumstances, either the user is not aware of the feature package in the equipment or the systems are just too complex for the average phone user to bother with.

To understand how these features and systems have dominated the telecommunication industry, it is first necessary to understand that account representatives (for BOCs) and information technology sales engineers (for CPE) are the best trained and best compensated individuals in their field. The industry is highly competitive and turnover is rapid. Burnout usually occurs within 5 years. These people live in a fast-paced, highly mobile environment. They are usually paid on a base salary plus commission schedule and are given increasingly higher market share goals to stay competitive.

Depending on the system or service involved, markup ranges from 200 percent to as high as 15 times cost. Occasionally, predatory pricing takes over whereby, to gain market share, a competitor will value a service or feature below cost to gain a toehold for future profitability.

To make their product appear better to the subscriber, terminology is modified periodically. Acronyms are abundant and it is considered to be intellectually stimulating to let go with a stream of "telco jargon" synonyms in the company of friends and customers. Unfortunately, it is extremely confusing to the unenlightened, and embarrassment is the punishment for not being part of the elitist insider group. The goal of such nonsense is either to cripple the opponent in the management information services (MIS) field or overwhelm the client. I have used it myself, and it works almost every time.

When confronted with such tactics, simply ask the person to explain what you do not understand. Telephone system feature groups are relatively understandable when broken down into the functions they serve. Before a potential customer initiates a "feature hunt," it is best to survey the user group to determine what they expect and need the telecommunication system to do for them, before committing to a particular feature package. You may find out, as many of my clients did, that dial tone, fax, and modem capabilities are all that is required.

6.2 Transmission Methods*

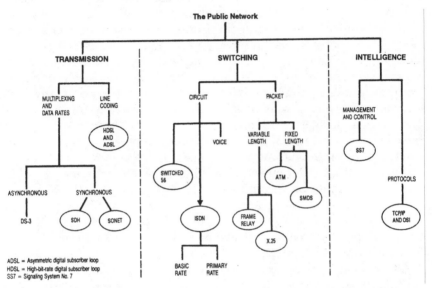

Figure 6.1 The three dimensions of public networks. The technologies that make up the world's public networks can be divided into three basic classifications: transmission facilities, switching technology, and intelligence features. (*Data Communications, December 1992.*)

In order to gain an understanding of how voice and data signals are processed, it will be necessary to view the evolution of telecommunication transmission.

Basically, there are two ways in which information of any type can be transmitted over telecommunication mediums: analog or digital. Analog means that the amplitude of the transmitted signal varies over a continuous range. Oscillating signals are normally transmitted and the frequency of the oscillation can also vary over a continuous range. Both the sound you hear and the light you see are analog signals spread over a range of frequencies. Sound, as any hi-fi enthusi-

---

*Datapro Research Corp., "An Overview of the Telephone Room Environment," MT50-110, Delran, N.J., September 1987, pp. 107–108.

ast knows, consists of a spread of frequencies from about 30 to 15,000 Hz (cycles per second), or, for people with very good ears, 20,000 Hz.

Sound cannot be heard below 30 Hz, and it cannot be heard above 20,000 Hz. If it is desired to transmit high-fidelity music along a cable to a given location, a continuous range of frequencies from 30 to 20,000 Hz would be sent. The current on the cable would vary continuously in the same way as air pressure does in the sound you hear.

The telephone company, in the interests of economy, transmits a range of frequencies that may vary from about 300 to 3400 Hz only. This is enough to make a person's voice recognizable and intelligible. When telephone signals travel over lengthy channels, they are packed together, or multiplexed, so that one channel can carry as many such signals as possible. To do this, a voice signal may be raised in frequency from 300–3100 Hz to 60,300–63,100 Hz. Another voice might have been raised to 64,300–67,100 Hz, for example. In this way, they can travel together without interfering with one another, but both are still transmitted in an analog form, that is, as a continuous signal in a continuous range of frequencies.

Digital transmission means that a stream of on/off pulses is sent like data traveling in computer circuits. The pulses are referred to as *bits.* It is possible today to transmit at a speed of many millions or even billions of bits per second.

A transmission path can be designed to carry either analog or digital signals. This fact applies to all types of transmission paths: wire pairs, high-capacity coaxial cables, optical fibers, microwave radio links, satellites, and other transmission media.

Until recently, most telephone channels into the home were analog channels, capable of transmitting a certain range of frequencies. To send computer data over them, the digital bit stream has to be converted into an analog signal using a special device known as a *modem,* a contraction of the words modulate and demodulate. This converts the data into a continuous range of frequencies—the same range as the telephone voice. In this way, it is possible to use any of the world's analog channels for sending digital data.

On the other hand, where digital channels have been constructed, it is possible to transmit the human voice over them by converting it into a digital form, with a device called a *codec* (a contraction of the words *code* and *decode*). Similarly, any analog signal can be digitized for transmission in this manner. It is possible to convert hi-fi music, television pictures, temperature readings, the output of a copying machine, or any other analog signal into a bit stream. The bit rate needed is dependent on the bandwidth, or range of frequencies, of the analog signal, as well as on the number of different amplitude levels desired.

Most of the world's telephone plant grew up using analog transmission. Much of this, however, has been replaced with both digital switching and transmission equipment. This is an enormous task, and most public telecommunications will not be fully digital until the late 1990s. The conversion from analog to digital started in that portion of the network that carries traffic between major cities, known as the *trunk network*. As the digitization process continues, the next layer of the network to be digitized is the switching offices of major cities, then out into the rural areas, until finally all subscribers are connected by a digital circuit to the network. At the same time, many corporations are converting their private telecommunication networks to digital. For a large public network operator such as a BOC this process represents massive investment, and, therefore, those areas that produce the highest revenue are converted first. Small telecommunication carriers that offer a more limited service are often able to convert their networks to digital operation faster. The net result is that, as both switching and transmission networks are digitized, a wide diversity of digital services becomes available, first to large business users and then to domestic subscribers.

6.2.1 Nonvoice traffic

For half a century, telephone technology has dominated telecommunications. There have been 1000 times as many telephone subscribers as other types of telecommunication users. Consequent economics of scale have dictated that telegraph and data traffic should be converted to a form in which they can travel over the telephone system.

Since the 1970s new types of common carriers have emerged and built their own nontelephone networks. Separate data transmission networks for computer users are available in all major industrial countries. Some of these operate by attaching special equipment to the telephone networks. Others employ new transmission networks, physically separate from the telephone networks.

It cannot be stressed too strongly that computer users have fundamentally different characteristics and requirements than users of plain old telephone service (POTS). A particularly important difference is the peak-to-average transmission rate requirement, which is 1 in telephone conversations but can be 1000 or higher in human-computer dialogs.

6.2.2 Telephone connections

A subscriber who picks up the telephone and hears a dial tone is in contact with the local central office directly, without any trunk being involved. If one subscriber dials another subscriber in a different city, the central office will switch the subscriber loop to a toll-connecting trunk, thus connecting the call to a toll office. The toll offices are interconnected with intertoll trunks, and this toll network establishes the connection between towns. There may be several intermediate switching offices in the path that is set up between the two toll offices. There are about 2000 toll and intermediate switching offices and about 25,000 central offices in the toll network of the United States. About 100 million local loops are connected to the central offices.

The long-distance trunks are all four-wire or equivalent. The term four-wire implies that there are two wires carrying the signals in one direction and two wires carrying them in the other direction. There are many repeaters (amplifiers) on these circuits, amplifying the entire group of lines in both directions. In reality the circuit may not consist of wire pairs but of higher-capacity transmission mediums; however, the historical term *four-wire* is still used and implies separate circuits in each direction.

In normal telephone service, the local loops are two-wire circuits, on which a single telephone call can be transmitted in both directions over the same pair of wires. The toll-connecting trunk can be either a two-wire or four-wire circuit. When a two-wire line is connected to a four-wire line, a special circuit is required to join them. For special purposes, a four-wire local loop can be used. Some data transmission machines require a four-wire rather than a two-wire connection.

6.2.3 Simplex, half-duplex, and full-duplex lines

Data transmission lines are classified as simplex, half-duplex, and full-duplex. In North America, these terms have the following meanings:

Simplex lines transmit in one direction only.

Half-duplex lines can transmit in either direction but in only one direction at once.

Full-duplex lines transmit in both directions simultaneously. One full-duplex line is thus equivalent to two simplex or half-duplex lines used in opposite directions.

The above are meanings in current usage throughout most of the world's computer industry. Unfortunately, however, the International Telecommunications Union (ITU) defines the first two terms differently, as follows:

Simplex (circuit): a circuit permitting the transmission of signals in either direction but not simultaneously.

Half-duplex (circuit): a circuit designed for duplex operations but which, because of the nature of the terminal equipment, can be operated alternately only.

Simplex and half-duplex are thus used differently by European telecommunication engineers and computer manufacturers (especially American ones) using the same facilities. The ITU and Consultative Committee on International Telegraph and Telephone (CCITT) are not consistent and sometimes use the North American wording.

Simplex lines (American meaning) are not generally used in data transmission because even if the data are being sent in only one direction, control signals are normally sent back to the transmitting machine to tell it that the receiving machine is ready or is receiving the data correctly. Commonly, error signals (positive or negative acknowledgment) are sent back so that there can be retransmission of messages damaged by communication line errors. Some data transmission links use half-duplex lines. This allows control signals to be sent and two-way conversational transmission to occur.

A four-wire circuit can always be used in full-duplex fashion (if half-duplex terminals are connected to it). A two-wire circuit is often used in half-duplex fashion. A two-wire circuit can be used in a full-duplex fashion if its amplifiers are designated to permit simultaneous transmission in both ways. Two-wire local loops can be used for full-duplex transmission. To do so, one needs a special modem that splits the band of frequencies into two parts, each for transmitting in one direction. The maximum speed available would be less than with a similar four-wire line. Sometimes the frequency band on a two-wire line is split to give greater speed in one direction than in the other. In a videotex system, for example, a specially adapted television set connected to a telephone line would send data to a remote computer at a speed of 1200 b/s, and the computer would respond at a speed of 75 b/s. The specifications of modems that connect computers and other data devices to the line state whether they operate in a half-duplex or full-duplex manner and whether they need a four-wire or a two-wire circuit.

Much computer-to-computer communication today is interactive, with messages passing both ways at once; therefore, full-duplex lines are normally used. The terms *half-duplex* and *full-duplex* are normally used when referring to computers and other devices connected to an analog telephone circuit using a modem. A computer or other device connected to a digital circuit does not require a modem, as the data output is in a digital format. A device connected to a digital circuit will normally always operate at full-duplex, 64,000 b/s passing both ways simultaneously.*

6.3 Public Network Services†

THE FULL SERVICE NETWORK
Evolution of services and enabling technologies

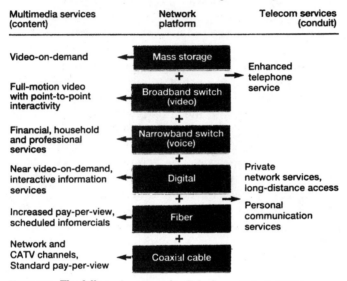

Figure 6.2 The full-service network. (*Telephony, Feb. 1, 1993.*)

6.3.1 An overview of centrex services

A few years ago centrex was seen as a product in obsolescence because of advances in PBX equipment functions. Today, centrex service providers offer a viable product with sophisticated PBX-like features. The existing installed base has been narrowed to the health, education, government, manufacturing, insurance, and banking industries. Centrex systems are concentrated most heavily in government agencies. Table 6.1 lists important centrex features.

---

*Datapro Research Corp., "An Overview of the Telephone Room Environment," MT50-110, Delran, N.J., September 1987, pp. 107–108.
†Datapro Research Corp., "An Overview of Centrex Service," MT20-200, Delran, N.J., November 1990, pp. 101–106.

TABLE 6.1   Centrex Features by Generic Software

Features	1AESS			5ESS					DMS100				
	All	1AE8A	1AE9	All	5E2	5E3	5E4	5E5	All	BCS19	BCS20	BCS21	BCS22
Abbreviated dialing	•			•					•				
Account codes	•					•			•				
Add-on conference—all calls	•			•					•				
Add-on consultation hold-incoming only	NP			•					•				
Advanced private line termination	•			•					•				
Attendant busy line/trunk verification*	•						•		•				
Attendant call-through test capability*	•							•	•				
Attendant camp-on*	•			•					•				
Attendant controlled conference	•			•					•				
Attendant control of facilities	•			•					•				
Attendant delay announcements	•			•					•				
Attendant direct station selection with busy lamp field (DSS/BLF)*	•			•					•				
Attendant ID on incoming calls*	•			•					•				
Attendant through dialing*	•			•					•				
Attendant transfer of incoming calls	•			•					•				
Authorization codes	•					•			•				
Automatic callback calling	•				•				•				
Automatic call distribution								5E6/7				•	BCS26
Automatic circuit assurance	•						•						
Automatic identified outward dialing	•						•		•				
Automatic message link	•						•						
Automatic route selection (ARS)—basic	•			•					•				
Automatic route selection (ARS)—deluxe	•			•					•				
Autovon access	•			•					•				
Basic private line termination	•			•					•				
Business group line	•			•					•				
Call diversion	•			•					•				
Call forwarding—busy line	•			•					•				
Call forwarding—don't answer	•			•					•				
Call forwarding—don't answer—incoming only	•			•					•				
Call forwarding—don't answer—with variable timing	•			•					•				
Call forwarding—incoming only	•			•					NP				
Call forwarding—intragroup only	•			•					•				
Call forwarding over private facilities	•			•				•	•				
Call forwarding—variable	•			•					•				
Call hold	•								•				
Call park	NA			NA					•				
Call pickup	•			•					•				
Call transfer attendant	•								•				
Call transfer—individual—all calls	•			•					•				

All of these features may not be available in some areas even when the serving office is equipped with the required generic program. Feature availability is based on tariff approval and individual telco policies.

*Attendant console required.

**Capability available with message link feature.

NA—Not available.

NP—Not planned.

TABLE 6.1   Centrex Features by Generic Software (Continued)

Features	1AESS			5ESS					DMS100				
	All	1AE8A	1AE9	All	5E2	5E3	5E4	5E5	All	BCS19	BCS20	BCS21	BCS22
Call transfer—individual—incoming only	•			•					•				
Call transfer—internal only	•			•					•				
Call waiting—incoming only	•			•					•				
Call waiting—lamps for attendant*	•			•					•				
Call waiting—originating	•			•					•				
Call waiting—terminating	•			•					•				
Cancel call waiting	•	•		•					•				
CCSA access	•								•				BCS23
Centralized attendant service	•					•			•				
Centralized station message detail recording	•							•	•				BCS29
Central office local area network	•				•				•				
Centrex complex	•				•				•				
Centrex customer change feature	•				•				•				
Centrex system control	•				•				•		•		
Circuit switched digital capability	NA								•				
56K bps	NA						•		•				
64K bps	NA						•		•				•
Circular hunting	•				•				•				
Citywide (area wide) Centrex			1AE10					•	•				BCS29
Code calling	•				•				•				
Code restriction	•				•				•				
Common control switching arrangement	•	•			•				•				
Consultation hold—all calls	•				•				•				
Consultation hold—incoming	NP				•				•				
Critical interdigital timing	•				•				•				BCS26
Customer access treatment (CAT) code restrictions	•				•				•				
Customer administration and control	•					•							BCS23
Customer changeable—speed calling	•				•				•				
Customer dialed account recording	•					•							
Customer station rearrangement (facilities)	•				•				•				•
Customer station rearrangement (lines/features)	•				•				•				
Customer traffic record feature	•					•							BCS23
Delay announcements	•								•				
Deluxe queuing	•						•		•				
Dial access to private facilities	•				•				•				
Dial call waiting	•				•				•				
Dial dictation access	•				•				•				
Dial station conference	•				•				•				
Dial transfer to tandem tie lines	•				•				•				
Direct inward dialing (DID)	•				•				•				
Direct outward dialing (DOD)	•				•				•				
Direct trunk group selection*	•				•				•				
Directed call pickup with barge-in	•				•				•				
Directed call pickup—non-barge-in	•				•				•				
Distinctive ringing and distinctive call waiting tone	•				•				•				
Electronic directory**	•				•				•				
Electronic set interface	NA						•		•				•
EPSCS access	•				•				•				

All of these features may not be available in some areas even when the serving office is equipped with the required generic program. Feature availability is based on tariff approval and individual telco policies.

*Attendant console required.

**Capability available with message link feature.

NA—Not available.

NP—Not planned.

TABLE 6.1   Centrex Features by Generic Software (Continued)

Features	1AESS			5ESS					DMS100				
	All	1AE8A	1AE9	All	5E2	5E3	5E4	5E5	All	BCS19	BCS20	BCS21	BCS22
ETS/ETN	•					•			•				
ETS access	•			•					•				
Expanded ETS dialing plan (EEDP)	•					•			NA				
Facilities administration and control	•					•							BCS23
Facilities assurance reports	•					•			•				BCS23
Facilities restriction level	•			•									
Flexible incoming call restriction	•							5E6	•				
Flexible route selection	•			•					•				
Fully restricted station line	•			•					•				
Group numbering plan	•			•					•				
Individual billing of directory number (IBDN)—for WATS	•			•					•				
Intercom dialing	•			•					•				
In-WATS calling	•								•				
In-WATS—simulated facilities group	•			•									BCS23 BCS26
Local area signaling services			•					•	•				
Loudspeaker paging	•			•					•				
Main-satellite service	•								•				
Meet-me conferencing	NA							•	•				
Message desk**	•						•		•				
Message detail recording	•					•			•				
Message link	•						•		•				
Modemless intraswitched data	•			•					•				
Multiline variety package	•			•					•				
Multiple position hunt with queuing	•							•	•				
Music-on-hold	•						•	•	•				
Music on queue	•							•	•				
Night service	•			•					•				
Outgoing trunk queuing on WATS	•				•				•				
OutWATS	•			•					•				
OutWATS—simulated facility group	•			•					•				
Power failure service	•		•	•									
Privacy/data protection	•			•				•	•				
Radio paging access	•			•				•	•				
Released link operation	•								•				
Remote activation of call forwarding			•	•				•	•				BCS23
Remote call forwarding	•			•					•				
Remote trunk verification by station	•							•	•				
Satellite attendant transfer	•			•					•				
Selective customer control of facilities	•						•					•	
Semirestricted (originating and terminating)	•			•								•	
Simplified message desk interface	•						•		•				
Simulated facility groups for in, out, and no intercom calls	•			•					•				
Simultaneous voice/data	•			•					•				
Six-way conferencing	•			•					•				
Special intercept announcements	•			•					•				
Speed calling	•			•					•				
Speed calling 10XXX	•		•	•									BCS24
Station call-through test	•							•					
Station message detail recording	•					•			•				

All of these features may not be available in some areas even when the serving office is equipped with the required generic program. Feature availability is based on tariff approval and individual telco policies.

*Attendant console required.

**Capability available with message link feature.

NA—Not available.

NP—Not planned.

TABLE 6.1   Centrex Features by Generic Software (Continued)

Features	1AESS			5ESS					DMS100				
	All	1AE8A	1AE9	All	5E2	5E3	5E4	5E5	All	BCS19	BCS20	BCS21	BCS22
Station message detail recording to premises—ETS	•				•				•				
Station message detail recording to premises—non-ETS	NA					•			•				
Switched modem pooling			•			•			•				
Tandem tie trunks	•				•				•				
Telecommunications enhanced real estate	•				•				•				
Three-way calling—all calls	•				•				•				
Three-way calling—incoming only	•				•				•				
Tie trunk access	•				•				•				
Time of day routing	•				•				•				
Toll diversion to attendant	•				•				•				
Traffic data to customer	•						•						BCS23
Traveling class marks	•					•			•				
Trunk answer any station	•				•				•				
Trunk dial transfer	•				•				•				
Trunk group busy indication*	•				•				•				
Uniform call distribution	•						•		•				
Uniform call distribution with queuing	•						•		•				
Uniform numbering	•					•			•				
Voice store-and-forward**	•							•	•				
WATS administration data	•		•						•				

All of these features may not be available in some areas even when the serving office is equipped with the required generic program. Feature availability is based on tariff approval and individual telco policies.

*Attendant console required.
**Capability available with message link feature.
NA—Not available.
NP—Not planned.

Many users stay with centrex because of the time and money required to maintain a PBX. The top three suppliers of centrex service are Bell Atlantic, Ameritech, and Pacific Telesis. BellSouth and Nynex are next. Regional Bell Holding Companies (RBHCs) comprise 93 percent of the market, GTE 2.4 percent, and United Telephone and Cincinnati Bell, 4.5 percent. Many firms upgrade to centrex because of its sophisticated features. The most popular enhanced features are:

- Customer moves and changes
- SMDR/CDR (call accounting)
- Voice mail
- Automatic call distribution (ACD)
- Integrated services digital network (ISDN)

- Automatic route selection (ARS)
- Uniform call directing (UCD)

Centrex is an excellent network switch for large multinode network users. It can function as an end point in a network or as a tandem-switching point within a very large network. Centrex switches operate in complex networks such as common-control switching arrangements (CCSA), electronic tandem switching (ETS), and the Federal Telecommunications System (FTS). It is an extremely flexible product that easily handles the most complex network configurations.

6.3.2 Digital switching systems

AT&T's 5ESS and Northern Telecom's DMS-100 are the two digital switching systems most widely used by the Bell operating companies. These digital switches provide computer-controlled time-division switching and support a distributed architecture that uses a host module and many microprocessor-controlled switching modules. Digital central offices can handle direct interface of digital carrier systems and T1 carrier to multiplex 24 digitized voice channels over two-pair wire at 1.544 Mb/s.

Digital centrex service, in conjunction with customer premises equipment, provides other features including protocol conversion, modem pooling, and electronic directory. While centrex supports standard trunking facilities, such as wide-area telephone service (WATS), foreign exchange (FX), tie lines, interexchange carrier terminations, and other specialized products, flexible route selection and automatic route selection features provide the end user with a least-cost routing function, allowing optimal network use. Digital switching systems are the basis for ISDN.

6.3.3 ISDN

As generic upgrades to these digital switches become available, users can select increased centrex capabilities such as access to ISDN features on an analog 1AESS switching system equipped with a remote module and either a T1/DS1 or DS3 (digital carrier types) interface from a 5E4ESS (digital switch) host, depending on the distance from the 5ESS central office (CO). Northern Telecom has approximately 55 ISDN sites throughout the United States. ISDN enables users to use the same interface for different types of terminals and applications; to transport terminal equipment from one location to another; and to connect private networks, specialized services, exchange carrier, and interexchange carrier facilities. It greatly reduces call setup time, lowering end-user expenses associated with long-distance calls and facsimile transmission. ISDN group IV facsimile machines can transmit a single document in seconds.

In addition, ISDN will provide a standard signaling arrangement for all services, enabling manufacturers to standardize hardware. All of these enhancements are the direct result of ISDN protocol applications as defined by the CCITT such as Q.931, CCS7 call control ISDN user part (ISUP), and transaction capabilities application part (TCAP).

6.3.4 Data communications

Centrex was never thought of as a viable data communication product until Bell Atlantic introduced the central office–local-area network (CO-LAN) feature in 1985. Either an analog or digital CO can serve CO-LAN users. In the Bell Atlantic region, for instance, a stand-alone data switch located in the local CO provides CO-LAN service. Its modular design provides for maximum port utilization and growth capacity. The data switch connects to customer premises data terminals and a host computer via private line facilities and local access loops. The data switch supports RS-232C and RS-449 interfaces and is protocol-transparent to the user. It handles asynchronous and synchronous transmission speeds ranging from 1.2 kb/s to 19.2 and 56 kb/s, respectively. Some of the other RBHCs provide CO-based LAN service in a similar CO configuration.

The BOCs made CO-LAN–type services available primarily to lower the end-user cost for data transmission, provide for centrex additions and rearrangements, and provide multiple lines, as well as to minimize customer floor space, power, and environmental requirements.

6.3.5 Features

SMDR. One of the most popular basic feature enhancements the BOCs made to attract centrex service subscribers is station message detail recording to premises (SMDR-P). SMDR-P provides users with length of calls, destination of calls, and a variety of other call record data. This feature is especially useful in organizations where individual department billing takes place.

Voice messaging. Voice messaging in a centrex environment provides complete telephone message coverage of a subscriber's unanswered calls. The voice mail system can be located at the customer's premises or in the central office. The centrex CPE voice mail packages today support a wide range of voice ports and voice storage configurations.

Station rearrangement. Centrex station rearrangement (CSR) is tariffed under different names by each of the seven RBHCs, but works similarly. The function provides direct access into the local telephone company's (telco's) database and allows changes and verifications to stations within the centrex system. The user can activate and deactivate features and change particular service options on a per-line basis; move line telephone numbers; and search for all lines that do or do not have particular features, service options, or assignments, and display the information for one or more lines. The usual CPE hardware requirements for centrex station rearrangement include terminals that conform to RS-232C interface requirements, operate at 300 b/s, operate in full-duplex mode, use even parity, are able to input data in uppercase, and provide hard-copy output. CSR is password-protected, and access is over a dedicated dial-up link.

Uniform call distribution. The uniform call distribution feature allows calls coming in on a group of lines to be assigned stations as smoothly as possible so that a higher number of calls are completed. Incoming calls that are holding for the longest time are usually queued for service first.

Automatic route selection. Automatic route selection in a centrex environment provides automatic routing of outgoing calls over alternative customer facilities based on the dialed long-distance number. The station user dials either a network access code or a special ARS access code followed by the number. Users can program their phone calls to be rerouted to another destination or to be held in a queue until their line is free.

The fiercely competitive marketing of centrex services versus the CPE PBX/hybrid/key-telephone system gives the end user more bargaining leverage and creates a lucrative revenue source for manufacturers. Centrex service enhancements for the small and large business user include CO-based LAN, message desk features, SMDR, access to all other features provided by the particular telco serving office, and attractive rate packages.

Responding to changing communication needs, many vendors offer voice mail systems, ACD, and key telephone equipment that operate compatibly with and enhance centrex service features. The key telephone equipment ranges from electronic multibutton telephones to complete stand-alone CPE systems that operate in conjunction with centrex.* We will now review these systems.

6.4   Premises Equipment*

One of the advantages for end users resulting from the enactment of Computer Inquiry II and divestiture is the freedom of choice, not only in equipment, but also in service and equipment suppliers. One of the biggest decisions an organization now has to make is between choosing centrex service or a PBX for handling communication functions. In the present environment, this might include both voice and data services. In this section, we will discuss the key items to be considered in making a decision between centrex and a PBX.

When centrex service was introduced in the early 1960s, the targeted market consisted of customers requiring very large end-user systems. Currently, centrex is available to users requiring anywhere from one to thousands of lines, and enhanced features are available to small and large system users. Centrex service is marketed by the nation's 22 Bell operating companies under various names.

PBX systems offer the latest digital technology and an array of features that should satisfy the requirements of any business. The real issue in choosing a PBX or centrex is between owning or leasing a switch located on your premises, using your floor space, or leasing the use of a switch located in a telephone company building. Station equipment, wiring, installation, and maintenance will not be covered because they will be generally equivalent in either system. As centrex service has resurged in the marketplace since divestiture, manufacturers seized the opportunity and are providing station equipment designed specifically for use with centrex service. The result is that the same variety of station equipment available for PBXs is also available for centrex service.

When choosing between centrex and a PBX, organizations should compare the two on many different points before making a decision. Prospective users of either alternative should note that there are advantages and disadvantages in each choice, and the decision

---

*Datapro Research Corp., "An Overview of Centrex Service," MT20-200, Delran, N.J., November 1990, pp. 101–106.

should be made on the basis of individual application. In large organizations with wide-ranging communication requirements, it may be best to devise a plan that would take advantage of both centrex and PBXs. Studies indicate that evaluating either alternative based on user satisfaction does not provide a clear decision criterion or solution.

6.4.1 Capital investment and recurring costs

The purchase of a PBX system requires a capital investment that would not have to be made if you were leasing a system, and with the removal of the investment tax credit, leasing has become more attractive. The purchase of a PBX also requires the establishment of a depreciation account in order to replace the equipment at the end of its economic life. Depreciation and equipment replacement on centrex service are the responsibility of the telco, and the end user is relieved of these concerns. A typical example of how most large corporations are depreciating their PBX equipment is as follows: 15 percent in the first year, 22 percent in the second year, and 21 percent over the next three years.

Cost of floor space, energy, and insurance. Valuable floor space on your premises must be provided for a PBX, as well as the required environmental conditions and insurance to protect against loss. If you cannot or choose not to put your telephone equipment in an office area, you will need a separate equipment room to house the equipment. Some of the advantages of an equipment room are controlled environment, security, dedicated space for expansion, and prevention of accidents. Depending on the size of the PBX and its environmental specifications, you may have to plan on a separate air-conditioning system for the telephone room, as the temperature and overall environmental conditions must be maintained year round.

Since clean, uninterrupted electrical power is critical to the smooth operation of a telephone room, extreme care must be taken when planning power requirements. It is generally the customer's responsibility to provide all electrical connections, and consideration for future growth is a key part of telephone room design. The power source should be isolated from other power systems in the center. Power specifications can be obtained from the equipment vendor.

With centrex service, the expenses for maintaining an equipment room are the responsibility of the telephone company. Electrical power failures can shut down CPE systems, leaving the user with only trunk lines to predesignated telephones (power failure feature). COs, on the other hand, are equipped with engine-powered generators that can supply the CO operating power requirements for extended periods of time. Users with centrex service may lose their building power, with the loss of lighting and air-conditioning, but the centrex lines will be working.

Mileage charges. The distance that your premises are located from the CO providing your telephone service is an important consideration in some areas, because mileage charges for every line could be involved. Mileage charges might apply only to the trunk lines required for a PBX, but could apply to every line on a centrex system. Centrex may not be available in all areas, and even where available, some features may not be available, depending on the vintage of the telco switch and—in the case of an analog or digital electronic switching system (ESS)—the vintage of the software feature package known as the *generic program*. A PBX switch can provide all of its features no matter where it is geographically located. However, the latest vintages of ESSs provide centrex features fully competitive with the latest PBX equipment. Several manufacturers provide modular CO switches that supply centrex service through a module located on the customer premises, but consent decree restrictions have prohibited BOCs from providing centrex service and CPE as a total package (this is changing).

Subscriber line charge (SLC). The SLC will vary in its application and amount from state to state. Some state tariffs list the SLC as a separate rate element applied per centrex line; others apply the SLC on a trunk-equivalency-per-station-line basis; and other tariffs do not list the SLC, but bundle it in with other rate elements. A PBX system will incur the SLC only on trunks from the local CO.

Maintenance. A centrex switch is maintained by the telco on a 24-h basis, with the cost included in the various centrex rates. The end user is responsible for the maintenance of a PBX, except if the PBX is leased with maintenance included. A lease for CPE may or may not include maintenance, and if included, the maintenance hours may be restricted on both a clock and calendar basis. A proposed lease for CPE should be carefully studied to determine whether maintenance will be provided and what limitations apply. Software updates on a centrex switch are automatically performed by the telephone company, while software updates on a PBX may be the responsibility of the end user, depending on the provisions in the lease. A time factor should be considered here, as there may be a delay in the application of software updates on a centrex switch because of its size and complexity and the schedule of the telco construction budget. Telco plant modernization projects usually depend on the location of the CO and the number of centrex customer lines being served by the switch. PBX software updates can be performed as they become available. Total system shutdowns can and do happen on CPE systems, but centrex service has an excellent record and total system shutdowns are rare.

Price comparison. The local telco is the sole supplier of centrex service. There are numerous vendors that provide PBX systems, and a range of pricing on PBXs can be obtained. When doing a comparison between centrex and PBX, use switching equipment prices only and exclude station equipment and wiring, as these costs should be virtually the same in either system. Assuming the purchase of a PBX, the total price of the PBX equipment (excluding station equipment) can be calculated to equal $\$X$ per station line. The total monthly price of the centrex service (excluding station equipment) for an equivalent system is used for comparison.

The cost of ongoing charges (number changes, restriction level changes, etc.) to your telephone system to meet the needs of your business should be carefully considered. If you must contract to have these changes done, they can be costly; however, features available on both centrex and CPE systems allow the end user to make the changes.

6.4.2 Regulatory considerations

CPE is completely deregulated and only needs FCC registration by the manufacturer to be legal for connection to the telephone network. In some states where centrex service is regulated, the telephone company may offer *rate stability* on centrex service, but the offer guarantees only that the BOC will not raise prices and leaves open the possibility of rates being increased by order of the state commission. In many states the telco's single largest centrex customer is the state government. Any political influence this might have on the regulatory agency would seem to benefit centrex customers in general, unless the state has made separate tariff agreements with the utility.

6.4.3 Technology

Staying on the cutting edge of technology is of vital concern to any business today. With CPE under your direct control, you can implement upgrades and modifications as they become available. Although the BOCs consider centrex their flagship service and are vigorously competing in the centrex/PBX market, centrex upgrades and modifications may take more time than the end user can tolerate. Equipment obsolescence is a factor to be considered by the end user with CPE, while end users having centrex service are relieved of this concern, as the BOCs routinely upgrade their switches, albeit perhaps not as quickly as some customers wish.

Data communication requirements can be provided by either system. Centrex offers CO-LAN, which provides a painless way to satisfy your low-speed LAN requirements, provided the transmission speeds are sufficient for your needs. It appears that ISDN may improve on the speed situation, where available.

6.4.4 An overview of key and hybrid systems

The key and hybrid system market consists of business telephone users in small- to medium-sized organizations (under 100 lines). Large organizations use key and hybrid systems in branch offices and for departmental applications, often behind PBX and centrex systems. The market has grown steadily since the AT&T divestiture in 1984, when users began to seek economical alternatives to the electromechanical equipment they had been leasing from AT&T.

The key systems market has always been characterized by a large number of vendors and a wide range of products, from basic two-line voice-only systems to sophisticated digital hybrids that support simultaneous voice and data communications.

Electromechanical systems. Electromechanical systems have been with us for some time, but in this day of microcircuits and telecommunication satellites, such technology is old-fashioned. The principal components of the 1A2-type key system are the key service unit (KSU), the key telephone units (KTUs), the power supply, and the terminal equipment.

The KSU can be wall- or desk-mounted or contained in a floor-standing cabinet. The KSU houses the switching electronics that connects central office lines, and individual stations. The CO lines, station lines, and other equipment units meet at the KSU. KTU circuit packs are mounted within the KSU and are integrated by internal KSU wiring. Main distribution frames (MDFs) are intermediate connection points that make installation of cables easier. Without them, all the wires would be terminated within the KSU itself.

The relay-operated KTUs contain a large number of electrical contacts, which require a wire for each contact. The KTU performs CO line connections, intercom functions, paging, and station connections. The number and types of KTUs are limited by the construction of the KSU. There are separate KTUs for the central office lines, station lines, and intercom circuits. Special-purpose KTUs can be installed to enhance a key system. Basic KTUs are grouped in the center block as common equipment, including a tone and ringing generator, an interrupter that mechanically regulates lamp flashing, and dialed number detectors and translators.

The power supply can be included within the KSU. More often, it is housed in a small cabinet mounted away from the KSU, near a commercial ac power source for convenience.

Electromechanical key systems use a wide variety of station equipment. Besides the easily recognized 6- or 10-button desk key telephones, much larger instruments with 30 or more keys are available.

Some systems support single-line telephones. An assortment of auxiliary units can be added—lamps that indicate when a station is busy, loud bells indoors or out that alert the user to an incoming call, external paging systems, automatic dialers, answering machines, a music-on-hold (MOH) source, and speakerphones that allow hands-free conversation.

Extensive wiring and the number of electrical contacts genuinely distinguish electromechanical key systems from their electronic counterparts. On the line KTUs, for example, three of four relays control hold, ringing, and other functions. When there are several KTUs in place, the total number of contacts grows accordingly. In addition to the KTU relay contacts, each telephone line button has several mechanically operated contacts. Each of these requires a wire, and as the number of line buttons increases, so does the size of the wire bundle. If, for instance, each line button requires 8 wires and there are 9 buttons, the total number of wires required is 72, or 36 pairs. For each line key, one wire pair is dedicated for a speech path and another for hold, and a third is used for illuminating the line button lamp. Additional features, such as hands-free operation, usually require at least one other pair of wires.

Electronic systems.   Many of the lessons learned in designing modular electromechanical key systems have been applied to electronic systems. The central equipment cabinet, also called a KSU for consistency, houses numerous printed circuit boards (PCBs), which are often called KTUs, again for consistency. Electronic KTUs are quite different, however, from their electromechanical counterparts. Instead of large relays and discrete electrical components, electronic system KTUs use integrated circuits for voice amplification and switching. Many modern electronic systems also use solid-state crosspoints instead of metallic relays for circuit switching. A wired backplate integrates the various subsystems, as in electromechanical key systems.

Like the electromechanical KTUs, system circuit boards perform specific functions. Sophisticated electronics has dramatically reduced the number of contacts. While a 1A2 key system has several contact pairs for each key, an electronic system has only one pair that is closed momentarily. Unlike electromechanical keys, which have to be held in place by a latching mechanism to maintain contact connections, the electronic system has specialized circuits that "latch" a selected line. This simplified design, in conjunction with widespread use of microprocessors for controlling functions, greatly simplifies wiring requirements. Usually only two or three pairs of wires carry all the control data and speech information.

Electronic station equipment is far more advanced than the typical electromechanical key phone. While customers are usually locked into a line of proprietary station equipment with the purchase of an electronic system, the tradeoff can be worthwhile, since most of the sets are microprocessor-controlled and have several feature keys that make telephone use easier. Built-in speakers and/or speaker/microphones are often available as standard equipment.

Hybrid systems.   Hybrid electronic key systems combine the best of both electronic key and PBX features into one compact package. A hybrid system usually contains a processor-based central control unit, containing one or two task-sharing microprocessors to increase call handling capacity and to provide a wide range of telephone management capabilities. Most hybrids are compatible with multibutton proprietary key telephones and 500 (rotary) or 2500 DTMF (Dual-Tone Multifrequency) type telephones for system operation and feature access. Hybrid systems' port-oriented architectures allow a variety of custom configurations to suit specific business applications.

Peripherals, such as maintenance terminals and SMDR printers, connect to the system via standard RS-232C interfaces. Many hybrids are compatible with PBX data-recording devices for cost accounting and traffic engineering studies.

What distinguishes a hybrid from an electronic key system? A commonly used but somewhat inaccurate definition is that a hybrid functions as a key system but includes some enhanced features usually associated with a PBX. More specifically, the FCC classifies a system *key* or *hybrid,* based on central office line access from a single-line telephone. If a single-line telephone can access only one line, the system is classified as a key system. If the single-line phone has access to a pooled group of lines, the system is considered a hybrid. Some systems are dually registered because software options allow them to be configured for either pooled or single-line access from a single-line telephone. Since the hybrid connects to trunks, and the key system to lines, access costs vary, depending on local telephone company tariffs.

Non-KSU systems. Another development in key system technology has been the introduction of key systems that do not require a KSU. Instead of being housed in a centrally located KSU, the circuit packs that control basic operations and features are located within each telephone. Because of their modular design, these systems are easy for most users to install themselves, and the price per station is much lower than that for traditional key systems that use a KSU. Non-KSU systems form the very low end of the key system market, generally offering fewer features and lower line/station capacities than electronic or hybrid systems. They can, however, provide a cost-effective communication solution for businesses that require basic telephone service for one to five telephones.

Centrex-compatible equipment. The rapid development of one area of the key system market is directly related to the recent resurgence of centrex. Today, the largest market for enhanced centrex service is the under-35-line business user segment. Centrex requires end-user-provided equipment. To meet this demand, several vendors offer key telephone equipment with software and hardware designed to simplify feature activation for centrex users. Buttons can be programmed to perform the steps, such as switchhook flash and number sequences, that are necessary to activate centrex features. Conversely, centrex service offers the key system user most of the features of a computerized PBX, including voice/data transmission and user-performed additions, moves, and changes.

Technology issues. Since AT&T Technologies stopped manufacturing the 1A2, there have been some backlog problems with spare parts. Because of marketplace demand, however, present manufacturers of electromechanical key systems have been progressively picking up the slack.

Installation charges are usually lower for electronic key systems, which require less time and fewer materials for installation. The ability of electronic systems to operate with fewer wire pairs means that less copper wiring is needed, which should result in slightly lower costs. In addition, thicker cable usually requires more time to install and two technicians are often needed for long pulls. Since labor costs can range anywhere from $35 to $75 per hour, the time factor becomes significant in determining costs.

Reconfiguration or moves of electromechanical systems usually require considerably more time. Electronic systems can often be reconfigured by entering commands from a console or reprogramming the system controller software. Using the same approach, a user can quickly activate new features and update restrictions on a moment's notice.

Hybrid versus key.   The discrepancy between the classification of key/hybrid systems by the FCC and the state regulatory agencies has not been resolved; until it is, costs will vary from one state to another. Low-cost key system CO line installation charges may not apply to a hybrid in some areas because of local tariff restrictions. These restrictions would otherwise give hybrids an advantage over a small PBX system. This situation also applies to any electronic key system that provides automatic CO line access, instead of individual direct CO line access on each extension in the system. Overall, the electronic key and hybrid systems appear to be the favored choice of the industry and consumers.

Selection guidelines.   Selecting a business telephone system requires careful planning. The first step is to determine present needs and make growth predictions for the next 5 years. The most basic consideration is the number and types of stations required.

Feature selection can be a real challenge, especially in evaluating software-driven electronic systems that offer numerous features. All features chosen should be functional and should make using the system easier. No one wants to use a complicated feature. That is one of the advantages of a key system—it is simple, for example, to put a call on hold. No multiple key command codes are required, and the system identifies which line is being held. Some systems not only show which lines are on hold, they specifically indicate which lines at each station were placed on hold by that station user.

In selecting a telephone system, it is important to consult its intended users. Personnel who perform central answering generally have valuable insight into the system's operational requirements. Features that are helpful and widely used in an existing system should be retained in the new one. Additional feature requirements should be identified. A new system will never be well-accepted if plans are not coordinated with the everyday activities of the company.

After the initial planning is done, a number of systems should be examined. Select from among the systems that provide enough line, station, and intercom capacity to accommodate the organization's projected needs over the next 5 years. System capacities, growth capabilities, and the vendor's reputation in the industry should be considered. The list of potential candidates can be narrowed further by choosing specific features. For instance, does the system provide a battery backup system as a standard feature or an option? Is this an important consideration? Can any station in the system be programmed for attendant operation, or is a special proprietary set required for this purpose? Is the automatic dialing capacity adequate? Will the system be shared with another organization?

Once the system has been chosen and its size established, an area should be allocated for the central equipment. A floor plan showing the physical arrangement of the system is a valuable management tool. The plan should show every user's location, identified by name, intercom number, lines terminated at the station, and hardware to be installed. It is not practical to indicate features in a floor plan; a separate list is more useful for that.

It is important to specify all requirements in a clearly written request for proposal (RFP). To reduce the field of potential vendors to two or three prime contenders, it probably will be necessary to ask some additional questions. Start by getting a customer list from each vendor. Interview the companies listed, asking about system reliability and quality of support. Pay particular attention to the types of businesses each vendor serves. Note also the product lines each emphasizes, and ask whether full service will be provided. Service charges not covered in a maintenance contract can be costly. There are firms that limit the extent of their commitment to service only; some sell and install systems, others only sell, leaving installation and maintenance up to the user.

Proceed to the contract stage only after all decision makers are reasonably assured that a particular product is suitable and that the reputation of the service company indicates long-term maintenance reliability.*

---

*Datapro Research Corp., "An Overview of Centrex Service," MT20-200, Delran, N.J., November 1990, pp. 101–106.

6.5   Cellular Radio, Microwave, and Satellite Options†

6.5.1   Cellular radio (telephone)

There has been a constant and growing demand for mobile communications since the early 1900s. Work to provide and improve mobile telephone service has been ongoing since the 1930s. By 1947, researchers at Bell Laboratories had uncovered the key concepts of cellular radio—by subdividing a large geographical coverage area into smaller sections, or *cells,* and by permitting the assignment of the same channels to multiple, nonadjacent cells (a concept called *frequency reuse*), traffic capacity on the channels could be significantly increased.

But additional technologies were necessary to administer and connect the multiple cells. For example, in the early days of mobile telephone, trunking to connect two or more channels to one another required two quartz crystals and a positive channel selector switch for each new operating frequency. The development of low-cost frequency synthesizers facilitated economic operation of a larger number of channels. Integrated circuit technology and the availability of low-cost microprocessors paved the way for the complex radio and logic circuitry used in modern mobile telephones. Electronic switching systems with stored program control can now provide the call processing and control functions necessary to operate a cellular radio service.

In 1974, the FCC allocated 40 MHz of spectrum (825 to 845 MHz and 870 to 890 MHz) for cellular radio service. These spectrum bands were formerly allocated for television channels 70 to 83.

At the same time, the FCC solicited applications for systems that could demonstrate the efficient use of the allocated spectrum. In 1975, AT&T became the FCC's first cellular radio applicant and was granted a license to operate a developmental cellular radio service in Chicago. AT&T subsequently formed Advanced Mobile Phone Service (AMPS), a mobile phone subsidiary, which began cellular radio field tests in Newark, N.J., in 1976. AMPS completed installation of the Chicago system in 1978. Shortly after AT&T's application was received, American Radio Telephone Service (ARTS) applied for and was granted FCC authorization to install a developmental system in the Baltimore/Washington, D.C., area. Motorola developed the hardware for this system.

---

†Datapro Research Corp., "An Overview of Infrared Communications," CA70-010, Delran, N.J., July 1991, pp. 201–211.

The AMPS and ARTS tests indicated that cellular radio can provide service of a technical quality at least equal to standard wireline telephone service. The tests also proved that cellular radio held enormous potential. When offered to users on a commercial basis in the later phases of testing, both services were highly successful. This prompted many telecommunication companies to develop cellular radio systems for their own commercial markets.

Technology basics.   Mobile telephone service operates much like the wireline telephone service in homes and offices—it uses full-duplex mode, permitting two-way simultaneous transmission, with each mobile unit assigned its own private telephone number. To draw a clearer picture we can compare it to its close companion, mobile radio service, which operates in half-duplex mode, and which is two-way but not simultaneous; only one party on the call can transmit at a time, usually by holding down a button or transmit key, and (unless scrambled) the transmission can be picked up by any listener tuned to that channel. The main difference is that a citizens' band (CB) or any other mobile radio today is generally designed for use by a specified community of users, although calls from the public telephone network can be manually "patched in." Mobile telephones are designed to interconnect with the public telephone network and with one another.

Components of a cellular radio system.   The four basic components of a cellular radio system are the mobile telephone switching office, the cell site containing a controller and radio transceiver, system interconnections, and the mobile or portable telephone units.

The mobile telephone switching office.   The mobile telephone switching office, a digital telephone exchange, is the system's heart. It controls the switching between the public switched telephone network and the cell sites for all wireline-to-mobile and mobile-to-wireline calls as well as for mobile-to-mobile calls. The switching office also processes mobile unit status data received from the cell-site controllers, switches calls to other cells, processes diagnostic information, and compiles billing statistics. The switch communicates with the cell-site controllers through two 240-b/s data links operating in a full-duplex mode.

When the switching office receives a call for a mobile phone user from the public telephone network, it deciphers the telephone number dialed by the wireline user and alerts the controllers at the cell sites to page the corresponding mobile unit. When a mobile phone user places a call, the switching office accepts the dialing data from the cell-site controller and dials the desired number.

The cell site. There is one cell-site controller in each cell. Operating under the direction of the switching office, the cell-site controller manages each of the radio channels at the site, supervises calls, turns the radio transmitter and receivers on and off, injects data onto the control and user channels, and performs diagnostic tests on the cell-site equipment.

Each cell contains one radio transmitter and two radio receivers; both receivers are tuned to the same frequency, but the one receiving the stronger radio signal is continuously selected. Each cell has at least one radio channel that transmits control data to and receives control data from mobile phone units. This control data tells the mobile unit that a call is coming from the switching office or, conversely, tells the controller that a mobile phone user wishes to place a call. To complete the connection, the controller uses the control channel to tell the mobile unit which user channel has been assigned to the call.

Completing each cell's radio-frequency system are transmitter combiners to connect multiple radios, receiver multicouplers, coaxial lines, and one or more antennas. The antennas generally reside on a free-standing tower or existing structure that is 100 to 150 ft high. In the initial system each cell has an omnidirectional antenna, but as a system expands, directional antennas that radiate more power in some directions than in others may be used to increase traffic capacity within the cell.

System interconnections. Leased four-wire telephone circuits are generally used to connect the switching office to each of the cell sites, although more systems are now using fiber-optic connections to improve transmission quality. For each of the cell's user channels, there is one dedicated 4-wire trunk; in addition, there must be at least one 4-wire circuit for a control channel.

Although leased lines are likely to be the most efficient interconnection when the number of lines is small, perhaps 12 or less, other alternatives are available. T1 digital lines, which handle the equivalent of 24 four-wire telephone circuits on a single line, allow the switching system to send multiplexed calls directly from a digital data stream without conversion back to analog voice signals. Translation back to individual analog circuits then must take place at the cell site.

Digital microwave radios are used by cellular communication companies to link cell sites. Architects, when designing cell sites that are 5 to 25 mi apart from each other and have a clear line of sight, choose microwave radios to link the sites. The microwave radios are usually the most efficient means of handling analog transmission and the multiplexing of data over four-wire telephone circuits (or the equivalent of one or more T1's on a single link). The microwave radios are suited for the cellular environment because they are easy to install and can be moved and reconfigured quickly as the market dictates.

Microwave transmission provides for multiplexing many four-wire telephone circuits, usually the equivalent of one or more T1's, on a single link. Like T1 transmission, microwave requires translation at the cell site to break the signal into the individual four-wire channels. The microwave equipment pays for itself through reduced leased-line costs and improved transmission-line reliability. Payback periods typically range from 1 to 3 years.

Mobile and portable telephone units. Mobile and portable units are basically the same, except that the portable telephones have a lower power output capability and do not have as efficient an antenna system as the mobile units. Mobile units are generally hard-wired into a car, while portable units can be carried.

A mobile phone unit generally consists of a control unit, a transceiver, and an antenna. The control unit contains all the user interfaces, including a handset and various user controls and indicators (e.g., the recall, clear, send, end, and store buttons) along with the 12-button dial pad. The transceiver uses a frequency synthesizer to tune into any designated cellular system channel. A logic unit, contained in the transceiver, interprets customer actions and system commands, and manages the transceiver and control units. The antenna can be installed on the trunk, roof, or rear windshield of a car and is usually a coil antenna.

Cellular services

Roaming. It is impossible to clearly define a coverage area for a given cell site. When a user moves to the outer limits of what should geographically be the serving cell, an adjacent cell might best handle the call; e.g., when ambient noise is stronger in the serving cell than in the adjacent cell. In this case, the system automatically passes the user to the adjacent cell.

Roaming, the ability to use cellular phones outside a home service area, has been made possible through agreements between cellular service providers in the United States, Canada, and Europe. These roaming arrangements between providers also include billing agreements. Roaming customers are billed for all calls through their home service provider, instead of getting separate bills from each system. Roaming costs vary, but usually include a daily usage cost plus a minute usage cost.

Data transmission. In general, cellular telephones were not designed for data transmission, therefore RJ-11 adapter connections and modulating/demodulating devices have been used. These adapters permit two-wire modular connections and ringing, dial tone, and pulse features. But, even with this adapter, the data handling capabilities of cellular radio were not yet perfect. The biggest problem was the handoff, during which a 0.05- to 0.2-s interruption takes place. While this interruption is imperceptible to the human ear, it can easily disrupt a 1200-b/s data transmission, necessitating substantial error detection and retransmission support.

One method used to compensate for data communication errors is to furnish the computing end of the circuit with a microcom networking protocol (MNP) modem and/or communication software.

Today, fax and modem support are readily available to the cellular market, with increased feature availability for higher-speed and better-resolution products on the drafting board.

6.5.2 Short-haul microwave communication systems

In microwave communications, activity remains concentrated on the 18- and 23-GHz frequency ranges. Most vendors have shifted their emphasis on microwave as a disaster recovery medium to microwave as a LAN extension. Vendors are redefining and enhancing their equipment to support 10-Mb/s token-ring and Ethernet traffic and to include network management capabilities.

Because of advances in microwave technology, the additional frequencies approved for use by the FCC in the last several years, and the changes divestiture has wrought, most of the growth potential in microwave communications is in the market segment commonly called *short haul.*

Technology basics. Fully understanding the basic principles of short-haul microwave technology requires an understanding of analog and digital communications.

Analog electronic transmissions rely on voltage changes related to the volume or image changes of an original sound or image. An analog signal varies in amplitude or frequency. Digital signals are basically a bit stream of ones and zeros composed of discontinuous pulses; information is contained in the durations, periods, and/or amplitudes. Digital signals are hard to distort; therefore, digital communication is perceived as the ideal means of data transmission. Communication via microwave has advantages over fiber-optic cable because, unlike fiber or cable, a microwave user does not need a continuous right of way.

Digital microwave communication refers to transmission in high bandwidth. The FCC recommends that businesses and common carriers transmit digital data in the 2-, 4-, 6-, 8-, 11-, 18-, and 23-GHz frequency ranges. Today, the most popular range in which digital microwave radios transmit is between 18 and 23 GHz.

In the 1990s, advanced bandwidth compression techniques, made possible by recent improvements in integrated circuit fabrication, will reduce digital microwave's extravagant use of scarce radio spectrum. Improvements in integrated circuit technology will also make feasible the introduction of high-frequency microwave equipment, thus opening up the sparsely used radio spectrum above 23 GHz for practical business use.

Components.   Microwave radio is a form of alternating-current technology. An electrical current smoothly increases in strength, from nothing at the beginning of a cycle to full strength one quarter of the way through the cycle. It then recedes to nothing at the halfway point of the cycle.

At this point, the current begins to flow in the opposite direction. When considering a microwave transmitter's operation, do not think of it as a collection of integrated circuits, printed circuit boards, and electronic power supplies. Rather, think of a microwave transmitter as merely a simple alternating-current generator, similar to the mechanical alternator in an automobile. This AC generator is capable of reliably producing frequencies of 18 to 23 GHz, the equivalent of 18 billion to 23 billion revolutions per second.

These higher-frequency bands have wider bandwidth allocation to allow the transmission of video for videoconferencing or of data for LAN extensions.

Antenna characteristics.   At high frequencies, such as those used in a short-haul arrangement, a 2-, 4-, or 6-ft-diameter antenna can be used. It is easier and cheaper (a savings of up to $1500) to install these systems on a roof, instead of erecting a tower. These small antennas have ranges of 2 to 15 mi and can support multiple T1 links while also supporting extra bandwidth for future expansion. The cost of a single T1 link for both ends is about $31,000 to $60,000.

The antenna's size will directly affect its gain and beamwidth. Gain is proportional to the effective antenna area in terms of wavelength. Higher antenna gain allows a longer distance between repeaters because the fade margin will increase with antenna gain, providing higher system reliability. The beamwidth decreases with increasing gain, thereby making the antenna more directional.

For longer path lengths, the lower frequencies are usually more attractive. Systems operating at 2 GHz can span distances of up to 40 mi, without repeaters. Transmission at lower frequencies is also much less vulnerable to the effects of rain attenuation. The engineering requisites at the lower frequencies are more complex, however, and larger antennas are required. In addition, the antennas must be mounted higher in the air to obtain a clear line-of-sight path over the greater distance involved. The combination of these factors pushes up system costs considerably.

The sheer size of these systems makes installation a major undertaking. Transmitters for long-haul systems are powered by tubes, called *waveguides*, that require indoor installation. Thus, in order to connect the transmitter to the outdoor antenna, longer waveguides are often necessary, resulting in more signal loss than with short haul.

Parabolic antennas are the most popular type in short-haul, private microwave systems. Parabolic antennas feature a curved dish reflector and a source of radiation at the focal point, called the *feed*. They are classified as *standard, high performance,* or *ultrahigh performance*. High-performance and ultrahigh-performance parabolic antennas are usually available with shrouding and a radome. Shrouding is primarily used in cities or other heavily congested areas where frequency coordination would be difficult.

Maintenance and diagnostics. Current digital microwave systems have inboard diagnostics, thus simplifying maintenance. Diagnostic routines can include alarms, bit error rate alarms, and status indicators.

Repeaters. Repeaters, used to extend the lengths of radio signal paths or maneuver them around obstacles, can be active or passive. Active repeater systems are usually configured with two antennas, an interface unit, and two radio-frequency units. Passive repeaters are configured to reflect the radio signal in another direction.

Advantages of microwave communication systems. Microwave technology has its advantages. When a company acquires short-haul equipment, it is replacing costly telephone company leased lines and protecting itself against future rate increases. Payback periods of less than 2 years are common.

Microwave systems are often installed very quickly. This is a paramount advantage when circumstances require a system to be on line at once. Another advantage is cost: microwave systems are often considerably less expensive than other mediums. Part of this lower cost results from the fact that microwave systems do not require a right of way, as cable-based systems do. Short-haul equipment can provide a path that telephone companies cannot, either because they cannot build them fast enough or because they do not have adequate facilities.

Many companies are discovering that microwave systems offer a readily installed, highly reliable method of restoring service when the primary system fails. To their even greater delight, however, microwave also augments and enhances those primary systems under normal operating circumstances.

The trade press has been full of stories about disasters that have crippled communication systems; many managers look on microwave as the only restorative solution. A hurricane can knock out phone service, and a fire or earthquake can disrupt telephone service for hotels, business offices, airlines, police departments, fire departments, and hospitals.

Of course, users who install microwave restoration systems quickly learn that they need not wait for an emergency to capitalize on the backup system's increased capacity. Also, special events often require that communication systems be given extra capacity to accommodate the increased demand.

Most business communication traffic takes place between points in close proximity to one another. These short hops are best served by short-haul microwave communication systems.

Telephone companies provide users with finite capacity. If users need more bandwidth, they must buy it. With short-haul systems, extra bandwidth is already available. This high bandwidth can handle videoconferencing, database transfers, and computer-aided design/computer-aided manufacturing (CAD/CAM) traffic.

Short-haul microwave limitations. Short-haul microwave is a powerful solution, but it must be applied with an understanding of its limitations, including the following:

- The high microwave frequencies, combined with their low output power, make rain attenuation effects significant. Attenuation is the decrease in magnitude of a transmission signal; in this case, heavy rainfall causes attenuation between the microwave transmitter and receiver. The effects of rain attenuation increase with the path length and rain's intensity.

- Interference from TV and FM radio stations is extremely low, mainly because they transmit in the 60- to 500-MHz range. Interference from other microwave systems is a greater problem; therefore, users must allocate time to get frequency surveys and an FCC license.

- There must always be an unobstructed path between the transmitting and receiving antennas (a clear line of sight). In some cases, towers may be required at one or both ends to obtain the required path.

- There is a limit on the length of cable from the indoor data connection point, known as the interface or indoor unit, to the antenna. Most systems are limited to either 500 or 1000 ft between the two devices.*

6.5.3 Infrared communications

Infrared technology uses the atmosphere as its transmission medium; it is a practical and cost-effective solution in many communication situations. Infrared systems eliminate the need for expensive cabling and the installation challenges associated with hard-wired systems. There is no FCC licensing requirement for infrared transmission, and no rights of way are needed. On the other hand, infrared communication has a number of limitations, notably line-of-sight requirements, range restrictions, and sensitivity to weather conditions.

In addition to outdoor applications, infrared is used indoors to connect PCs to LANs with wireless links and in factories to link robots.

Technological aspects of infrared communications. Infrared communication is a type of electromagnetic emission that fills a niche for short-range cable-free transmission. Available since the late 1960s, it is a cost-effective and easily deployed method of transmitting voice, data, and video.

Infrared transmission occurs at considerably higher frequencies than conventional microwave. Infrared signal propagation is, therefore, much more directional than microwave; this results in a number of advantages to the user, but it does cause some disadvantages. Infrared signals are loosely referred to as light signals because they are very close to the visible range of the spectrum. Infrared, typically occurring at 8300 Å (0.83 μm or 830 nm), is not visible to the human eye, since the eye's response is between 6900 Å (red) and 4300 Å (violet).

Electromagnetic signals may be propagated in three different ways. First, they may be passed through an electrical conductor. This mode of transmission becomes impractical at higher frequencies because of the tendency of the signal to migrate to the outer surface of the con-

---

*Datapro Research Corp., "An Overview of Short-Haul Microwave Communications Systems," MT20-580-101, Delran, N.J., May 1990, pp. 101–110.

ductor, a condition known as *skin effect*. The second transmission mode is via a waveguide, the channel that allows the signal to be radiated into a confined space and then guided by the walls of the physical medium. The final transmission mode is to radiate the signal into the atmosphere or into space. Copper twisted-pair transmission is representative of the first mode; fiber cable is representative of the second mode; microwave and infrared transmission are representative of the third mode.

Infrared is a short-range alternative to microwave, radio, cable, or fiber-optic facilities, especially in harsh or difficult environments. Atmospheric infrared links are ideal for electrically noisy environments such as power stations, reactors, and automated factories. Since the data are transmitted via line of sight through the atmosphere, no permits, cables, wires, fibers, ducts, or conduits are required. Thus, crossing streets, alleys, railroad rights of way, or from building to building can be done with ease, provided no obstructions exist. At the application level, infrared communication systems can be considered complementary to microwave.

Transmission. The objective of a free-space communication system is to superimpose information on a transmitter at a given location, send that signal to a receiving location, and detect and isolate the original information so that it can be passed to a receiving user. The transmitter must generate and concentrate a microwave or optical/infrared signal and direct it to an antenna system or optical receiver. This concentration renders the system less likely to be affected by interference or obstruction. A similar signal concentration at the receiver side, through a directional antenna or an optical collector with a small field of view, masks competing signals and other forms of interference. Like microwave technology, an infrared transmitter-receiver pair is mounted in a stable fashion to maintain alignment; sighting scopes are built into each unit. Infrared uses the same type of light sources and detectors as fiber-optic components.

To transmit information, it must be superimposed on a carrier signal. This process is called *modulation.* The total range of frequencies used for transmission is called the information channel's *bandwidth.* The bandwidth of a channel determines its information capacity. A series of mathematical relationships determines the maximum information capacity of a channel per unit of bandwidth; the typical value is 2 b/s per hertz (cycle per second) of bandwidth, with modern modems achieving slightly higher figures.

There are several general effects on transmission that relate directly to the carrier. Perhaps the most important is the capacity of the channel to carry information. A bandwidth of 100 MHz on a carrier frequency of 1000 MHz (1 GHz) would be a 10 percent variation on the carrier—a wide swing to handle for receiver technology. The same 100 MHz would be only 1 percent at 10 GHz, the K_u band of satellite communications, and a minuscule 0.00003 percent in the infrared range (about 330,000 GHz). Clearly, it is easier to provide high bandwidth at high carrier frequencies; light has a higher frequency than microwave, allowing it to support large bandwidths.

Practical bandwidths on infrared systems can be as high as 45 Mb/s, though the more typical figure is 1.544 Mb/s. Bandwidth on microwave systems can be as high as 45 Mb/s for digital systems and 25 MHz for analog systems. The combined bandwidth of the basic band can be higher (for example, 500 MHz for satellite links; fiber systems currently have bandwidths upward of 2.4 Gb/s).

System architecture. Most infrared systems are point to point in nature, with the optical path providing an alternative to copper wire, microwave, or other traditional transmission mediums for voice, data, or video. Point-to-multipoint infrared data systems, in which a central node broadcasts to multiple slaves on one channel while the slaves compete on a time-division multiple-access (TDMA) basis for return access, have been tried experimentally. The power of an infrared emitter is very low compared to that of a microwave transmitter, and the only reason reception is practical at all is that the beam is so concentrated. Typical systems have a beam width of only 1 m at normal operating distances, and spreading the beam to cover a large area in which multiple receivers are located would reduce the number of photons captured by a receiver to below the minimum needed for detection.

Because the distances involved in optical communications are so small, there is no propagation delay, but some links are subject to dropouts due to fading effects caused by fog, rain, or obstruction. Data communication links that employ infrared optical paths can easily deal with the low bit-error rates they generally provide, but they may be affected by the relatively long fade intervals. Users may need to adapt their protocols to ensure that a link that has temporarily faded will not be declared down.

Hardware components. It is with the issue of modulation that the parallels between microwave and optical communications begin to break down. In microwave systems, modulation is generally applied through electronic means; frequency or phase shifts in the carrier caused by the information are generated by changing electrical characteristics in the device that generates the carrier. When it comes to infrared systems, very large scale integration (VLSI) devices with direct optical couplings are just now being developed, but the available technology cannot process light (photons) in the same way that it processes microwave signals carried as electrons.

The light output of an optical laser, however, can be varied by changing the current through the emitter diode; this is also true of light-emitting diodes (LEDs). Optical modulation can also be provided by placing a modulation component in the light beam path. Polarization and other optical effects may be caused by electrical effects in some materials, making them suitable for optical modulation; these materials are generally not used for normal communications.

System components. An optical transmission system consists of a light emitter and a light collector, each sensitive to the same optical band. Infrared in the range of 8000 to 10,000 Å (0.8 to 1.0 μm) is simply another form of light and is processed the same way. Emitter technology may be based on coherent light generated by laser diodes or on noncoherent light generated by LEDs. (Coherent light is of a single frequency and is polarized in one direction without appreciable divergence; noncoherent light, or random-phase light, is dispersed throughout different phases.) Receivers are generally photodiodes sensitive to infrared frequencies. A simplex system employs a single emitter/collector pair, while a full-duplex system has two pairs of elements, one for each channel.

Transmitters. Laser systems emit a very narrow beam of light, which is polarized and limited sharply in its bandwidth. The light intensity of a laser can be managed by controlling the power applied. The ease of modulation translates into wide variations in signal state, representing the zeros and ones of data and making it easy to detect and demodulate the signal in a robust fashion at the opposite station. Laser optical systems have theoretical ranges of up to 10 mi under good conditions.

The light levels generated by LEDs are very small in comparison with those generated by lasers, and the losses associated with modulating LEDs are higher. There is a general relationship between the current applied to an LED and the intensity of its light, but the curve is not as linear as that of a laser. In addition, LEDs have a relatively slow switching time, making it difficult to pulse them; typical LEDs with rise time of about 1 μs can support data rates in the order of 1 Mb/s. Outdoor applications generally employ laser diode technology; indoor systems use LEDs.

Neither the output of a laser diode nor that of an LED presents any physical risk. The lasers used in communication systems radiate power at low levels, and the reduction in average power is often accomplished by pulsing the laser. This results in high power levels during the pulse, which facilitates signal detection; low pulse duration, however, eliminates exposure risks.

Terrestrial atmospheric laser communication systems on the market today use laser diodes operating in the range of 100 to 1000 mW, a power level too low to cause damage to tissue except in extreme cases of exposure (infrared transmission systems using optical fiber employ lasers operating at relatively higher power, since the fiber confines the beam and eliminates risk of burns or eye damage). Nonetheless, one should not look directly into a laser diode emitter at a distance of less than a few feet. All laser diode systems are required to carry warning decals cautioning users not to look into the aperture at close range. Most laser systems are rated as class IIIb by the National Center for Devices and Radiological Health, meaning that they are not dangerous unless viewed directly or via reflection (Document 21 CFR 1040, "Performance Standards for Laser Products," as amended).

Typically, an atmospheric system puts out several hundred times less light energy than a regular flashlight. LED systems present no harmful effects at all; neither the FCC nor the Bureau of Radiological Health places any restrictions on LED products, and they are as absolutely safe as a remote control device for the home television.

Receivers. Optical and infrared receivers can generally be classified as direct photon detectors or optical interference detectors. In the former technology, a photodiode in the receiver traps photons from the transmitter beam and converts them to electrons, which can then be amplified and processed to extract the signal; this technology is very common today. Interference systems use the same principle as radio and television reception. The incoming light beam is optically mixed with a reference light signal. The interference field is then detected directly as the information channel; this technology may become more prominent in the next few years.

Direct detection is an easy and inexpensive method, but there are some limitations. The photon detectors will operate very well with LED or other noncoherent sources but are more sensitive to background noise. The sensitivity of a photodiode photon detector peaks in the desired infrared range, but in this region there are other light sources that may interfere with reception; for example, incandescent light for indoor applications, or daylight for external applications (the signal emission from fluorescent lights does not interfere, since its output at 9000 Å is minimal). Placing an infrared color filter in the collector for the photodiode can help to reduce the noise from other light sources, but it will also attenuate (weaken) the real signal.

Cost of technology. An entry-level infrared system (two transmitters and two receivers) can be purchased for as little as $5000, and systems capable of megabit-range communications can be installed for less than $20,000. This is far lower than the price of most microwave systems of equal capacity. If the path lengths are less than 1 mi, infrared is a strong alternative and should be considered. In areas where rain and fog are limited, reliable communication at distances of 2 to 3 mi may be possible. For paths under 2 mi, the operation range will be approximately twice the minimum visibility, while for greater ranges the distances converge: 1.2 times maximum visibility at 5 mi and 1 to 1 beyond 7 mi. Models that handle voice and data tend to cost somewhat more than systems intended for video only. Some manufacturers also offer the equipment on short- or long-term rental.

On a single application basis, there are almost no disadvantages associated with infrared technology, provided the constraints on path length and interference are met. The equipment is reliable, is easily installed and moved, requires no licensing, and presents no hazards in normal installation and use. An infrared system will generally have a payback period of 1 year. It interacts well with analog video equipment and with digital voice and data communication equipment, and it provides what is probably the most secure of all atmospheric communication. The systems are generally compatible with encryptors, should extra security measures become necessary.*

6.5.4 Satellite communications†

Satellite communication transmission technology continues to offer capacity and new service opportunities for video, data, and voice networks. Satellites provide the capability for the simultaneous transmission of several video broadcasts, thousands of telephone conversations, and millions of bits per second of data, all through the same

---

*Datapro Research Corp., "An Overview of Infrared Communications," CA70-010, Delran, N.J., July 1991, pp. 201–211.

†Datapro Research Corp., "Satellite Communications Technology Briefing," MT20-620, Delran, N.J., July 1989, pp. 101–120.

satellite. Earlier use of large earth stations for long-distance point-to-point communication paths supported the migration of traditional terrestrial communications to high-capacity, low-cost satellite channels. Today, corporations are turning to private satellite networks because they are proving to be effective in industries such as retail, banking, insurance, automotive, utilities, and others. Corporate executives are recognizing the strategic benefit of a relatively inexpensive, quickly implemented satellite communication system that provides faster inventory control, better employee training, real-time monitoring of operations, and many other applications.

Satellite transmission is used extensively in many areas of the world and has become an integral part of the telecommunication infrastructure of most countries. Not only have those countries with the most advanced telecommunication networks embraced this technology, but less developed countries with widely dispersed populations or rugged terrain have found satellite systems particularly appropriate for their telecommunication needs. New satellite applications emphasize the ease of implementing the broadcast and private network capabilities of satellites.

Satellite communication systems.   A satellite is a communication transmission device which receives a signal from a ground station, amplifies it, and broadcasts it to all earth stations that are able to see the satellite and receive its transmissions. No end-user transmission either originates or terminates at the satellite, although the satellite does send and receive signals for monitoring and correction of on-board problems or signals indicating its position in orbit. A satellite transmission begins at a single earth station, passes through the satellite, and ends at one or more earth stations. The satellite itself is an active relay, much like the relays used in terrestrial microwave communications.

A satellite communication involves three basic elements: the space segment, the signal element, and the ground segment. The space segment involves the mechanics of the satellite's orbit, the means of launching it into that orbit, and the design of the satellite itself. The signal element involves the frequency spectrum used for communicating by satellite, the effects of distance on communication, the sources of signal interference, and the modulation schemes and protocols used to ensure proper transmission and reception. The ground segment includes the placement and construction of earth stations, the types of antennas used for different applications, and multiplexing and multiple-access schemes that allow fair and efficient access to satellite channels. The space segment, the signal element, and the ground segment are relatively complex and technically diverse and will not be described further in this text. Should the reader be interested in gaining further information on the technical aspects of satellite communication, an abundance of sources are available at most public libraries.

Advantages and restrictions of satellite communications. Communication satellites have unique attributes that distinguish them from other communication technologies. Some attributes provide advantages that make satellites practical and attractive for certain profitable applications. Others are inherently restrictive, making satellites impractical or impossible for other applications.

The advantages of satellites include:

- *Stable costs.* The cost of transmission by satellite is the same regardless of the distance between the sending and receiving stations. Additionally, all satellite signals are broadcast. The cost of a satellite transmission, therefore, remains the same regardless of the number of stations receiving that transmission.

- *High bandwidth.* Satellite signals are at very high bandwidths, capable of carrying large amounts of data.

- *Low error rates.* Bit errors in a digital satellite signal occur almost completely at random. Therefore, statistical systems for error detection and correction can be applied efficiently and reliably.

Among the restrictions to satellite use are:

- *Signal delay.* The great distance from the ground to a satellite in geosynchronous orbit means that any one-way transmission over a satellite link has an inherent propagation delay of roughly 250 ms (a quarter of a second). This delay creates a noticeable effect in voice communications and makes the use of satellite links extremely inefficient with data communication protocols that have not been adapted for use over a satellite circuit.

- *Earth station size.* The combination of the FCC-mandated 2° orbital spacing of satellites, the low power of satellite signals in some frequency bands, and the great distances the signals must travel produces an extremely weak signal at the receiving earth station. Until new, higher-powered satellites are in place, these factors will tend to keep the earth station diameter large, limiting the ease of installation.

- *Interference.* Satellite signals operating at K_u or K_a frequencies are very susceptible to interference from bad weather, especially rain or fog. Satellite networks operating at C band are susceptible to interference from terrestrial microwave signals. Bad weather interference can provide sporadic unpredictable performance in the K bands for a few minutes to a few hours. Terrestrial interference in C band limits the deployment of earth stations in major metropolitan areas where users are concentrated more heavily.

These advantages and limitations of satellite systems heavily influence the decisions regarding the use and types of satellite systems selected for private networks. Users with satellite-compatible network requirements, e.g., networks with geographically dispersed locations and large bandwidth requirements within the network, will be interested in the economic advantages of satellites over terrestrial networks.

Satellite voice and data transmission. Frequency-division multiplexing (FDM) is a technique used to multiplex numerous voice or data signals onto a single satellite transponder. All satellite carriers make extensive use of FDM transmission techniques to send telephony signals on their respective satellite networks.

With FDM, the waveform of each telephone signal is filtered to limit its bandwidth to a range of audio frequencies between 300 and 3400 Hz. The waveform is then converted to a single-sideband AM (SSB/AM) waveform. Twelve individual SSB/AM signals are then multiplexed together into a composite baseband signal. Each group is composed of telephone signals located at 4-kHz intervals throughout the composite baseband signal called a *group*. Several groups are remultiplexed to form a *supergroup* that can contain from 12 to 3600 separate voice channels. This supergroup is frequency-modulated into a 70-MHz intermediate-frequency (IF) carrier, which in turn is translated to microwave frequencies before being uplinked to the satellite.

Single-channel-per-carrier frequency modulation (SCPC/FM) is a satellite transmission format that assigns a single frequency-modulated radio-frequency (RF) carrier to each audio or data signal. These SCPC signals are located at spaced intervals throughout the transponder's frequency range. Both domestic and international satellite systems use SCPC/FM to transmit radio network and telephony traffic.

Time-division multiplex (TDM) is another method widely used to transmit voice and/or data services on a single transponder. While FDM signals are assigned to separate frequency segments within the transponder, each TDM transmission accesses the full channel bandwidth. To permit multiple signals to be relayed by a single transponder, each audio or data signal must sequentially modulate pulsed subcarriers that are spaced in time so that no two users can ever occupy identical time frames.

TDM earth stations typically employ a modulator/demodulator (modem) to transmit and receive TDM signals. The modem transmits information in high-bit-rate bursts. Buffer circuits are used to store both incoming and outgoing information bursts. Stored input is released at timed intervals; received signals are then released as a continuous bit stream that is spread over those time periods when the earth station is not in the receiving mode.

Because TDM transmissions permit only one signal to occupy the available full channel bandwidth at any moment, unwanted intermodulation distortion products are not created. Network operators can therefore transmit at full power without concern about generating unwanted crosstalk between signals sharing the same frequency spectrum.

Satellite video transmission.   Television cameras slice a picture into several hundred separate horizontal lines. Each individual line is made up of a number of dots, each with its own level of brightness. In the United States, the 525-line frame system, switched at 30 frames per second, is known as the National Television Standards Committee (NTSC) standard. The British modified this standard to a 625-line frame system switched at 25 frames per second. In Germany, Telefunken found that the NTSC standard resulted in uneven color reproduction as a result of phase errors introduced during the propagation of NTSC signals. To cancel out the phase errors, the color signal was inverted by 180° on alternate lines. This scheme, phase alteration by line (PAL), is a video standard used by more than 40 countries.

The United States, Great Britain, and Germany adopted AM systems. The French adopted an FM system to transmit the color information, and the phase error problem was eliminated. Under the French scheme, color information is separated into red and blue and sent by two international carriers. A memory circuit within the TV set remembers the signals for regular color reproduction. This system was named SECAM, which means sequence with memory.

High-definition television (HDTV) has recently received a lot of attention as a result of the interest in very powerful direct broadcast satellites and the imperfections of NTSC when projected onto a large TV screen. HDTV requires about 1125 lines and therefore more bandwidth than the 36-MHz transponder channels that were mentioned earlier.

All of these standards create a problem for earth station operators that want to receive international TV transmissions. To overcome these incompatibilities, the British Independent Authority developed a video transmission standard that would use a universal transmission mode to give satellite services a quality that would exceed anything now in use. The technique, called *multiples analog component* (MAC), separates the monochrome and color components of the video signals during the transmission process.

Application areas

Business television. An emerging application concerning public network television and videoteleconferencing (VTC) is referred to as *business TV*. The forerunner of this service, Public Satellite Network, has made meaningful strides with contracts secured with various Fortune 500 firms. The offering, in essence a closed-circuit television show, is encountering much more favorable response in the business community than it would have a few years ago, when the tendency would have been to reject it as an entertainment medium.

Hotel/motel programming. Somewhat similar to the business television concept is satellite-delivered programming to the hotel and motel industry. Some of the services are on a pay-per-view schedule while others are incorporated into the room charge and thus appear to be free. Nearly every hotel/motel system employs TV receive-only earth stations, and use is therefore restricted to broadcast. Two notable exceptions are the satellite programs of Hilton and Holiday Inn. Each Hilton hotel owns its on-premises equipment, including steerable earth stations. Thus there are no substantial restrictions on the type of use or on the selection of satellites from which signals may be received. In addition to entertainment, individually tailored VTC can be delivered, primarily on a customer-specific contract level.

Education and training. Another broadcast-oriented application on an upward growth curve is satellite-delivered education. There are three formats under which this use appears: foreign/U.S. college sharing; college course work for remotely located student bodies; and college/private company exchange.

Satellite news gathering (SNG). SNG via K_u band is growing rapidly. Through truck-mounted, transportable uplinks communicating back to stations equipped with fixed receive-only terminals, the service permits live reporting on an extremely fast basis. Not only can the number of stories be increased to meet deadlines, but they also can be professionally prepared through the editing facilities aboard the mobile on-the-scene vehicle. SNG terminals are small enough (less than 400 lb) to be transported via aircraft, with a two-person crew for extended coverage well beyond the limits of mobile microwave units.

Mobile communications. A forecasted high-growth application is the use of positioning service for air, ground, and water vehicles anywhere in the United States and its coastal waters. A hub computer spotting signal is transmitted via satellite to a receiver/transmitter unit in a moving vehicle, which will then respond when its individual number/identification is called out. This provides the central station with the data it needs to calculate the vehicle's precise location using an internal mapping feature.

Radio determination satellite service (RDSS). RDSS is a set of radio-telecommunication and computational techniques that enables users to precisely determine their geographical position and to relay this and similar digital information to any other user. The system works by maintaining and processing a continuous flow of information between a system control center, geosynchronous satellites, and user transceiver terminals. The RDSS transmit control center transmits a signal at 1618 MHz many times per second through a geosynchronous satellite. The population of user transceivers receives the signal, and individual transceivers transmit a response at 2492 MHz through at least two satellites. These two signals are received at the control center where the position location computation is made. The position coordinates are embedded in the continuous signal sent out to one of the satellites and the individual transceivers extract from this interrogation signal those coordinates that are addressed to their ID codes. The technique has application to aeronautical, marine, and land mobile needs.

Teleports. Teleports are facilities in which carriers or earth station operators can locate earth stations that provide access to satellites. The idea is to provide a high-capacity uplink/downlink earth station, at gateway cities, that could be shared by many users instead of having users build their own systems. The idea of multiple shared usage of satellite facilities which can access any space segment has been advocated as an advantage to end users. Terrestrial links from the city to the teleport are generally included in the service offering.

Although teleports have been discussed since the early 1980s, their economic impact has not been significant in the satellite field. Advocates indicate that the network of teleports in the United States is only beginning to take shape. The U.S. teleport industry is driven by corporate demand for data communications, by real-time international video, and by the television industry's demand for real-time international news gathering. International teleports are also emerging, generally at trade centers.*

6.6 System Integration†

Assessing the future of network management systems is more than an intellectual exercise. Network managers must make decisions that have substantial dollar and resource costs attached to them. These decisions should be based not only on today's knowledge, but on future resource availability as well.

In the early stages of deregulation, network managers and other telecommunication decision makers were happy simply to have a growing range of equipment and control tools from which to choose. Demand for integration began to increase substantially by the mid 1980s, and many surveys show that the primary concern of today's telecommunication managers is the net management system's inability to deal effectively with all of a typical network's components. Slowly, vendors and voluntary committees are establishing interface capabilities and standards which will ultimately avail the network manager of the tools needed to integrate, on a cost-effective basis, practically every component in the system.

The following are current techniques and probable future solutions to the network integration dilemma.

6.6.1 Bandwidth partitioning

In its most basic form, virtual networking is simply the partitioning of available bandwidth with simultaneous implementation of security measures across that partitioned bandwidth. To the user, the partitioned bandwidth appears as a separate circuit or a collection of circuits.

In private networks, the challenge is to bring virtual networking capabilities down to low levels while still permitting access and security controls to be maintained over what would now be two or more separate networks. The challenge does not rest in the partitioning process itself but in controlling access to partitioned bandwidth. This can be met only through a combination of technical advancements and security arrangements.

---

*Datapro Research Corp., "Satellite Communications Technology Briefing," MT20-620, Delran, N.J., July 1989, pp. 101–120.

†Datapro Research Corp., "Integrated Network Management," NM40-100, Delran, N. J., December 1989, pp. 101–106.

Initially the virtual network is defined at the port level; access to that network will be restricted according to a rule-based system that relates user access privileges to port mappings. While integrity between different virtual networks and different users is maintained, some problems exist with this method. Most important, unused bandwidth in one virtual network cannot be used by those with access privileges to another network. The result is that at least some of the dynamic routing/rerouting capabilities in today's net management systems will be lost.

The second generation of small-scale virtual networking products, which allocate bandwidth to individual virtual networks through a software-controlled process, solves that problem. The process combines dynamic routing with expert systems to allow individual virtual networks to expand or contract specific geographical and bandwidth requirements.

6.6.2 Public/private interaction

As recently as a few years ago, network management systems were structured almost entirely as client premises products rather than network-compatible products. With client premises products, the only interaction with the public network was in using individual trunks provided by the public network vendors. With network-compatible products, interaction occurs between the control features within both the client premises and public network equipment.

Digital access and cross-connect system compatibility (where client premises T1 systems can drop and pick up individual T1 channels within the public network) blurred this distinction somewhat. The ongoing development of integrated services digital network capabilities and the resulting ISDN-compatible claims have blurred the distinction even further. Recent petitions by some Bell operating companies to remove the information provision restrictions imposed by divestiture indicate they are willing to provide further interaction with private networking systems.

The most feasible (and potentially the most profitable) combination of public and private network control technology in the next 2 years is the extension of network compatibility to the point where the private network can exert a form of software control over portions of the public network. This is not a new concept—some carriers have been offering forms of software-defined networks (SDNs) for at least 5 years.

Further, the most basic goal of the ISDN user or manager is to control parts of the public network in much the same way a private network is controlled. From a viewpoint of cost and system management, however, a combination of public network SDN and private network management capabilities makes the most sense in the short term. The private network sector is demanding products that will expand the ability of the network manager to view and configure the network.

The ability to exert software and database control over portions of the public network at the switch level will allow the private network manager to create or delete additional network capacity as bandwidth demand warrants by adding or deleting circuits within the public network. Effectively, the network manager will be able to create virtual subnetworks or temporary additions to the primary network, which will have a considerable impact on network management efficiency.

Geographical areas that currently do not warrant fixed network facilities will be integrated into the master private network by adding virtual subnetworks derived from the public network.

Finally, the availability of expanded networks to individual companies will mean a combined increase in both information flow and application availability.

Developments and changes in network management systems are becoming increasingly market-driven. As private network managers become able to define the new features that their systems require to function properly, these features will be developed by individual vendors. That development process is expected to accelerate more rapidly due to changing regulatory and vendor environments.

6.6.3 Multivendor interface options

The lack of integration between different vendors' software and hardware is the primary barrier to the development of a true network management system. Control code and interface information within individual net management systems, or between components within a single vendor's network, is typically proprietary to that vendor.

Individual systems simply do not communicate well with one another. In fact, they cannot communicate unless vendors are willing either to exchange proprietary control, interface, and alarm information, or to agree on a single set of standards for that information.

Although a growing number of systems and components can interface through third-party monitoring systems such as IBM's NetView, such interfacing is typically limited to passing status or alarm messages to the host.

The result is that no network management system today can truly manage the components of multiple vendors. This is the result of vendors' having no incentive to exchange proprietary information with their competitors. There are, in fact, a wide range of options for constructing some form of interface between individual systems.

The applications program interface (API) takes a particular application within the network management process, performs a software protocol conversion on it, and hands it off to another set of controls. The main advantage to this method is that the user can tailor the API to fit a particular situation. A disadvantage is that large numbers of applications require large numbers of APIs.

The asynchronous ASCII terminal interface is another option. This is the most commonly used of all interfaces. The vendor simply enables the user to view or control the system or component through an ASCII terminal. The extent of the user's ability to view or control is typically limited by the vendor, and enhancements to existing capabilities may be difficult or impossible.

A third interface option is reverse engineering. Control codes and other organizational factors are broken down into individual components and rewritten in a language compatible with the primary system. This method is efficient in that all aspects of the integration process can be addressed. Reverse engineering can be tremendously expensive, though, and even minor system changes that require reverse engineering can mean that the entire process must be repeated.

Software bridging is another interface option. A software bridge passes an application from one system to another, with the passage being transparent to the user. This tool is very efficient from an engineering viewpoint, but designing a bridge requires substantial proprietary information from each system.

Standard message protocols are a fifth option. Examples include CCITT X.400 and the International Standards Organization's (ISO) common management information protocol.

The main advantage of these protocols is that they allow users and vendors to focus on a known set of standards.

Another option is third-party "umbrella" systems, in which a single set of networking standards is used across a series of networks and equipment. This method has the most promise but requires agreement from a large number of vendors to be successful.

Wrap boxes are the final interface option. These are devices that either bypass or bridge individual components within a system so that the performance of those components can be monitored to some extent. However, rarely is either complete monitoring or control possible.*

---

*Datapro Research Corp., "Integrated Network Management," NM40-100, Delran, N. J., December 1989, pp. 101–106.

7

Data Processing Applications

7.1 Industry Overview

By their nature, information technologies transcend network boundaries. The value of both voice and data communications is proportionate to the growing number of terminals that can communicate and the development of commerce in equipment and systems. The value of communications is enhanced when it is able to cross geographic boundaries. Therefore, agreements on protocols are required. The Consultative Committee on International Telegraph and Telephone (CCITT) has been tasked with this responsibility. The CCITT, in conjunction with the International Telecommunications Union (ITU), prepares recommendations for communication standards, including data communication over the public networks. The recommendations are just that, recommendations. Compliance is voluntary, although manufacturers have considerable incentive to adapt their equipment to the recommendations unless they are large enough to establish and implement their own standards.

7.2 ISO and IEEE Models*

The most important work on LAN standards is being done by two organizations: the International Standards Organization (ISO) and the Institute of Electrical and Electronics Engineers (IEEE). ISO has published an almost universally accepted model for describing networking systems. Known as the *open system interconnection* (OSI)

---

*Rowland Archer, *A Practical Guide to Local Area Networks,* Osborne McGraw-Hill, New York, 1986, pp. 7–8.

model, it describes the general pieces that a network should contain. Each piece is slowly evolving into one or more standards, starting with the lowest levels. It is analogous to a committee's deciding that a house will be built from a foundation, framing, plumbing, wiring, a roof, and the like, and then looking at each piece in turn and refining it; the foundation will be poured concrete, wiring will be 110 or 220 V, pipes will be 1 or 2 inches in diameter, and so forth.

The OSI model consists of seven layers. The layers represent both the hardware and software of a local-area network (LAN), including servers, workstations, and connections. The same layers are used to describe the hardware and software in a workstation or server. The lowest layers are the best defined and contain some real standards, such as RS-232C for asynchronous communications (familiar to most microcomputer users from the *serial port*). The upper layers are the least defined, and represent the interface to applications using the LAN.

The layers of the ISO open system interconnection model are:

Layer 7: Application

Layer 6: Presentation

Layer 5: Session

Layer 4: Transport

Layer 3: Network

Layer 2: Data link

Layer 1: Physical

Layer 1, physical, is the lowest layer. It represents the network connection: the cables and hardware interfaces to the workstations and servers on the LAN. This layer describes mechanical connectors, electrical impedances, frequencies, and voltages. Compatibility at this layer is vital.

Layer 2, data link, defines how information is put onto the LAN and how transmission errors are detected. If an application program wants to send the word *hello* to the print server, that word needs to be packaged so that the LAN will know to send it to a particular print server. The data includes some information that will help the layer 2 software on the print server determine that an error has occurred— that, for example, it has gotten the word *jello* instead of *hello*.

Layer 3, network, addresses the routing and relaying of information from one network to another. If you need to send mail to someone on another LAN, software at this layer needs to determine how to find that other LAN. This layer also guarantees that if one workstation sends another workstation two messages in a row, the first message received will be the first one sent.

Layer 4, transport, contains the logic needed to split a long message into pieces of the size that the lower layers can handle. It also reassembles the smaller pieces into the original message at the other end of the wire. Layer 4 is also responsible for reliably transmitting information from one network node to another. For example, if an error is detected by layer 2, layer 4 must request the sender to send the message again.

Layer 5, session, handles the connection of two network nodes for the purpose of exchanging data. If a session is disrupted, this layer is responsible for restarting the session when the disrupting influence is gone. For example, say a user is sending 20 data files to another and the network fails after only 10 are transmitted; once the network is working again, proper session control will restart the transmission where it left off.

Layer 6, presentation, addresses such issues as the way data are displayed on the workstation monitor and how character sets used by a workstation are translated to/from those used for data transmission.

Layer 7, application, provides an interface for application programs, using the network. In layer 7 reside the programs that use the network, rather than programs that make up the network.

The last two layers, 6 and 7, are not governed by any significant standards. The lack of a complete set of official standards for all the layers is the reason that network components from one vendor may not work with components from another vendor, even though both are, for example, "Ethernet compatible." Since both are designed for Ethernet, it is very unlikely that either will "go up in smoke" when connected to the same Ethernet cable, but there is no guarantee that one vendor's workstation will work with another vendor's print server.

Local-area network standards evolve in much the same way as other communication standards did. Manufacturers experimented with communication and access techniques, developed numerous proprietary techniques, and gradually demonstrated the feasibility of local network access methods. Most manufacturers developed their own proprietary protocols, limiting the compatibility between the network and existing equipment, and compatibility with other manufacturers.

In 1980 the IEEE Computer Society appointed a committee to work on project 802 for the development of local-area network standards. The 802 committee's overall objectives, which have largely been met, were to establish standards for the physical interconnection of devices to local networks and the data link between devices. The following requirements have become established standards for local-area networks:

- Application is light industrial and commercial functions, excluding heavy traffic and home use.
- The distance between terminals is up to 2 km.
- The network configuration supports up to 200 devices on a single cable.
- Data rates can be from 1 to 20 Mb/s.
- The network system ensures that there will be no more than one undetected error per year, and that the failure of any single device on the network will not degrade the entire network.
- Existing data communication standards should be incorporated into the IEEE standard as much as possible.

The 802 committee completed its work on bus standards in 1982 and put them out for comment. The IEEE adopted the standards in 1983 and a few weeks later ISO announced that it would initiate processes to standardize the 802 format.

These standards are published in five parts:

802.1	Document containing the reference model, tutorial, and glossary
802.2	Link Layer Protocol Standard
802.3	Contention Bus Standard
802.4	Token Bus Standard
802.5	Token Ring Standard
802.6	Metropolitan Area Network

7.3 Wide-Area Networks

Of particular interest in data communications are the X series and V series of CCITT recommendations. The X recommendations relate to data communication within digital networks processed over the public network. The V series relates to data transmission over the voice circuit or analog public networks. Generally they relate to the following:

Number	*Title*
X.1	International user classes of service in public data networks
X.2	International user facilities in public data networks
X.4	General structure of signals of International Alphabet No. 5 code for data transmission over public data networks
X.20	Interface between data terminal equipment and data-circuit-terminating equipment for start-stop transmission services on public data networks

X.21   General-purpose interface between data terminal equipment and data-circuit-terminating equipment for synchronous operation on public data networks

X.24   List of definitions of interchange circuits between data terminal equipment and data-circuit-terminating equipment on public data networks

X.25   Interface between data terminal equipment and data-circuit-terminating equipment for terminals operating in the packet mode on public data networks

X.26   Electrical characteristics for unbalanced double-current interchange circuits for general use with integrated circuit equipment in the field of data communications

X.27   Electrical characteristics for balanced double-current interchange circuits for general use with integrated circuit equipment in the field of data communications

X.30   Standardization of basic model page-printing machine in International Alphabet No. 5

X.31   Characteristics for start-stop data terminal equipment using International Alphabet No. 5

X.32   Answer-back units for 200-baud start-stop machines in accordance with International Alphabet No. 5

X.33   Standardization of an international text for the measurement of the margin of start-stop machines in accordance with International Alphabet No. 5

X.40   Standardization of frequency-shift-modulated transmission systems for the provision of telegraph and data channels by frequency division of a primary group

X.50   Fundamental parameters of multiplexing schemes for the international interface between synchronous data networks

X.51   Fundamental parameters of a multiplexing scheme for the international interface between synchronous data networks using 10-bit envelope structure

X.60   Common channel signaling for synchronous data applications— data user part

X.70   Terminal and transit control signaling for start-stop services on international circuits between asynchronous data networks

X.71   Decentralized terminal and transit control signaling systems on international circuits between synchronous data networks

X.92   Hypothetical reference connections for public synchronous data networks

X.95   Network parameters in public data networks

X.96   Call process signals in public data networks

V.1	Equivalence between binary notation symbols and the significant conditions of a two-condition code
V.2	Power levels for data transmission over telephone lines
V.3	International Alphabet No. 5 for transmission of data and messages
V.4	General structure of signals of the 7-unit code for data and message transmission
V.10	Use of the telex network for data transmission at the modulation rate of 50 baud
V.13	Answer-back unit simulators
V.15	Use of acoustic couplers for data transmission
V.21	200-baud modem standardized for use in the general switched telephone network
V.22	Standardization of modulation rates and data signaling rates for synchronous data transmission in the general switched telephone network
V.23	200–1200-baud modem standardized for use on the general switched telephone network
V.24	Functions and electrical characteristics of circuits at the interface between data terminal equipment and data-circuit-terminating equipment
V.25	Automatic calling and/or answering on the general switched telephone network
V.26	2400-b/s modem for use on four-wire leased point-to-point circuits
V.26B	2400-b/s modem for use on the general switched telephone network
V.27	Modem for data signaling rates up to 4800 b/s over leased circuits
V.28	Electrical characteristics for interface circuits
V.30	Parallel data transmission systems for universal use on the general switched telephone network
V.31	Electrical characteristics for contact-closure-type interface circuits
V.35	Transmission of 48-kb/s data using 60- to 108-kHz group bank circuits
V.40	Error indication with electromechanical equipment
V.41	Code-independent error control system
V.50	Standard limits for transmission quality of data transmission
V.51	Organization of the maintenance of international-telephone-type circuits used for data transmission
V.52	Characteristics of distortion and error rate measuring apparatus for data transmission

V.53   Limits for the maintenance of telephone-type circuits used for data transmission

V.56   Comprehensive tests for modems that use their own interface circuits

V.57   Comprehensive test set for high transmission rates

The above recommendations represent many of the requirements imposed for transmitting data over the public switched network. These standards must be met when data transfer utilizes this network.

Private networks occur in buildings and in campus environments. These communication systems do not have to meet the public standards and in many cases are quite different in architecture, protocol, and topology.

7.4   Data Speed†

In general, the higher the frequency of the carrier signal, the greater its information-carrying capacity. In telecommunication literature, the term *bandwidth* is often used to refer to the capacity of communications. A channel's bandwidth is the difference between the highest and the lowest frequencies that are carried over the channel. The higher the bandwidth, the more information can be carried. For example, a telephone channel supporting voice communication transmits frequencies ranging from about 300 to 3100 Hz. So the range of frequencies, or bandwidth, support is $3100 - 300 = 2800$ Hz, or about 3 kHz. The transmission mediums used with local-area networks support bandwidths much larger than this.

A channel's bandwidth has a direct relationship to its data rate, or the number of bits per second that can be carried over it. Since local-area networks deal mainly with data transmission, we will find bits per second to be a more useful measure of a channel's capacity than bandwidth.

Another term that is used to express channel capacity is baud. Baud is a measurement of the signaling speed of a channel; a certain communication channel is said to have a speed of X baud. This refers to the number of times in each second the line condition changes. Suppose that we are using amplitude modulation and that one amplitude value is used to represent binary 0 and another to represent

†James Martin, *Local Area Networks,* Prentice Hall, Englewood Cliffs, N.J., 1989, pp. 21–30.

binary 1. In this case, the line's signaling speed in baud is the same as the line's data rate in bits per second. Suppose, however, that we use four different amplitudes to represent the binary value 00, 01, 10, or 11. In this case the data rate in bits per second will be twice the signaling speed in baud. If the signals are coded into eight possible states, then one line condition represents a tribit, and the data rate in bits per second is 3 times the signaling speed in baud.

7.5 Topology

Network topology, or architecture, is the pattern of interconnection of the terminals. The major topologies employed in local-area networks are star, bus ring, and branching tree schemes.

7.5.1 Star

Star networks (Fig. 7.1) are perhaps the most familiar topology. The telephone system is configured as a star, with lines radiating from a central switching machine to individual terminals. The central switch can be a minicomputer or PBX. With minor variations, all circuit-switched networks use this form, in which circuits from the terminals are brought into a central switching system. Messages or packets are

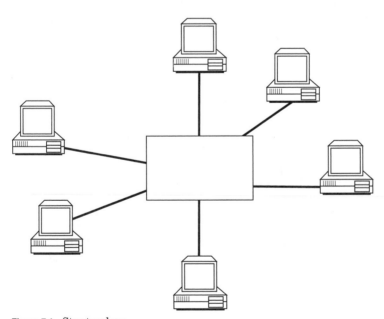

Figure 7.1   Star topology.

sent to a terminal only through the central switch, or controller, which is assumed to be the one and only intelligence resource with command functions over a circuit inaccessible to other terminals. In a star network, access is allocated by the central controller or switch. It is possible for other terminals to be star-connected without a central switch or controller. In this configuration, the star is the electrical and logical equivalent of a bus network. Additionally, a hierarchical topology can be incorporated in which, besides the master central switch, there are nodes, such as intelligent concentrators, which might themselves act as star structures. This approach leads to sub-networking, but is always based on centralized principles. These star configurations are the oldest and least reliable type of topology.

7.5.2 Bus

The bus network topology (Fig. 7.2) is used in the multidrop configuration (which gives the ability to add additional stations to the network). It is neither centralized nor a loop. This is the most common configuration in Ethernet-type LANs. Terminals are connected to a central circuit, or bus, and connect with one another for access. Messages or packets are broadcast simultaneously to all terminals on the bus. Either access is allocated by a control node or the nodes are treated as equals and can contend for access. It is important that the open ends of a bus be terminated in the characteristic impedance to prevent reflections of data signals. In waterpipe terminology, a termination is equivalent to an infinitely long pipe connected to the source so that the flow is continuous. The bus network's major problem is maintenance. Typically, the system is coax-based and requires a great amount of planning for physically balancing the electrical potentials. The Ethernet protocol is contention-mode and collision-detection based.

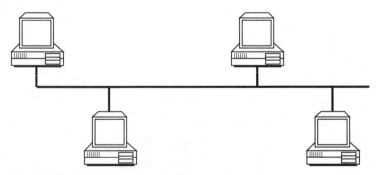

Figure 7.2 Bus topology.

7.5.3 Ring

The ring network (Fig. 7.3) is, in effect, a bus with the outer ends joined. Terminals are normally inserted in the network rather than braided as they are in a bus, so that all messages are regenerated by each node. Control of a ring network is usually assigned to one station and is accomplished by passing a token from node to node. A control node must be designated to monitor the network for a lost or mutilated token, and to recreate the token when necessary. Alternatively, all nodes can contain the control logic, with the first one discovering the difficulty taking independent corrective action.

7.5.4 Branching tree

The branching tree (Fig. 7.4) is similar to bus topology in that all data messages are broadcast simultaneously throughout the network. However, instead of one bus with two open ends, the branching tree can have multiple legs and as many open ends as there are branches. Care must be used in connecting the branches to avoid impedance irregularities. This is accomplished with devices known as *splitters* in cable television terminology, or by connecting legs through an amplifying device called a *repeater*. This ensures that at branching points the electrical signal is divided between legs without the signal being

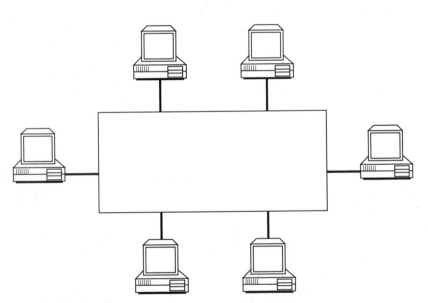

Figure 7.3 Ring topology.

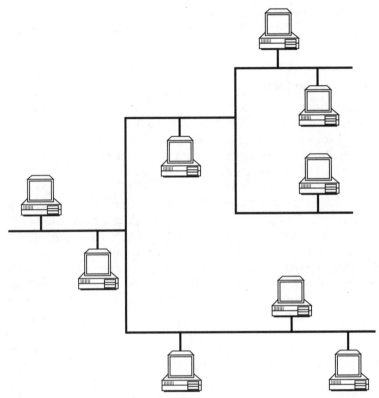

Figure 7.4 Branching tree topology.

reflected back to the source. The branching tree topology is most commonly used in broadband networks employing a contention access method, where it allows distributed control as in a bus network.

7.6 Protocols and Protocol Conversion

The protocol establishes compatibility between devices by solving the problem of the sharing of one connectivity medium (the physical link) among several devices; each machine knows exactly what to expect from the others. The protocol also establishes the line discipline, that is, whether the session is half or full duplex and what the signals mean on the circuit. Another protocol function is to set up and terminate connections. For example, if the connection is to be billed, the protocol signals the billing equipment to start at the moment the connection is established. The protocol can assist in error detection and correction. By enabling message blocks to arrive out of sequence, the receiving terminal can recognize the error and deliver the packets in the proper sequence. The protocol is the main defense against mutilated or out-of-sequence messages.

7.6.1 Cable access protocols

Each topology supports at least one type of access, but most topologies can support more than one type. The physical layout does not influence the protocol as much as one might think. There are, fundamentally, four cable access protocols, but each can have dialects and there are also hybrid possibilities.

Slot. The slot method is exemplified by the Cambridge ring. It is based on a type of time-division multiplexing. A constant number of fixed-length slots continuously circulate around the ring, and a full/empty indicator within the slot header signals the real state.

In this solution, any station ready to transmit occupies the first empty slot by setting the full/empty indicator to full and placing its date in the slot. Sender and receiver workstations share the ring bandwidth until they have completely transmitted their packets. When the sender receives the occupied slot, it must change the full/empty indicator to empty. This guarantees equal sharing of the bandwidth among all stands. In a LAN the slots are usually short; a message may have to be transmitted within several slots.

Between two consecutive access possibilities of a certain workstation, the slot travels exactly once around the ring in the empty state. All time periods of one cycle, during which the slot is empty, are put together in one continuous empty slot interval.

Token passing. The application of token passing is exemplified by round-robin (token bus) and token rings. The token is an attribute of the session connection which defines unambiguously, dynamically, and possibly exclusively whether or not a particular workstation has the right to access a given service. This is a necessary, but not always a sufficient, condition. Examples of tokens include send data, parameter reselection, terminate session, and synchronization. There are also user-defined tokens. A token will always be in one of the following states:

- Assigned or not assigned to a workstation
- Assignable or not assignable
- Available or not available

When a token is assigned to a workstation, the user has the right to use the service. If it is not assigned, the user does not have the right to use the service at that particular time, but may acquire the right without entering the reselection. If a token is not assignable, the user does not have the right to use the service; reselection is required to change to the assigned or not assigned state. If the token is not available, the user does not have the right to use the service over the lifetime of the session connection.

It follows that the management of the token status must be guaranteed. The management activity consists of controlling the system manner by which the system is initialized. When the system is initialized, a designated station generates a free token which travels until a station ready to transmit changes it to a busy and puts its packet onto the ring. The packet can, in principle, be of arbitrary length.

The sending workstation is responsible for removing its own packet from the ring. At the end of its transmission, it passes the access permission to the next station by generating a new free token. Possible queues are serviced in a cyclical manner by, for example, a rotation switch that either empties the queue or admits only a limited number of packets at each access. The former is known as *exhaustive* and the latter as *nonexhaustive* service.

In token passing, there is no control station and no master/slave relationship, and it is possible to build a priority mechanism so that some stations have priority over others.

Contention. The most favored protocol is carrier-sending multiple access with collision detection (CSMA/CD), most commonly used with Ethernet LANs. While particularly suitable for baseband, collision detection seems to be very difficult on broadband. The capacity of the segment is important; at 10 Mb/s, a single bit has a length of about 25 m. (Other typical baseband solutions are frequency duplication, Manchester encoding, nonreturn to zero, and differential Manchester.)

With CSMA/CA (collision avoidance) protocols, every station wanting to transmit a packet must listen to the bus in order to establish whether any transmission is in progress. If it is, the station defers its transmission until the end of the current transmission. But packet collisions cannot be completely avoided because of the nonzero propagation delay on the bus. If a collision is detected, transmission is aborted and the station reschedules its packet. The retransmission interval is dynamically adjusted to the actual traffic load.

CSMA/CD can be applied with coaxial cable, flat wire, or twisted wire. Cluster One, for instance, uses a CSMA/CD. The carrier line is activated whenever the cable is in use. Stations desiring to use the cable check for presence of the carrier signal can access the cable only when it appears to be idle. As in the coaxial-based Ethernet, this is a noncentralized access control mechanism with the same priority for all stations. An alternative possibility with CSMA is collision avoidance, used by some LANs such as Omninet.

In both Cluster One and Ethernet, collisions are detected by software during the first several bytes of packet transfer. In the Cluster One case, byte collisions occurring during retransmission are not checked in the same early byte sense, but they are detected by bad check sums at the end of the colliding packets. Both colliding packets will be retransmitted.

Ordered access bus. The protocol proper to the ordered access bus solution is known as multilevel multiple access (MLMA). It is frame-based and sees to it that information transmission occurs in variable-length frames. A controller provides start flags at the time intervals, which signal the beginning of a frame. The frame is divided into two parts, a request slot and an arbitrary number of packets, with every station attached to the bus owning one bit within the request slot. By setting its private bit, a station indicates that it wants to transmit a packet within the frame. The transmission sequence is given by a priority assignment known to all stations.

In line with this protocol, the bus is modeled as a single server facility. Packets from all stations, which have been newly generated and have not yet been scheduled, form a distributed queue; they cannot be transmitted within the current frame, but must instead wait until a new frame starts. However, to assure that all stations know the entries made in the request slot, the scheduling time may have to be insignificantly longer than the pure transmission time of the request slot.

The MLMA method is somewhat similar in concept to the contention resolution method for computer interrupt systems. There seem to be some basic assumptions in its use, the more restrictive one being that the distance between two stations transmitting in succession is uniformly distributed throughout the maximum bus length. Should the distance be uneven, slight time transmission differences would occur, giving the closest station the advantage.

The real-world applications of information technologies contain several LAN systems with different protocols. In order for these systems to talk to each other (share information), devices have been developed to change the language of one system to that of another. These devices bridge the tie using conversion software and can be used in private systems or over the public network. They are commonly referred to as *protocol converters.*

7.6.2 Protocol converters

A protocol converter is defined as any device that translates a binary data stream from one format to another according to a fixed algorithm. Although code converters are perhaps the simplest form of protocol converter, primary concern centers on converters that not only convert code but also simultaneously shift speed, change from synchronous to asynchronous, buffer data, perform error control, acknowledge polls, and reframe or packetize data. The most common type of protocol converter is the one that takes the data from a dumb ASCII asynchronous cathode-ray tube (CRT) terminal and packages it into bisynchronous or Synchronous Data Link Control/Systems Network Architecture (SDLC/SNA) blocks so that the terminal can communicate with an IBM or other synchronous-frame format computer.

The reasons for using a protocol converter may be divided into two categories: economic and technical. In economic terms, if an asynchronous CRT plus single or multichannel protocol converter is less expensive than synchronous terminals and their controllers, then the asynchronous CRT should be used. Gradually, however, the declining cost of 3270 compatible terminals has made this argument weaker. The technical reasons for using protocol conversion, however, are much more persuasive. Protocol converters can solve some major network design and integration problems.

Since port selections usually can switch only asynchronous data under CRT terminal operation control, it is best to remain asynchronous until after the port selector. This arrangement also makes it possible for a single terminal to switch itself between asynchronous and synchronous applications. The same principle applies in data-voice networks, PBXs, and dial-up security systems.

The protocol converter answers all host polls so no timeouts occur and no framing or polling characters need go long distance. This arrangement makes it unnecessary to change from bisynchronous to SDLC protocol or, with SDLC protocol, to go from 7 to 128 blocks before acknowledgment, which can cause buffering problems and decrease efficiency.

Some protocol converters can look like more than one controller at a time, so the asynchronous terminal operator can request one application program, use it to extract data, switch to another application briefly, and return to the original program. Alternatively, several single-controller converters can be used with a manual switch or port selector to select sessions.

Rather than alternate between synchronous applications, a protocol converter can switch a terminal from a synchronous computer to an asynchronous one. The terminal may be asynchronous or synchronous, depending on the converter chosen.

Several systems are available. Device selection should be discussed with qualified individuals on each system for which bridging is required.

7.7   Vendor Connectivity Specifications

The vendors of data processing equipment specify various operational limitations for their hardware as it pertains to the physical link between devices. Each manufacturer specifies different standards and, at times, uses different nomenclature to describe like performance requirements.

Due to the proprietary nature of each manufacturer's specific equipment and protocols, it will be necessary for the planner to acquire certain information from each vendor, survey the existing cable plant to determine its electrical performance capability, and match the spacing of equipment to satisfy the performance criteria for the devices.

In general, vendors will divulge the following performance limits for their equipment:

- Resistance
- Decibel loss
- Cable medium (coax, twisted pair, optical fiber)

The physical connection of the medium may be unique or allow for connection using standard telephone modular connectors. In either case, pin-out assignment integrity must be maintained throughout the system to assure signaling, alarm, and control feature delivery.

In general, most data systems can tolerate a 7-Ω loop, with a $\pm$ 3-dB loss between the LAN interface and user terminal, with a maximum physical loop not to exceed 90 Ω from each terminal to LAN server. Larger systems can extend the physical loop to 300 Ω, but will still only tolerate a 3-dB loss.

When using existing telco wire, the following average values can be used for calculating loop characteristics:

Gage	Resistance, Ω/kft	Loss, dB/kft
26	83	3
24	52	3
22	32	3
19	16	3

Add to this approximately 3 Ω per 66-type connector in the loop and 0.5 Ω for high-density connector blocks. Baluns will add approximately 3 Ω.

To assure circuit performance, it is strongly recommended that each circuit be inspected, reterminated at junctions, and tested for actual electrical characteristics, prior to assignment. To improve performance, the circuit can be "double conductored," using two wires in each direction instead of one, to halve the resistance.

If the circuit proves to be within resistance design limits, but will not pass data at high speeds, there may be too many connector block terminations in the path. This situation can be corrected by either direct splicing or retermination on high-density blocks.

As a last resort, the existing connectivity may be conditioned with a carrier to bring resistance values to 0 between devices. While expensive, many times it is cheaper to introduce additional electronics rather than replace the embedded wiring system.

8

Management and Control Systems

8.1 System Administration Requirements

With the local telephone utilities getting out of the intrabuilding wiring business, the general public has been placed in the position of having to manage all telecommunication and information wiring (connectivity) and associated equipment within residences and commercial buildings.

This chapter discusses a broad range of flowcharts and forms developed to assist the individual in addressing the specifics of a flexible and comprehensive management and control system. These documents and recommendations are designed to work in conjunction with the information presented throughout this work.

It is recommended that all equipment inventory forms contain certain common entries. Each form should ask for the type of equipment, date of installation, department and building location, and the universal information outlet (UIO) identification, and whether the equipment is owned or leased. Each form will require its own entries unique to the type of equipment.

It is recommended that all serving closet terminal record forms contain certain basic components. There is a requirement for information on the origin and end termination of all facilities and equipment serving and leaving the serving closet terminals. If the serving closet feeds the end-user terminal, the UIO and room number are also required.

The network access point (NAP) record will require, in addition to the information listed above, the circuit or telephone number of the facility on each cable pair or circuit. The type of facility (voice, data, video, alarm, etc.) is a helpful piece of information on this form. The UIO number should also be on the NAP form. This tends to tie together the circuit or telephone number to the end-user termination.

The form used for the PBX equipment room requires information relating to the PBX port, trunk, and station number.

It is recommended that the cable record be laid out differently, as it requires information on the size of the cables and their pair count. It should also describe the type of cable (primary, secondary, etc.). The circuit/telephone and UIO number will be referenced for each cable pair.

The final recommendation deals with the numbering plan developed in this chapter. The numbering plan used for cables will describe the type of cable and the floors the cable runs between.

The numbering plan for serving closets will describe the type of closet, the floor of the closet, and, if there is more than one closet of its type on that floor, the number of the closet.

The UIO number plan will reflect the floor serving closet feeding the outlet, plus the individual number assigned to that outlet.

Trouble reporting and repair analysis systems are designed to record certain required information on a daily trouble report log. At the end of each month, the data from the daily trouble forms will be tallied on a monthly trouble report summary. This form will allow the user to look at an entire month's trouble report and determine whether patterns on repeat troubles are developing.

Along with the daily trouble reporting log, it is suggested that the trouble report work sheet be used on all facility troubles. This form allows the individual doing the trouble analysis to carry a copy of all the cross-connect locations of the facility in trouble.

There are also three flowcharts that will assist in determining whether the trouble is in the terminal equipment or on the facility (network wiring). On single-line modular sets, the repair technician will even be able to isolate the problem to one of several replaceable components.

The final control document included in this chapter is a service request work sheet form provided as part of the service order process. This form should be used to request additions, disconnects, or rearrangement of facilities or terminals within a building or network.

8.2 Facility Inventory System

Maintaining accurate records on the amounts, types, and locations of equipment in buildings is becoming increasingly important as ownership of telecommunication wiring and equipment increases. In order to create and maintain a manual equipment inventory, which can later evolve into a software database, certain information must be collected and recorded.

The type of information that is required in any equipment inventory will vary with respect to individual equipment being inventoried. Data equipment will require a different format on an inventory document than single-line telephone sets or key system equipment.

Most forms will have certain common entries such as type of equipment, date of installation, department and building location, the universal information outlet identification, and whether the equipment is owned or leased.

8.2.1 Data equipment forms (Fig. 8.1)

In addition to the entries listed above, the data equipment forms will require certain additional items such as the circuit number appearing on the data terminal, the patch panel jack, and row information and controller port assignment.

8.2.2 Single-line telephone set forms (Fig. 8.2)

The single-line set form will require only a few entries. In addition to the common entries found on all forms, the telephone number terminating on the set is all that will be needed on this document.

8.2.3 Key telephone system inventory forms (Fig. 8.3)

Key system inventory records will be divided into two forms. The first will deal with the common equipment and will ask for some of the same information as the other forms (type of equipment, date of installation, and building and department location). Some of the net items required on this form are key system number (KS#), common equipment location, and the maximum capacity of lines and stations provided on the common equipment. The form will also show the number of lines and stations already in service.

The second key system form will provide information on the type of terminal, the intercom number assigned to that set, any associated equipment working with each set, the telephone lines being picked up on these terminals, and the universal information outlet that the terminal is plugged into.

8.2.4 PBX equipment inventory forms

PBX equipment inventory forms will be provided by the vendor furnishing the PBX system. These forms should include the same common entries found on all the previously discussed forms. It is also

Data Equipment Records

| Circuit number | Patch panel # | | Term type | Universal information outlet # | Date of installation | Bldg. # | Dept. | Owned or leased |
	Row #	Jack #						

Figure 8.1 (*a*) Data equipment records.

Data Terminal Type

Code	Make	Model	Description	Bit rate
001	IBM	3260A	BISYNC terminal	2.35 MBS
002	IBM	3270A	SDLC terminal	2.35 MBS
003	IBM	3770C	Printer	4.00 KBS
004	COMPAC	4XA	PC with 2 FD	1.00 MBS
005	IBM	AT	PC with 20 MHD	1.00 MBS
006	AT&T	6300	PC with 10 MHD	1.00 MBS
007	IBM	3277X	Cluster controller 16 sta.	2.35 MBS
008	CODEX	48D	Sync modem	4.8 KBS
009				
010				
011				
012				
013				
014				
015				
016				
017				
018				
019				
020				
021				
022				

Figure 8.1 (Continued) (*b*) Data terminal type.

Single-Line Terminal Records

Phone or ext. #	Universal information outlet #	Voice terminal type	Date of installation	Owned or leased	Dept.	Bldg.	Floor #

Figure 8.2 (a) Single-line terminal records.

Voice Terminal Type

Code	Make	Model	Description	Dial type	Ringer
AA	WECO	500	Desk single line	Pulse	Bell
AB	WECO	2500	Desk single line	TT	Bell
AC	III	2500	Desk single line	TT	Tone
AD	WECO	532	Wall single line	Pulse	Bell
AE	WECO	564	6 Button desk	Pulse	Bell
AF	WECO	2565	6 Button desk	TT	Bell
AG	WECO	2575	10 Button desk	Pulse	Bell
AH	WECO	2585	20 Button desk	Pulse	Bell
AI					
AJ					
AK					
AL					
AM					
AN					
AO					
AP					
AQ					
AR					
AS					
AT					
AU					
AV					
AW					

Figure 8.2 (Continued)   (*b*) Voice terminal type.

Key System Common Equipment

| Date of installation | Key system # | Vendor | | Bldg. | Dept. | Common equipment location | Number of lines | | Number of sets | |
		Model					Max	Installed	Max	Installed

Figure 8.3 (a) Key system common equipment.

Key System Terminal Records

	Key System #					Building #				Floor #				Room #											
	10	11	12	13	14	15	16	17	18	19	20	21	22	23	24	25	26	27	28	29	30	31	32	33
Intercom #	AH	AG	AG	AE	AE																			
Equpt. type																								
Telephone #																								
442-XXXX	PB	P																						
442-XXXX	PB	P	P	P	P																			
442-XXXX	PB	P	P	P	P																			
Univ. info. outlet #																								
Associated equipment																								

Figure 8.3 (Continued) (*b*) Key system terminal records.

common practice for the electronic PBX and key systems to have internal programming software and, depending on system size, software-driven inventory systems.

8.3 Cable Numbering Records

The cable numbering plan (Fig. 8.4) is divided into three components:

8.3.1 Type of cable

A two-character alphanumeric identification will be used to represent both the type of cable (primary, secondary, interbuilding, etc.) and the floor the cable originates on. A sample of this would be P01. The P represents the type of cable (primary distribution cable), and the 01 signifies that the cable originated on the first floor of the building.

8.3.2 Terminating location

A two-digit number from 01 to 99 will represent the floor the cable will terminate on.

8.3.3 Cable numbering

If more than one cable goes from the same origin to the same terminating location, they will be assigned a number from 1 to 9, predicated on the chronological sequence of placement.

A sample cable designation of O01-09-01 would represent a primary distribution cable going from the first floor to the ninth floor, and it is designated as cable number 1 in this cable path.

Cable Numbering Plan

	Type of cable ①	Place of origin (floor)	Terminating location (floor)	Number of cable
Example	P	01	09	1

① P = primary distribution cable
S = secondary distribution cable
T = tertiary distribution cable
I = interbuilding cable
A = agency distribution cable
N = network access point

Figure 8.4 Cable numbering plan.

Serving Closet Numbering Plan

	Serving closet code ①	Floor number	Closet number
Example	A	03	A

① A = apparatus closet
S = satellite closet
E = equipment room
N = network access point

Figure 8.5 Serving closet numbering plan.

8.4 Building Closet Numbering

The serving closet numbering plan (Fig. 8.5) will allow the user to identify everything needed to be known about a serving closet. This form is divided into three components:

8.4.1 Serving closet code

This is an alphabetic character signifying the type of closet. For example, A represents apparatus closet, E represents equipment room, and N represents the NAP.

8.4.2 Floor number

This is a two-digit code from 01 to 99 reflecting the floor the closet is located on.

8.4.3 Closet identification

If there is more than one closet of this type on any floor, each closet will be identified by an alphabetic character. For example, the first apparatus closet on a floor will be shown as A, the second as B, and so on.

A sample serving closet designation of A-05-B would indicate an apparatus closet on the fifth floor and the second apparatus closet on that floor.

8.5 Universal Information Outlet Numbering Plan

The universal information outlet numbering plan (Fig. 8.6) consists of seven alphanumeric characters. The following is a breakdown of the three main components of the plan:

Universal Information Outlet Numbering Plan

	Floor number	① Type closet	② Closet # on floor	Individual outlet on floor
Example	03	A	A	123

① A = apparatus closet
   S = satellite closet
   E = equipment room
   N = network access point

② Number of this type of closet on a floor

Figure 8.6 Universal information outlet numbering plan.

8.5.1 Floor number

A two-digit floor identification ranging from 01 to 99.

8.5.2 Type of serving closet

A two-alphabetic-character code. The first character represents the type of closet (A = apparatus; S = satellite; N = NAP). The second character represents the closet identification (example: the first apparatus closet on the floor will be shown as A, the second as B, and so on).

8.5.3 Universal information outlet number

A random, three-digit number assigned to a particular end-user location served from a particular closet. In a sample UIO number, 01-AA-123 would indicate that the end-user location is served from the first apparatus closet located on the first floor of the building. The last three digits indicate which station location the UIO represents.

8.6 Building Terminal Records

The heart of any building distribution system is the location where the building cables are terminated and then cross-connected to the utility company network access cables. This location is referred to as the *main building terminal,* the *1.1 terminal,* the *network access point,* and more recently the *minimum point of presence* (MPOP) and the *demarcation point.* (Isn't this a terrific example of buzzword arrogance!) These termination points can be at either the main building terminal location or any of the apparatus or satellite closets. If the building also has a PBX room or closet, individual records for the terminations within the equipment space will also be required.

There are certain major components common to all of these areas. The first item required is a record of all cable and pairs coming into the termination area. This record will include where the cables come from (MPOP, apparatus closet, etc.), the type of cable (primary, secondary, etc.), and where the pair is cross-connected to either an outgoing cable to another termination point (such as an apparatus or satellite closet), or to the end user. The end user will be identified by the UIO number.

8.6.1 Network access point (NAP), MPOP, or demarc records (Fig. 8.7)

This record, no matter which terminology you use, is the first point of record keeping in a distribution system. Before the local telephone utility will terminate a new line in a building, it will ask what pin on the registered jack (RJ21X) the new line should be placed on. The owner of the building should be able to look at a set of records and know what the next available pin is on this registered jack. It is recommended that records on the MPOP be kept by the telephone or circuit number appearing on the pin or cable pair, depending on how the circuit is tagged.

The MPOP records contain information on the block number (A to Z) and the pair (01-XYZ) associated with each telephone or circuit number. It will also contain such other information as:

- Circuit type (voice, data, etc.)
- Cross-connect location (distribution block, equipment block, etc.)
- Cable and pair termination at cross-connect location
- Universal information outlet number

8.6.2 Apparatus and satellite closet terminal records (Fig. 8.8)

The information needed for the apparatus closet and the satellite closet terminal records will be the same. These forms will contain the following information:

- Incoming cable from MPOP, another apparatus, or satellite closet
- Incoming cable and pair designation
- Cross-connect location (interagency cable, tie cable, data patch panel, etc.)
- Outgoing cable and pair number

Network Access Point Record

Network interface		Circuit or telephone #	Circuit type	Cross-connect location	Outgoing cable		U.I.O. #	Misc.
Block #	Pair				Cable	Pair		

Figure 8.7 Network access point record.

Apparatus Closet Cable Record

Incoming cable		Cross-connect location	Outgoing cable		Termination location		Cross-connect to equipment		Miscellaneous
Cable	Pair		Cable	Pair	U.I.O. #	Room #	Block #	Pair	

Figure 8.8 Apparatus closet cable record.

- Termination location by universal information outlet and room number

- Cross-connect to equipment and identification

8.6.3   Equipment room records (Fig. 8.9)

Office buildings may have one or more equipment rooms to house customer voice and data switching and controlling equipment. The records to keep track of what is working on the cables connecting the equipment room with the MPOP and other serving closets will contain the following information:

- Incoming cable—cable and pair number

- Trunk or telephone number

- Outgoing cable and pair number

- UIO number of end-user station

8.6.4   Data patch panel records

This form will keep a record of the row and jack number of the termination from the UIO, the row and jack number of the cable and pair going to the next termination point, any end closet location, the circuit number, and the type of circuit (LAN, modem, etc.).

8.7   Cable Assignment Records
(Figs. 8.10, 8.11, 8.12)

Labeling cables and keeping records on them is essential in maintaining any distribution system. The records should include the items listed below:

- *Number*—cable number taken from the cable numbering plan.

- *Place of origin*—the location at which the cable starts (MPOP, equipment room, etc.). This number will be taken from the serving closet numbering plan.

- *Termination location*—the location where the cable ends (apparatus or satellite closet, equipment room, or building number).

- *Size*—total physical capacity of the cable.

- *Count*—the sequential count of the physical facilities within the cable from facility 01 to the total number of good facilities within the sheath.

- *Type of cable*—primary, secondary distribution, or interbuilding cable.

Equipment Room Record

Room # _____   Floor # _____

Incoming cable		Trunk or telephone #	Station #	U.I.O. #	Outgoing cable #		Miscellaneous
Cable	Pair				Cable	Pair	

Figure 8.9 Equipment room record.

8.17

Satellite Closet Cable Record

Incoming cable		Cross-connect location	Outgoing cable		Termination location		Cross-connect to equipment		Miscellaneous
Cable	Pair		Cable	Pair	U.I.O. #	Room #	Block #	Pair	

Figure 8.10 Satellite closet cable record.

Interbuilding Cable Record

Cable #	Place of origin	Termination location	Size	Count	Type of cable		

Pair #	Circuit/ telephone #	U.I.O. #	Pair #	Circuit/ telephone #	U.I.O. #	Pair #	Circuit/ telephone #	U.I.O. #

Figure 8.11 Interbuilding cable record.

Distribution Cable Records

Cable #	Place of origin	Termination location	Size	Count	Type of cable		

Pair #	Circuit/ telephone #	U.I.O. #	Pair #	Circuit/ telephone #	U.I.O. #	Pair #	Circuit/ telephone #	U.I.O. #

Figure 8.12 Distribution cable records.

The second half of this form will reflect the facility assignments and will contain the circuit or telephone number and the UIO number assigned to each circuit.

8.8 Location Records

These records provide a visual reference to the wiring location in buildings and interbuilding transitory facilities. This document can be drawn to include basic data necessary for building riser duct location and inventory of the wiring plan's end-to-end architecture and will complement the forms described previously. The basic components of the location record are detailed below.

8.8.1 Cable location records
(Figs. 2.1, 2.2, and 4.14)

These records will graphically display the following information:

- Site boundaries and building locations
- Geographical and street reference information
- MPOP, closet, and equipment room locations
- Cable identifications and paths

Individual station information is not included in the location record. This information is contained on previously described forms and would be difficult to draw because of the no-scale configuration of the record. Additionally, this record is intended as a visual diagram of the backbone cable system only and is to be used for rapid physical reference during maintenance on the system.

8.8.2 Campus location records

The campus environment may include aerial, buried, and underground conduit facilities. The figures mentioned in Sec. 8.8.1 show the method of identifying the routes of these various cables, but other location techniques will apply for maintenance of these interbuilding facilities.

8.8.3 Aerial facilities

Occasionally cables using pole line construction will be placed, usually on a temporary basis at construction sites, on power company joint poles. The power company will maintain the pole line (poles only), and, as the facilities will be telephone-company-owned in most situations and temporary, the cable will be maintained by others. In the event the client wishes to maintain a permanent record for these facilities, the information displayed on the campus distribution record will be sufficient.

8.8.4 Buried cable

In addition to the information contained on the campus distribution record, geographic, scaled drawings will have to be maintained for buried-cable facilities. This is necessary for locating the cable in the field for maintenance, subsequent digging activity, and interception for future extensions. The most economical and efficient means for accomplishing this is to retain a copy of the as-built construction drawings and then to reference the contract number as shown on the campus distribution drawing.

8.8.5 Conduit location records

The conduit location record method is the same as that for buried cable. Additionally, duct assignment is critical. The improper assignment of ducts will result in cables blocking access to vacant ducts, requiring a very costly cable rearrangement for new conduit construction. The rules for multiple duct assignment are contained on the attachment to the campus location records.

8.9 Facility Modifications (Fig. 8.13)

Facility modifications will be required for moves, changes, and/or additions to the information system. A service request work sheet should be generated whenever there is a need to add, disconnect, or rearrange an existing line or circuit within a building or network. This form is very similar to the trouble report work sheet with only minor changes such as the headings and the *date service requested* entry.

The person filling out the service request work sheet should complete the following entries on the form:

- Request from (name and department)
- Request date
- Building and room number where service is required
- Contact number—telephone number including any extension or station where service-initiation contact may be reached
- Date service is required

Service Order Work Sheet

Request from: _____ NameContact # _____ Ext. # _____

_____ Work completed: _____ Date

Request: _____ _____

_____ Completion reported to: _____ Name

Work done by: _____ _____

Building # _____ _____ Date service required _____

Description of request:

Circuit or telephone # _____ Circuit type _____

Terminal type _____ Universal information outlet # _____

Distribution map

Distribution system location	Cross-connects	Existing		New		Comments
		Cable	Pair	Cable	Pair	
Network interface	In					
	Out					
Primary distribution cable	In					
	Out					
Apparatus closet	In					
	Out					
Apparatus closet	In					
	Out					
Satellite closet	In					
	Out					
Satellite closet	In					
	Out					
Data patch panel	In					
	Out					
Building entrance terminal (campus)	In					
	Out					
Universal information outlet	Outlet #					

Figure 8.13 Service order work sheet.

- Description of request—include any information as to features and systems required that may be helpful to installation forces
- Circuit or telephone number
- Circuit type—voice, data, other
- Universal information outlet number

The information for the distribution map is to be filled out by the person taking the report. This work sheet will then be given to the person assigned to completing the service request. After the work is completed, fill in the name and telephone number of the person this service request has been cleared with.

8.10   Facility and Equipment Diagnostic System

At some point in time, it will become necessary to determine why a terminal does not work. Three flowcharts are furnished to assist in and simplify the diagnostic process.

8.10.1   Modular and nonmodular set flowchart

Figure 8.14 will take the user through a step-by-step diagnostic procedure to determine the source of the impairment on modular single-line sets and modular or nonmodular key telephone sets. This chart is designated to be used only for new installations and is not intended to cover older, nonmodular single-line telephone sets. Following the steps on this flowchart will enable the user to isolate the source of the problem to either the terminal equipment (e.g., telephone set, cords, or handset), the building distribution system, or network (utility company) facility.

8.10.2   Line troubleshooting procedure

If the steps in Fig. 8.14 show the trouble to be on the line, then go to Fig. 8.15, which will help the user to determine if the trouble is within the building distribution system or out on the local utility company's network.

8.10.3   Data troubleshooting procedure

Figure 8.16 is similar to Fig. 8.14, but it deals with isolating trouble on data terminals instead of telephone sets. Since data equipment usually requires vendor maintenance, there are fewer steps in isolat-

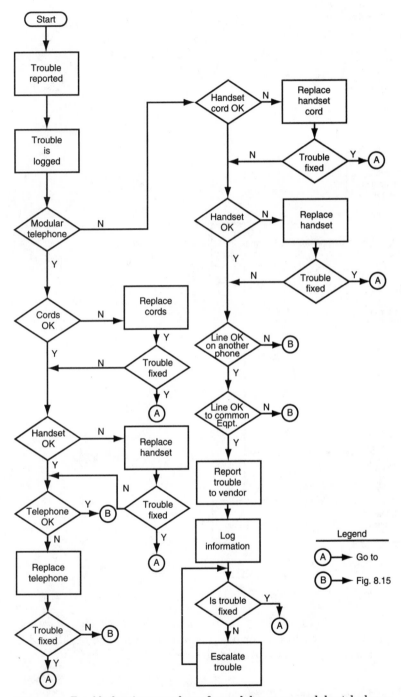

Figure 8.14 Troubleshooting procedures for modular or nonmodular telephones (KTS).

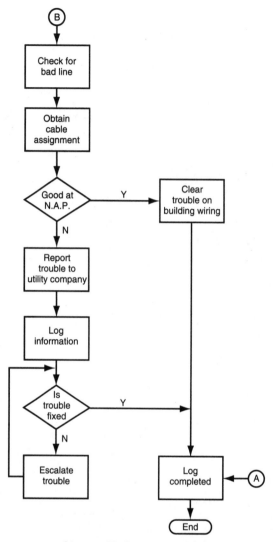

Figure 8.15 Line troubleshooting procedures.

ing the trouble between the terminal equipment and the building distribution system or utility-provided circuit.

8.10.4 Diagnostic equipment

If the using agency has determined that the trouble on a terminal is
isolated to the wiring in the building distribution system, certain

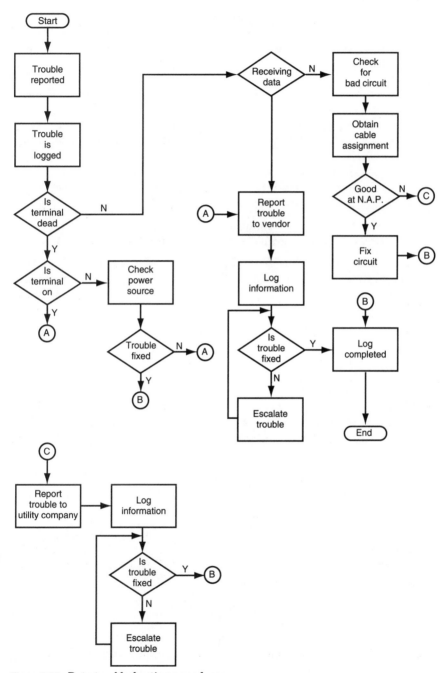

Figure 8.16 Data troubleshooting procedures.

basic types of test equipment will be required to assist in further isolation of the source of the trouble. The following is a generic list of basic test equipment which would be helpful in troubleshooting low-voltage interconnect facilities.

■ *Lineman's test set.* Called a *butt-in* within the industry, it looks very much like a cellular hand-held phone with patch cords. It should be equipped for touch-tone or dial pulse output, be totally line powered, and have built-in polarity guard, light-emitting diodes (LEDs) for polarity indication, and an on-line and monitor/talk switch. Modular output is desirable but if not available, obtain test set replacement cords to allow testing directly to modular outlets.

■ *Tone generator.* For producing modulated signals over a pair of wires to perform continuity and pair location tests.

■ *Modular phone jack tester.* Tests termination correctness of UIO and modular patch panel jacks. Usually uses LEDs to indicate test result and should be equipped to test modular jacks with as many as four pairs.

■ *Hand-held digital multimeter.* Full-range capabilities, overload protection, long battery life, and low-battery indicator.

■ *Ohmmeter.* Voltage, continuity, resistance, dB, and grounding measurement capability.

■ *Time-domain reflectometer (TDR).* Data test set which produces digital signals in synchronous and asynchronous modes, T-carrier emulation, and basic data test signal capability common to both digital, radio-frequency, and analog signals.

■ *Optical time-domain reflectometer (OTDR).* Synthesizes optical signals for full range of light wavelengths propagating over an optical fiber.

There is a wide range of devices in each of the above types of test equipment and a corresponding range of prices. It is best to check with your equipment manufacturer first before purchasing any test equipment beyond the basic ohmmeter.

8.11 Trouble Reporting and Repair Analysis Procedure

All trouble reports should be logged on the daily trouble report log. This log will be used to record all troubles on line facilities and terminal equipment reported with the wiring network.

The daily trouble report log will contain the following information (Fig. 8.17):

- Date and time report is received
- Trouble reported by (name, telephone number, work group identification, and UIO)
- Type of trouble reported (physical circuit or equipment)
- Type of trouble reported
- Person dispatched on trouble (time and date)
- Time trouble cleared (hour and date)
- Duration of outage
- Trouble found (description and location)
- Corrective action reported to (name and date)

If the troubleshooting procedures isolate the problem to the facilities within the building, use the trouble report work sheet to record the information needed to follow the distribution path throughout the building. This form will allow the person troubleshooting the facilities to carry a complete record of each element of the distribution system from the network interface up to the universal information outlet. As the need arises to replace any of the elements, the new assignment can be recorded on the form.

If the trouble reported has been isolated to a piece of vendor-provided equipment, use the vendor trouble report work sheet to keep track of vendor performance.

The data accumulated from the daily trouble reporting logs will be summarized at the end of each month, on the monthly trouble report summary (Fig. 8.18).

Categories found on the monthly report form will include the following information:

- Date—month, day, and time of trouble report
- Facility circuit/telephone number
- Voice terminal UIO number
- Data terminal UIO number
- Trouble reported
- Duration of outage
- Trouble found
- Department/building reporting trouble

Daily Trouble Report Log

| Report received | | Trouble reported by name, number, and dept. | Type of trouble | | Trouble reported | Trouble assigned to | Trouble cleared | | Duration of outage | | Cleared report to name, number | Trouble found |
Date	Time		Ckt. # or tele. #	Voice or data U.I.O. #			Date	Time	Hr.	Min.		

Figure 8.17 Daily trouble report log.

Monthly Trouble Report Summary

Date	Facility Circuit # telephone #	Voice terminal U.I.O. #	Data terminal U.I.O. #	Trouble reported	Duration of outage Hr.	Min.	Trouble found	Reporting trouble Dept.	Bldg.

Figure 8.18 (a) Monthly trouble report summary.

Trouble Reported Codes		Trouble Found Codes	
Code	Description	Code	Description
01	No dial tone	01	Open wire
02	Noise	02	Shorted
03	Errors	03	Bad patch cord
04	No data	04	Network trouble
05		05	
06		06	
07		07	
08		08	
09		09	
10		10	
11		11	
12		12	
13		13	
14		14	
15		15	
16		16	
17		17	
18		18	
19		19	
20		20	
21		21	
22		22	
23		23	
24		24	
25		25	
26		26	
27		27	
28		28	
29		29	
30		30	
31		31	

Figure 8.18 (Continued) (*b*) Trouble reported and trouble found codes.

An example of the type of entries that can be used in the columns marked *trouble reported* and *trouble found* appear on the forms. There are many other entries that can be placed on these forms, and the system manager can add whatever entries best fit the users' needs.

When all data are accumulated, the manager should review the monthly report and look for common trouble types and/or locations. At this point the manager can also multiply the hours spent on trouble and assess a monthly cost to maintain the existing, or status quo, facilities and use the data for evaluating repair or rehabilitation alternatives.

9

Planning Guidelines

9.1 A System Approach to Connectivity Planning

The purpose of this chapter is to familiarize the system manager with the standard techniques for an economic evaluation of whether to retain an existing system or replace it with an upgraded wiring network. Because sound business decisions are based on a relatively stable set of criteria, emphasis will be placed on the application of the specific variables associated with wiring systems and their relation to economic criteria. While the requirement for a formal study may not apply in all cases, it is recommended that the procedure be followed for all planning cost studies.

The economic study should give the manager whatever information is required for making a valid choice among alternatives involving different amounts of capital expenditures, expenses, and revenues, and the presentation of these results should be in a format that is as helpful as possible. Once presented with this information, the manager must be able to objectively evaluate the study. To ensure confidence in the final recommendation, each study should contain the following components:

- A definition of the objective of the study
- A description of the alternatives considered
- A description of the data used
- The study techniques employed
- The study results
- The plan recommended

9.1.1 Study objective

The first component of the economic study identifies the reason for the study by answering the following questions:

- What is the problem?
- Where is the problem?
- Why is there a problem?
- What is the extent of the problem?
- Why fix the problem?
- Why fix the problem now?
- Why fix the problem this way?

Once the need for and purpose of the study have been established, subsequent analysis will determine if the problem can be solved, and, if so, at what cost. The objective should also establish the limits on the scope of the study and its overall impact on the system.

9.1.2 Alternatives considered

Prior to selecting a course of action, it will be necessary to identify plans which are compatible with the environment and which perform the functions required by the study and meet the objective statement. Courses of action that are not compatible with the environment can be rejected outright. Once truly comparable alternatives are identified, the analyst will compare each system on an economic basis.

Each reasonable plan which is rejected for any reason should be backed up with adequate documentation explaining why it was rejected. In cases where there are too many viable courses of action to perform separate evaluations, the rationale should be stated, explaining why certain alternatives were selected for detailed analysis.

9.2 Study Techniques

The objective of economic comparison studies is to describe and predict the economic impact a particular decision will have on the business or department (including the do-nothing alternative). Alternatives can be analyzed and compared by a mathematical model—an inexpensive way to examine a large number of alternatives without actually making a commitment of resources.

Generally, economic and noneconomic considerations will not be served well in the same study—for example, maximizing service quality versus minimizing cost.

The most commonly used approaches to determine the economic advantage to any particular plan of action, or nonaction, is to develop one or more of the following comparative economic indicators for each alternative.

9.2.1 Internal rate of return

The internal rate of return views projects from the investor's point of view. Internal rate of return refers to the critical cost of money at which a project breaks even. However, an *incremental* internal rate of return must be calculated in order to decide among alternatives.

The *incremental internal rate of return* can be defined as the interest rate which causes the present worth of the incremental cash flow between two projects to be zero. If the incremental cash flow represents an investment followed by its recovery, the internal rate of return is then the break-even cost of money.

It is important to note that the internal rate of return approach can be applied only in deciding between two mutually exclusive alternatives. When more than two alternatives must be evaluated, it is necessary to make sure that all the pairs are compared, i.e., A to B, A to C, B to C, etc.

Further, this technique assumes competition for borrowed funds from various sources. It applies more to venture expenditure than the normalized funding architecture of the typical corporate or business environment.

9.2.2 Present worth of expenditure

Of more use to the system manager is the present worth of expenditure technique, which approaches an economic study from the customer's viewpoint. This method will select the alternative with the lowest revenue requirements because its objective is to find which alternative will meet the service objectives at the least cost. The decision is based on the lowest total present worth of expenditure over the life of the study. This method assumes that budgeting will be sufficient to fund any alternative selected.

9.2.3 Present worth of annual costs

The present worth of annual charges method also approaches an economic study from the customer's point of view. However, with this technique the actual expenditures are represented by a levelized equivalent annual amount which represents the amount which will be expended on a monthly billing arrangement. This method is time-sensitive in that it relates the useful life of the purchase and prorates that expense over the study period. This study may be useful when contemplating first-cost expenditures for common components with different service lines, such as a PBX with fiber versus cable.

9.2.4 Benefit-to-cost ratio

The benefit-to-cost ratio is a measurement of overall efficiency, defined as

$$\text{Efficiency} = \frac{\text{output}}{\text{input}} + \frac{\text{benefit}}{\text{cost}}$$

For a given net cash flow, the output is the positive net flow and the input is the negative net flow. The benefit can be defined as the present worth of the positive net cash flow, and the cost as the present worth of the negative net cash flows:

$$\text{Efficiency} = \frac{\text{output}}{\text{input}} + \frac{\text{PW (positive net cash flows)}}{\text{PW (negative net cash flows)}}$$

If the benefit/cost ratio calculated by this formula is greater than 1.0, the alternative is generating enough cash to recover the invested funds. The most efficient alternatives will generate ratios much higher than 1.0. Consequently, it will be crucial to constrain study parameters to allow truly equal comparisons.

9.2.5 Payback period

This is the most common decision-making economic study because of its relative simplicity and the ability it gives a manager to relate to time-period investment return criteria. The payback period is the time (in months or years) required for the cost savings and receipts from an investment to equal the amount of investment with no interest. The major drawback to this method is that it ignores all events which occur after the payback itself.

9.2.6 Break-even study

Often, an alternative will be more economical under one set of circumstances than under another. By altering one variable while holding all others stable, it is possible to assess a value to the variable which will make the competing investments economically equivalent. The manager can then estimate the likelihood of an occurrence of the sensitive variable or variables to the extent that it or they would alter the economic selection.

Some of the decision quantification models commonly associated with economically sensitive models are:

- Expected monetary value
- Conditional profit
- Loss minimization
- Managerial analysis

A detailed discussion of these effects on an economic study would fill many pages and be of questionable value to an analysis of wiring systems. This is because they are irrelevant to the capital funding system employed by most business enterprises and because the decision process required for reinforcement and replacement associated with designing buildings for ultimate wiring or circuit capacity is quite straightforward.

9.2.7 Study technique summary

A broad range of study alternatives has been described because it is necessary to view alternatives from more than one perspective. The internal rate of return, the present worth of expenditures, the capital requirements, the risk, the annual cash flows, etc. are all integral to the decision process. Since most if not all funding for cable replacement or reinforcement comes from the general funds of a business or department and not from market-competitive sources such as bond issues, much of the interest-sensitive criteria inherent in the study techniques mentioned above will not apply. However, since the system manager may be involved in the selection of various staffing proposals to deal with the purchase and subsequent maintenance of existing building wiring systems, it will be appropriate to view future requirements as an investor. Therefore, while not entirely within the scope of this work, it is recommended that projects involving force and/or investment increases incorporate those methods employed by most profit-sensitive investments to evaluate effective and economic plans of action. A detailed discussion of these methods will be found in *Engineering Economy.**

---

*American Telephone and Telegraph Co., *Engineering Economy,* 3d ed., McGraw-Hill, New York, 1977.

9.3 Economic Variables

Studies are usually quite specific in terms of equipment and hardware. However, data gathered from general average accounting systems will include all components associated with an account, not just the specific equipment and hardware expenditures. It will be necessary, in this case, to acquire subunit information that is valid for comparative or quantitative extrapolation. Also, concurrent projects may affect economy-of-scale impacts that could qualify study alternatives that would not be competitive in a stand-alone study.

Besides being confident that the data are reasonable, reliable, and consistent, the analyst must assure that all relevant economic factors have been considered. Some of the major economic factors to be considered are:

- Capital expenditures
- Salvage
- Inflation
- Rent
- One-time expenses
- Recurring expenses
- Maintenance expense
- Cost of removal
- Equity return
- Interest during construction

In addition, many times technological advance can negatively or positively impact the status quo environment. Worker and revenue effects must also be calculated for the alternatives for accurate cost comparisons.

9.4 Information Gathering

Often, information to be included in a study will come from numerous sources, not all having the same degree of reliability. To help in assuring the credibility of data to approval authority, it will be necessary to document the source of information and all assumptions associated with it.

Some sources of data are:

- Internal records of similar placements

- Engineers and architects
- Purchasing department records
- Vendor representatives
- Other system managers
- Accounting office
- Technical letters and journals
- Supply catalogs
- Engineering economy texts
- Operating telephone company organizations
- Advertisements
- Manufacturer representatives
- System reports
- Trade journals
- Professional comparison journals and publications

Keep in mind that factors which on the surface appear insignificant may, on detailed examination, have a substantial economic impact. Additionally, any economic factors which are eliminated from the study because they have a common effect on all alternatives should be specifically noted.

9.4.1 Per-unit cost estimating process

Per-unit costs represent current average costs to construct, repair, or rearrange major items of wire and cable. Aerial, buried, underground, and building cable costs are shown in Sec. 9.4.4. The prices represent average utility company costs per unit of plant. These figures may be used until the analyst acquires sufficient data to develop a personal set of costs. These costs will be used by the administrators of building wiring systems to prepare cost comparisons only. *Should the analyst desire to determine approximate budgetary allocations, then a 15 to 35 percent increase in the costs presented here must be included for vendor profit.* These costs should be considered valid only for the development of cost comparisons between equal alternatives and should be avoided for budget development.

9.4.2 Cost applications

After a plan of action is developed, the per-unit costs will be allocated to each type of facility proposed. Cable prices list two costs: one with splicing labor included and one without. Generally, cables within buildings will not require splicing and the *excluding splicing* cost will be used. Cable termination costs for labor plus material would then be added. The most common setting for splicing arrangements will be in the aerial, buried, and underground campus placement environment.

Once all items have been priced, a sum will be developed to represent the estimated cost to construct the project.

9.4.3 Additional costs

Any special condition can be accommodated by developing an additional value for it. These conditions include, but are not limited to, the following:

- Core bore in buildings
- Additional conduit (entrance to building)
- Additional labor for cable placement in congested facilities
- Rearrangement costs
- Manhole/cable-vault pumping
- Cutting concrete, sidewalks, or pavement
- Miscellaneous repairs
- Motor vehicle expense/travel
- Special equipment rental
- Contract administration
- Engineering and incidental expense
- Overhead expense

Once all costs are identified, an analysis may be prepared by using the methods outlined above. These studies may then be compiled in any manner consistent with internal approval authority. *Should the costs be developed for a budgetary process, the amount developed using these guidelines must be multiplied by 1.35 (vendor profit factor) to reflect the expected gross bid amount.*

9.4.4 Per-unit costs

The unit prices in Fig. 9.1 represent utility company averages and are commonly used by competing vendors. Again, vendor profit can be a major factor in determining the final product price. At times, profit can exceed 100 percent of the actual estimated time and material calculations, so be careful when attempting to forecast actual expenses on wiring bids. In balance, it must be recognized by the estimator that study and planning costs are a pure form of determining the most economical method of picking a system or project alternative. Budgets, on the other hand, will require inclusion of profit and common costs to meet the overall needs of the funding process. In this regard, pick high, as no one will ever be disappointed when the bids come in lower than the estimated amount.

The types of cable plant in Fig. 9.1 are listed by environment, unit to be estimated, unit cost, and labor hours per unit.

CAUTION: Per-unit cost estimating is an extremely precise art. Often components and/or activities are omitted due to error or statistical manipulation. In either case, the result is that a misleading decision-making recommendation is presented. This can be extremely embarrassing and, in the long run, expensive!

Figure 9.1 Per-unit cost estimating charts.

(a) Building cable construction.

CONSTRUCTION

Cable	Unit	Total unit cost (excluding splicing)	Total unit cost (including splicing)
25 pair—all gauge	ft	2.15	4.93
50 pair—all gauge	ft	2.28	5.44
100 pair—all gauge	ft	2.66	6.83
200 pair—all gauge	ft	3.92	9.87
300 pair—all gauge	ft	4.43	12.14
400 pair—all gauge	ft	5.06	14.67
600 pair—all gauge	ft	6.83	20.11
900 pair—all gauge	ft	7.34	26.06
1200 pair—all gauge	ft	8.60	32.76
1500 pair—all gauge	ft	9.87	39.47
1800 pair—all gauge	ft	11.01	46.05

NOTE: Splicing cost = 1 standard splice prorated on cable distance of 100 ft.

Ground wire	Unit	Total unit cost
Wire ground, B no. 6	ft	2.15

Cable stubs	Unit	Total unit cost (excluding splicing)	Total unit cost (including splicing)
10A1-50/10B1—50 stub 30 ft	Each	69.58	354.20
10A1-100/10B1—100 stub 30 ft	Each	82.23	442.75
10A1-200/10B1—200 stub 30 ft	Each	113.85	657.80
10A1-300/10B1—300 stub 30 ft	Each	139.15	847.55
10A1-400/10B1—400 stub 30 ft	Each	164.45	1049.95
10A1-600/10B1—600 stub 30 ft	Each	290.95	1518.00
10A1-900/10B1—900 stub 30 ft	Each	354.20	2099.90

(Continued)

Figure 9.1 (Continued) Per-unit cost estimating charts.

(a) (cont.) Building cable construction.

Cable stubs	Unit	Total unit cost (excluding splicing)	Total unit cost (including splicing)
50-pair stub 30 ft	Each	63.25	379.50
100-pair stub 30 ft	Each	75.90	493.35
200-pair stub 30 ft	Each	113.85	708.40
300-pair stub 30 ft	Each	139.15	910.80
400-pair stub 30 ft	Each	177.10	1138.50
600-pair stub 30 ft	Each	278.30	1593.90
900-pair stub 30 ft	Each	354.20	2226.40

NOTE: Splicing cost = 1 standard splice.

(b) Aerial cable construction, removal, and rearrangements.

CONSTRUCTION

Cable (strand or self-supporting, excluding building cable)	Unit	Total unit cost (excluding splicing)	Total unit cost (including splicing)
25 pair—26, 24, 22 gauge	ft	1.27	1.90
50 pair—26, 24, 22 gauge	ft	1.27	1.90
100 pair—26, 24, 22 gauge	ft	1.90	2.53
200 pair—26, 24, 22 gauge	ft	2.53	3.80
200 pair—19 gauge	ft	4.55	5.57
300 pair—26, 24, 22 gauge	ft	3.16	4.43
400 pair—26, 24 gauge	ft	3.16	4.48
400 pair—22 gauge	ft	4.43	6.33
600 pair—26 gauge	ft	4.43	6.33
600 pair—24 gauge	ft	4.48	7.59
600 pair—22 gauge	ft	6.96	8.86
900 pair—26 gauge	ft	5.69	8.86
900 pair—24 gauge	ft	6.96	10.12
900 pair—22 gauge	ft	9.49	12.65
1200 pair—26 gauge	ft	6.96	9.49
1200 pair—24 gauge	ft	10.12	12.65
1200 pair—22 gauge	ft	13.28	15.81

NOTE: Splicing cost = 1 standard splice prorated on cable distance of 600 ft.

Terminals (building)

50 pair terminal	Each	468.05
100 pair terminal	Each	721.05
200 pair terminal	Each	1201.75
300 pair terminal	Each	1771.00
400 pair terminal	Each	2277.00

NOTE: Splicing cost = 1 standard splice + termination splice.

Figure 9.1 (*Continued*) Per-unit cost estimating charts.

(*b*) (*cont.*) Aerial cable construction, removal, and rearrangements.

REMOVAL

Cable	Unit	Total unit cost (excluding splicing)	Total unit cost (including splicing)
All sizes and gauges	ft	0.63	1.20
Stubs			
50 to 900 pair	Each	18.98	657.80
Terminals			
All sizes and types	Each	18.98	303.60
Terminal cross-connects			
All sizes and types	Each	63.25	759.00
Connecting blocks			
All types	Each	18.98	

REARRANGEMENT

Aerial cable		
Moved	Cable	63.25
Transferred	Cable	101.20
Relocated	ft	1.27
Replace—400 pair and smaller	100 ft	177.10
Replace—400 pair and larger	100 ft	278.30
Building cable		
Relocated	Sheath ft	6.33
Building terminal		
Relocated	50 pair	88.55

(*c*) Buried cable construction, removal, and rearrangement.

CONSTRUCTION

Cable	Unit	Total unit cost (excluding trenching and splicing)	Total unit cost (including trenching and splicing)
25 pair—26, 24, 22, 19 gauge	ft	2.91	3.29
50 pair—26, 24 gauge	ft	2.66	3.16
50 pair—22, 19 gauge	ft	3.42	3.92
100 pair—26, 24, 22 gauge	ft	3.29	4.05
100 pair—19 gauge	ft	4.68	5.44
200 pair—26, 24 gauge	ft	3.80	4.93
200 pair—22 gauge	ft	4.68	5.82
200 pair—19 gauge	ft	6.58	7.72
300 pair—26, 24 gauge	ft	4.43	5.95
300 pair—22 gauge	ft	5.19	6.70
300 pair—19 gauge	ft	8.60	10.12
400 pair—26 gauge	ft	4.68	6.58
400 pair—24 gauge	ft	5.57	7.46
400 pair—22 gauge	ft	6.70	8.60

(*Continued*)

Figure 9.1 (*Continued*) Per-unit cost estimating charts.

(*c*) (*cont.*) Buried cable construction, removal, and rearrangement.

CONSTRUCTION

Cable	Unit	Total unit cost (excluding trenching and splicing)	Total unit cost (including trenching and splicing)
400 pair—19 gauge	ft	9.36	13.28
600 pair—26 gauge	ft	5.82	8.48
600 pair—24 gauge	ft	6.96	9.61
600 pair—22 gauge	ft	7.72	10.37
900 pair—26 gauge	ft	7.34	11.01
900 pair—24 gauge	ft	9.11	12.78
900 pair—22 gauge	ft	10.12	13.79

NOTE: Splicing cost = 1 standard splice prorated on cable distance of 300 ft.

Trenching	Unit		Total placing cost
Contract—all areas	ft	11.07	13.28
Joint trench (50% telco)	ft		7.59
Customer provided	ft		4.43

Plowing	ft		5.06

Boring			
Cable 2 in or less diameter	ft		12.65
Cable 2 in or larger diameter	ft		15.18

Splice boxes/pits	Unit	Total unit cost
Splice box 3' × 5' × 3'6"–4'6" 12" × 22" × 12"	Each	1948.10
Splice box 27" × 30" × 12"	Each	354.20
Splice pit	Each	543.95

Cabinets (outside type)	Unit	Total unit cost
29 type	Each	404.80

Closures		
Large closure (all)	Each	379.50
Pedestal-type closure	Each	215.05

Terminals	Unit	Total unit cost (excluding splicing)	Total unit cost (including splicing)
Distribution type (all)	Each	202.40	404.80

NOTE: Splicing cost = 1 standard splice (averaged all sizes).

Cross-connect type			
100 pair	Each	543.95	759.00
200 pair	Each	683.10	1012.00
300 pair	Each	796.95	493.35
400 pair	Each	961.40	1518.00
600 pair	Each	1277.65	2074.60

NOTE: Splicing cost = 2 standard splices.

Figure 9.1 (*Continued*) Per-unit cost estimating charts.

(*c*) (*cont.*) Buried cable construction, removal, and rearrangement.

REMOVAL

Terminals	Unit	Total unit cost (excluding splicing)	Total unit cost (including splicing)
Distribution type	Each	113.85	215.05
Cross-connect type	Each	240.35	506.00
Cabinets	*Unit*		*Total unit cost*
29 type	Each		240.35
Closures	*Unit*		*Total unit cost*
All types	Each		37.95

REARRANGEMENTS

Terminal replaced	Unit	Total unit cost (excluding splicing)	Total unit cost (including splicing)
Outside distribution type	Each	215.05	442.75
Buried cable			
Relocated	Cable	12.65	
Transferred	Cable	189.75	
Pedestal closure			
Replaced	Each	113.85	

(*d*) Pole line construction, removal, and rearrangement.

CONSTRUCTION

Poles	Unit	Total unit cost
22 ft and under	Each	290.95
25 ft	Each	417.45
30 ft	Each	468.05
35 ft	Each	556.60
40 ft	Each	600.88
45 ft	Each	790.63
Anchors		
$\frac{3}{4}$-in rod and smaller	Each	215.05
1-in rod and larger	Each	366.85
Guys		
6M and smaller	100 ft	240.35
10M and larger	100 ft	341.55

(*Continued*)

Figure 9.1 (*Continued*) Per-unit cost estimating charts.

(*d*) (*cont.*) Pole line construction, removal, and rearrangement.

REMOVAL

Poles	Unit	Total unit cost
22 ft and under	Each	37.95
25 ft	Each	63.25
30 ft	Each	88.55
35 ft	Each	126.50
40 ft	Each	151.80
45 ft	Each	215.05

Cross arms		
Cable and guard arms	Each	25.30
Extension arms (all types)	Each	25.30

REARRANGEMENT

Poles moved	Unit	Total unit cost
1 ft to 15 ft	Each	75.90
Over 15 ft	Each	290.95
Pole stub placed	Each	88.55

Cross arms (all types)	Unit	Total unit cost
Move/replace pins	Arm	25.30
Transfer/replace	Arm	37.95

Guy moved		
Move existing guy	Guy	37.95
Replace guy	Guy	50.60

Anchors		
Moved	Anchor	177.10
Replaced	Anchor	240.35

(*e*) Underground cable construction, removal, and rearrangement.

CONSTRUCTION

Cable	Unit	Total unit cost (excluding splicing)	Total unit cost (including splicing)
25 pair—26, 24, 22, 19 gauge	ft	0.89	1.52
50 pair—26, 24, 22, 19 gauge	ft	1.27	1.90
100 pair—26, 24, 22, 19 gauge	ft	1.64	2.53
200 pair—26, 24, 22, 19 gauge	ft	2.40	3.54
300 pair—26, 24 gauge	ft	2.28	3.54
300 pair—19, 22 gauge	ft	4.05	5.31
400 pair—26, 24 gauge	ft	2.78	4.30
400 pair—22 gauge	ft	4.30	5.95
600 pair—26 gauge	ft	3.29	5.31
600 pair—24 gauge	ft	4.43	6.45
600 pair—22 gauge	ft	6.07	8.10

Figure 9.1 (*Continued*)   Per-unit cost estimating charts.

(e) (*cont.*) Underground cable construction, removal, and rearrangement.

CONSTRUCTION

Cable	Unit	Total unit cost (excluding splicing)	Total unit cost (including splicing)
900 pair—26 gauge	ft	4.68	7.46
900 pair—24 gauge	ft	6.20	8.98
900 pair—22 gauge	ft	8.86	11.64

NOTE: Splicing cost = 1 standard splice prorated on cable distance of 600 ft.

Rodding duct		Total unit cost
Duct	Unit	0.10

Cable stubs	Unit	Total unit cost (excluding splicing)	Total unit cost (including splicing)
100 pair (Avg. 20 ft)	Each	25.30	607.20
200 pair (Avg. 20 ft)	Each	37.95	822.25
300 pair (Avg. 20 ft)	Each	50.60	1049.95
400 pair (Avg. 20 ft)	Each	56.93	1265.00
600 pair (Avg. 20 ft)	Each	82.23	1707.75
900 pair (Avg. 20 ft)	Each	113.85	1834.25
1200 pair (Avg. 20 ft)	Each	145.48	2327.60

NOTE: Splicing cost = 1 standard splice.

REMOVAL

Cable	Unit	Total unit cost (excluding splicing)	Total unit cost (including splicing)
All sizes and gauges	ft	0.51	1.14

Cable stubs			
All sizes and gauges (Avg. 20 ft)	Each	8.86	1075.25

REARRANGEMENT

Cable relocated	Unit	Total unit cost (excluding splicing)
	100 ft	0.38
	cable	56.93

9.5   Capital Expenditure Study Technique

For the purposes of this text, the capital expenditure study technique will be a rare occurrence in practice. This work deals primarily with cost-estimating various ultimate connectivity configurations, with funding assumed.

However, should competing capitalization projects be contemplated for the connectivity of a large private or public network where the physical link is a revenue-producing unit, the present worth of annual costs (PWAC) model lends itself to the cause. Capitalization projects usually concern themselves with infinite time frames and economic reinforcement of existing (embedded) plant configurations, using time-sensitive growth demand forecasts.

These studies compare economic variables peculiar to a given set of criteria established within each operational unit and apply them to all available facility reinforcement and potential facilities and components, recomputing infinite subsets of the data. The technique lends itself well to mainframe environments where potentially thousands of variables are updated and processed continuously.

Should the analyst decide to proceed, it is strongly recommended that he or she contact a Bell or General Telephone Company engineering representative to arrange a time-sharing or program purchase agreement. Be sure to include training as part of the package. While the studies are relatively simple, improper data entry will lead to inaccurate recommendations.

9.6   Maintenance Expense Study Technique

For dealing with the matter of maintaining an existing wiring system or replacing it, the following hybrid of payback period, present worth of annual cost, and benefit-to-cost ratio studies is presented as a model for expenditure analysis of maintaining existing facilities versus rehabilitation. For the purpose of uniformity, this type of study will be referred to as *status quo* versus *rehabilitation*.

9.6.1   Study assumptions

For conducting status quo versus rehabilitation studies, the following assumptions will be made:

- All new facilities will be sized for the ultimate requirements of each candidate site.
- Funding will be available for any alternative selected.

- Economic efficiencies will be the sole rationale for alternative selection.

- All plan service features will be equal.

- Equipment costs will not be considered in any alternative unless they are unequal for any plans.

Having satisfied this set of assumptions, and any others that may be appropriate for the study, the analyst can proceed to develop the study by the following procedure.

9.6.2   Selection of study alternatives

Several alternative courses of action will usually meet the ultimate objective of a project. They are normally mutually exclusive since the selection of one alternative rules out all others. To save study time, it is reasonable to eliminate marginal alternatives from further study by inspection and then analyze each of the remaining alternatives against the basic plan.

Selection of basic alternatives.   The basic plan is the alternative which represents the lowest first cost up to the time of cutover and meets established demand and service criteria. If there are several alternatives under study, the results are meaningful only if the status quo plan is compared with each of the proposed plans. It is important that the plans should include all revenues, first costs, net salvage, and cash expenses that are not common between all plans.

Incremental computations.   Each comparison allows for two alternatives. The first, or A plan, is the proposed alternative; the second, or B plan, is the basic alternative, or status quo. The data for each plan are composed individually and all dollar amounts are calculated as the incremental difference between plan A and plan B. The timing of either plan can thus be shifted without affecting the other alternative.

9.6.3   Data collection

Data must be supplied for both alternatives. The study is only as accurate as the input data. Data are grouped by category type. Data that are needed for a status quo versus rehabilitation study of distribution plant are listed below by input category. (*Note:* Eliminate any data that are common between plans.)

Revenue income. Revenue income is considered to be independent of plant condition. In most cases, system network connectivity yields little if any revenue, but may reduce recurring expense by providing increased operating processing speeds, thus eliminating on-line dial time.

Capital expense. The capital dollars required for rehabilitation, both construction and removal dollars, are required in this category. Estimated project costs are used for the proposed alternative. Also entered in this category is the net salvage value. The net salvage value is the value recovered from the salvage of plant at the end of its service life, minus the cost of removal. The value is entered as a dollar amount of first cost.

One-time noncapital expense. This category includes any maintenance and repair dollars required for cutover or maintenance. Since rehabilitation usually releases distribution circuits or pairs that are classified as bad, their recovery should be included in the study as the cost it would take to clear these facilities in the status quo plan. The average cost to clear a bad facility can be determined from the repair form mentioned in Chap. 8, which provides the number of circuits or pairs cleared and the labor hours required to clear those pairs over a period of time. Total cost is derived from the loaded labor rate times the labor hours per bad pair or circuit cleared. Any other one-time expenses incurred or anticipated should be documented and included in this category.

Operating expense. Entries in both alternatives should include costs based on total incidences for line and station transfers, idle assigned circuits, and rearrangement of dedicated circuits. The number of incidences can be obtained from the repair history forms. Their costs can be determined by multiplying the labor hours required per year by the loaded labor rate. Also entered in this category are the cable maintenance costs. The incidences and labor hours can be obtained in the same manner as above. Cost can be determined by taking the labor hours per case times the number of cases per year times the loaded labor rate. The average annual cost can be used for both plans to cutover.

Any other recurring expenses should be included in this category, even if common, to present a realistic cost of doing business until rehabilitation.

9.6.4 Program time conversion

The status quo versus rehabilitation study process is based on annual compounding and uses the following time convention:

- Capital expenditures on one-time expenses are made at the beginning of the appropriate study year.

- Revenue, operating expenses, depreciation, debt interest, and net salvage are reflected at the end of the appropriate study year.

The study always starts at time point zero, the first day of the first study year. The time between time points represents one study year. Therefore, a 25-year study would run from time point zero through time point 25. The study period extends from the beginning of the first year through the end of the last study year. The length of the study must be sufficient to reach a time where the differences between the alternatives will be insignificant when brought to present worth. The maximum study period for this type of study is 50 years.

9.6.5 Cable and support structure lives

It is imperative that equipment lives based on the actual situation involved in the study are used rather than the average service lives established for accounting purposes. The accounting average life is a composite of the estimated lives of a particular group of installations as they happen to exist at the time of the last depreciation study. Only by coincidence would the actual life be the same as the average life. Average life for building cable is 34 years, aerial cable is 34 years, and conduit is 65 years, based on industry averages. Technological advancements will render some types of cable facilities obsolete in much shorter time spans, and this should be factored into the study when appropriate.

In order to avoid cash flows beyond the study period, resulting in inaccurate profitability rates, the coterminated plant assumption must be used. This means that all plant will retire simultaneously regardless of the installation date. This may require the input of adjusted project lives and net salvage values in the later study years. However, if the study is carried out far enough in time (usually 25 years), these inaccuracies will be insignificant when brought to present worth.

9.6.6 Special study considerations

Typically, when noncoincidental retirement of existing plant occurs in one plan earlier than in another, salvage that can be realized from the plant, costs of maintaining the plant, and any expense that must be paid on it are relevant, since they are not common between plans.

The most reasonable principle to apply to this situation is to consider that retaining the plant is the same as purchasing the plant for its net salvage value. Charge to the retaining plan, as a first cost, the net salvage value at the earlier retirement date and show the later actual salvage as salvage. Charge also to the retaining plan the costs of maintenance and ad valorem taxes on the retired plant.

The capital repayment of original first cost of existing plant should not be considered in most engineering economic studies. The obligation is not charged by the timing of the removal of the plant except as it might affect the amount of net salvage. The most that can be affected is the time or manner in which remaining repayment is made. This is significant only if a very large replacement program is undertaken with a magnitude great enough to affect the average service life of the total group.

An additional special study consideration involves applying investment tax credit as an advantage to any particular plan. This credit should be included as a savings for the depreciation period.

Current economic conditions and projections indicate the need to establish an inflation rate for economic selection studies. This amount may be obtained from the gross national product (GNP) and economic indicator publications available through the federal government.

9.7 Sample Study Analysis

In this section, a simple sample problem is presented with one alternative. In the economic analysis of the status quo versus rehabilitation of the sample, it is assumed that the criteria and tasks associated with Sec. 9.6 have been satisfied.

We will compare placing a 400-pair, 24-gauge distribution cable between the network access point and apparatus closet A, and removing the existing 200-pair, 24-gauge cable (rehabilitation) while retaining the existing wiring facilities. It is expected that the existing 200-pair cable will exhaust in 5 years due to increased occupancy planned for the building.

Additionally, as a result of prepared documentation, we have discovered the recurring expenses in the table for plan A below, using the following definitions for recurring expense items:

PLAN A   Status Quo

Item	Occurrences (annualized)	Unit cost, $	Total cost per year, $
Transfers	5	19.86	99.30
Rearrangements	2	40.25	80.50
Reassign circuit	5	23.80	119.40
Clear defective circuit	10	50.15	501.50
No-trouble-found dispatch	15	18.00	270.00
Other (installation)	10	16.00	160.00
Administrative costs	50	18.00	900.00
Total annual costs			$2130.70

- *Transfer*—moving primary service from one location to another. Usually involves moving circuits between distribution cables at the network access point (NAP).

- *Rearrangement*—moving service in apparatus closets, satellite closets, and the NAP. Normally the end-user location remains the same.

- *Reassign dedicated pair*—moving idle pairs or circuits appearing at one location to another location to provide additional capacity at the new location.

- *Clear defective pair*—correcting a condition on a circuit or pair of wires which renders it unusable (i.e., shorts, grounds, crosses, etc.).

- *No trouble found*—cases where trouble was reported and the agency communication technician was dispatched, but no trouble was found and the service appeared normal.

- *Other*—any recurring cost. In this case, installation was included to indicate the normal cost of providing service for new end users because facilities were too congested to dedicate facilities to each location.

- *Administration costs*—reflects the costs associated with record keeping for the wiring systems studied.

The per-unit costs reflected above were developed by multiplying the hours spent performing the function by the loaded labor rate (salary plus overhead) of the agency communication technician performing the work.

The next step is to prepare the economic data for the reinforcement plan. In the process, it must be recognized that maintenance costs will be significantly reduced but not entirely eliminated. Also, all services will have to be reinstalled at the apparatus closet and costs will have to be included for these rearrangements. See table for plan B below.

PLAN B   Rehabilitation

Item	Occurrences (annualized)	Unit cost, $	Total cost per year, $
Transfer	0		0.00
Rearrangement	0		0.00
Rearrange dedicated circuit	0		0.00
Clear defective pair	0		0.00
No-trouble-found dispatch	5	12.00	60.00
Other (installation)	10	14.00	140.00
Administrative costs	25	18.00	450.00
Total annual cost			$650.00

An annual savings of $1480.70 can be realized by replacing the existing facilities. However, there are one-time expenses that must be included to allow for a fair analysis of the proposal.

One-time cost may include the following categories of work:

- Removal expense
- Installation labor
- Support structure modifications
- Rearrangements unique to rehabilitation alternatives
- Material and equipment costs, not common between plans

The salvage value of cable resources will be shown as a credit to the rehabilitation plan. Also, any recurring or one-time expenses that are common between plans will be eliminated from the study so as not to distort the decision process. When common costs are included, they will lessen the impact of real savings between alternatives.

Hence, the estimated cost for cable placement will be developed by using the per-unit cost evaluation. A cost of $3260.00 is estimated for all one-time expenditures associated with the rehabilitation alternative B. The next step is to develop a time diagram. As mentioned earlier, the existing 200-pair cable will exhaust in 5 years, so the status quo solution will reflect the expense of reinforcement with inflated dollars using future worth amounts.

The costs for a 5-year period are now developed, shown in the table for plan A versus plan B, below. This time frame was selected to reflect the next exhaustion of existing facilities and accurately compare short-term savings between the plans. All exhaustion beyond the 5-year period are considered to be common between plans. This represents a comparison of the cost-effectiveness of replacing a moderately high maintenance plant now or waiting until the normal exhaustion of facilities before replacement.

PLAN A VERSUS B   Five-Year Costs

	Costs				
	Year 1	Year 2	Year 3	Year 4	Year 5
Plan A					
Recur	$2130	$2305	$ 731	$ 760	$ 791
Installed first cost (IFC)			3527		
Cumulative	$2130	$4435	$8693	$9453	$10,244
Plan B					
Recur	$ 650	$ 703	$ 731	$ 760	$ 791
IFC	3260				
Cumulative	$3910	$4613	$5344	$6104	$ 6,895

Alternative B, rehabilitation, shows a 33 percent savings over the 5-year period, based on an inflation rate of 4 percent. However, plan B does represent a significant year 1 investment over alternative A, status quo. To assist in evaluation, the time diagram below is presented.

TIME DIAGRAM

	0	1	2	3	4	5
						Year

The time diagram shows that the expeditors for plan B, rehabilitation, break even in year 2 and begin paying back in years 3 through 5 and beyond. In this type of study, break-even and payback are the same.

9.8   Study Analysis and Budget Process*

The manager should be able to see that the study results are a logical conclusion based on verifiable and applicable data. Since the results

---

*American Telephone and Telegraph Co., *Engineering Economy*, 3d ed., McGraw-Hill, New York, 1977, pp. 414–420.

are transmitted by way of the presentation, the presentation may be the most important part of an economic study. To effectively and objectively prepare, a decision must first be made among the alternatives studied.*

The basis for this decision is often difficult to grasp. Each corporate body determines capital budget distribution in a different manner. Public agencies have devised elaborate and extremely rigid budgetary practices. The one common thread among all processes is politics. It is incumbent, therefore, on the manager to be "tuned into" the process, to know to whom the real power belongs, to not become emotionally involved with one's "own" project, and to be willing to compromise. For instance, if the indicia is short-term oriented, first-cost expenditure will normally rule. However, if the company is well-funded, long-term investment payback will normally dominate the decision process. Make sure you know the rules before charging forth with your project!

Depending on the term of financing, or capital commitment, that drives the budgetary process in your environment, you will normally be able to make a decision based on raw economic data. Normally, a 15 percent or greater advantage to one plan over another indicates a clear selection. However, when the difference between plans is less than 15 percent, a clear advantage does not exist and other means to determine selection can be imposed. Justification for noneconomic selection in these cases is based on the frailty of the study data in comparisons of estimated versus actual project costs.

In such cases, the best method to select one alternative over another is to evaluate three components in each recommendation:

- Degree of difficulty
- Progress toward goal
- Degree of acceptability/compatibility

Each category will be rated high, medium, or low. The manager can define the variables, but should be sensitive to the corporate direction when doing so. The ideal rating for each of the categories would be:

- Degree of difficulty—low
- Progress toward goal—high
- Degree of acceptability/compatibility—high

---

*American Telephone and Telegraph Co., *Engineering Economy,* 3d ed., McGraw-Hill, New York, 1977, pp. 414–420.

Keep in mind, the approval authorities will always lean toward the lowest cost, no matter how small the amount, so be prepared to explain the reason for going beyond that criterion with documentable estimates of actual cost data, or, if there is none available, vendor price spread information.

Next, discuss your definitions for the above categories with your peers or, if possible, the approval authority, before going to the mat. In all cases, do not get emotionally involved with "your" project, and avoid hidden agendas if possible!

9.8.1 Project Presentation*

The purpose of the economic study is to help managers decide which plan is most economical, and the same care and planning should be given to the presentation as to the analysis. It is the presentation which gives the conclusions of the analysis in tabular, narrative, or graphic form, and with enough details to establish the integrity of the study.

Although many different formats can be used to present a study, there are a few basic principles to follow. The most important one is that a study should be presented in different levels of detail, depending on which managers will use the study. It is important to keep in mind that different managers have different goals, and therefore have a need for different degrees of detail.

The presentation should include a written narrative progressing from the most general to more and more specific. The most general level is the conclusion. It may be as simple as, "of all the plans considered, plan B is the least costly to the firm over a 5-year period." Or perhaps the conclusion is to defer a decision.

Next, each conclusion needs to be qualified, but in just a simple statement. A short beginning paragraph can include the conclusions, any qualifications needed, and enough information to identify the project being considered. This one paragraph may be all that is required for some managers. Others will require greater and greater detail, possibly even including the working papers used in the study.

The narrative can be continued to explain the study conclusions in more detail for those managers who need it. Sufficient documentation should be included to support any assumptions and projections that were made in doing the study. This can include tables of data, as well

---

*American Telephone and Telegraph Co., *Engineering Economy,* 3d ed., McGraw-Hill, New York, 1977, pp. 414–420.

as narrative discussion, but graphical illustrations, as shown above, are often a more effective way to present the conclusions.*

9.9 Contract Bid Strategies

There are three basic means by which to receive contractor or vendor bid packages: invitation for bids, bid advertisement, and sole source.

The invitation process involves the sending of bid packages to from one to any number of potential, prequalified vendors and/or contractors. The operative word here is *prequalified.* Concerns involved must have already proved to your company or agency that they have the necessary skills, expertise, equipment, and financial resources to complete the work requested in a timely and satisfactory manner.

Advertising bids requires considerably more information in the bid specification package as to details, legal requirements, worker skills, work environment, payment schedule, and acceptance parameters.

The sole source scenario applies, in general, to *cost plus* type of work—that is, where a contractor or vendor is already embedded by means of a maintenance master contract or other ongoing situation. The rate at which the vendor will perform the work is already set, the skill and ability of the vendor's forces are already established, and time is of the essence.

In all but the sole source bid strategy, the manager must make sure that any information given to one source is given to all. The bid contact person must never give out information on other bidders, or other contending bid amounts, prior to contract award. It is appropriate, however, to give information on previous bids for the same type of work on jobs that have been completed. Remember, to avoid pesky lawsuits, be consistent, be unemotional, and do not let bidders buy you lunch, ever!

9.10 Bid Specifications (Figs. 9.2 and 9.3)

The development of bid specifications is, at best, a meager attempt to describe the work you want done. It is best to approach the processes using the KIS (keep it simple) method. The more precise, the better. As you will see in the sample bids in Figs. 9.2 and 9.3, it is possible to get so carried away with protecting yourself that you, and potential

---

*American Telephone and Telegraph Co., *Engineering Economy,* 3d ed., McGraw-Hill, New York, 1977, pp. 414–420.

bidders, could have serious trouble locating the description of work. It is also possible that such strict contractor qualifications could be asked for that only one vendor would be qualified. It is best to develop a set of statements that apply to a range of expenditure levels. To make requirements the same for all contract bid specifications will, in the long run, eliminate qualified firms and eventually lead to higher costs for your organization.

The basic bid package should include the following information:

- Transmittal letter

- Description of work

- Description of work environment

- Construction schedule

- Contractor qualifications

- Contract language (legal conditions)

- Rigid bid format

- Bid evaluation criteria

- Firm bid due date

- Firm bid award date

Since every bid will be different, it is important to develop each of the categories above to reflect the unique circumstances of your project. I have included two samples in their entirety to assist in the development of standards for different situations. However, in developing specifications, the only cardinal rule is, *do not let any vendor help you.* When we get to the next section, you will understand better, but allowing vendors to feed you specifications will lead to bid award protests in practically every instance.

9.11 Contractor Strategies (the Vendor's Point of View)

As a contractor, I, and my employees, depend on winning contracts to sustain our business. Regardless of size, this is the cardinal rule of contracting. Work comes to my business sporadically, and supply is a constant unknown. I am successful if, and only if, I continue to procure more work. All work does not have to be profitable, but all work is necessary. I must always satisfy the customer. I desire to monopolize the market to gain security and attain financial stability.

The bidding game, and it is a game, is open to innovation and creativity. I succeed when I outsmart the competition by being either better informed, better qualified, or better connected. I use the following techniques to gain contracts:

- *Gain knowledge.* I make every attempt to make direct contact with contract award personnel. If possible, I meet personally and try to develop a mutual-benefit relationship with the potential client. I find out as much information as I can about the contract from the design personnel. I find out what the potential customer thinks the bid amount will be, and I find out what the hidden agendas are. I find out the personal career ambitions of the contract administrator and attempt to make that person believe I can help him or her realize expectations by using my firm. I make as many "tasteful" contacts as possible in an attempt to make the potential client feel comfortable with my company.

- *Bid development.* In the information gathering process, I find out what the potential client has paid for like work in the past, who has performed this work, and whether the client is satisfied with the present contractor. Next, I will call two or three subcontractors and ask them to bid on portions of the work. I will also call those I know to be direct competition and either attempt to get cost information on segmented portions of the work or investigate the potential benefit of team bidding the project if the scope of work is sufficient to support such an effort.

Once I have completed the information gathering processes, I will carefully review the bid specifications. In so doing, I will discover if critical bid components have been omitted. I will, if such omissions have occurred, discretely investigate the reasons for their omission. If I determine that the omission was due to oversight, I will develop a cost to provide such services or materials separate from the bid I intend to submit, knowing that this will be an extra after-contract award. This cost will be quite high and will allow me to shave profit margins on the original bid.

I will then prepare the bid. Before submittal, I will make one more contact with the potential client and my competitors to determine if there have been any last minute revisions or changes in direction. I prepare the bid to be as adherent to the bid specification as possible.

After bid award, I seek to find ways to extend the contract by identifying the extras I already know about. As work progresses, I attempt to enhance the system I am working on by being knowledgeable about the client's needs and presenting alternatives that would make the client's work more efficient. I also attempt to gain information on future project work and begin positioning my company to become the sole source for that work. I attempt to become embedded!

There are also other strategies used by various competitors to undermine my company and its credibility. I am constantly mindful of these tactics and prepare business profiles and customer satisfaction ratings on competitors. Because I want to appear dignified in my client's eyes, I do not use this information except in defense and also present both the good and the bad when discussing my competition. If I cannot find any good, I will give them positive attributes common to all companies courageous enough to enter the market. I never waste energy by getting mad and I never appear to be getting even; I focus only on attaining more contracts for my company!

Of course there are many variations on this basic technique. There are as many styles as there are personalities. Some use only a fraction of these techniques and others use many more such tools; but the code is the same with all of us: get the work or die!

9.12 Bid Evaluation

In this process you are going to make one company very happy and the rest angry. Be prepared for this because it happens almost every time.

If you have presented a concise description of work, have given all vendors the same information, and have clearly defined the evaluation criteria, the process will be quite civilized and dignified. If the award process is contested, you will have to resort to the escape language in the contract. If the contesting vendor is not satisfied with that recourse, you will have to invoke your power as the award authority. However, if there has been an injustice, the best recourse would be to correct the deficiency and rebid the work.

Sometimes the contract bid manager experience can be quite humbling; but always be fair and learn from your mistakes. Some, but not all, of the vendors really are trying to trip you up.

STATE OF CALIFORNIA—STATE AND CONSUMER SERVICES AGENCY   PETE WILSON, *Governor*

DEPARTMENT OF GENERAL SERVICES
TELECOMMUNICATIONS DIVISION
601 SEQUOIA PACIFIC BOULEVARD
SACRAMENTO, CA 95814-0282
(916) 657-9903

April 28, 1992

Dear Interested Bidder:

<u>REQUEST FOR QUOTATION NUMBER R-92-015</u>
<u>CAPITOL FIBER LOOP MAINTENANCE SERVICES FOR STATE OF CALIFORNIA</u>
<u>DEPARTMENT OF GENERAL SERVICES, TELECOMMUNICATIONS DIVISION</u>

The Department of General Services, Telecommunications Division, is soliciting maintenance services for the Capitol Fiber Loop, Sacramento, California. The enclosed Bid Package consists of the following:

1. BID SPECIFICATIONS, BID NUMBER R-92-015

2. RIDER A, TERMS AND CONDITIONS

3. RIDER B, EQUIPMENT LIST

4. RIDER C, CONTRACTOR HOURLY RATES FOR NONCONTRACTED SERVICES

5. RIDER D, STATEMENT OF COMPLIANCE

6. RIDER E, CONFIDENTIALITY STATEMENT

7. RIDER F, MINORITY/WOMEN/DISABLED VETERAN BUSINESS PARTICIPATION/GOOD FAITH EFFORT REQUIREMENTS

8. RIDER G, DRUG-FREE WORK-PLACE CERTIFICATION

9. RIDER H, SMALL BUSINESS PREFERENCE FORM

10. BID PROPOSAL FORM AND BID PROPOSAL SUMMARY FOR CONTRACTED MAINTENANCE SERVICES

<u>READ RIDER "F" MINORITY/WOMEN/DISABLED VETERAN BUSINESS PARTICIPATION/GOOD FAITH EFFORT REQUIREMENTS CAREFULLY AND BEGIN THIS PROCESS NOW -- THE REQUIREMENTS OF THE GOOD FAITH EFFORT, WHICH INCLUDE ADVERTISING, TELEPHONE CONTACTS, SOLICITATIONS/RESPONSES, ETC., TAKES TIME -- START IMMEDIATELY!!!!!!!!</u>

<u>A MANDATORY WALK-THROUGH OF THE FACILITIES/EQUIPMENT WILL BE HELD ON MAY 6, 1992 AT 8:30 A.M. EACH BIDDER WHO WISHES TO SUBMIT A BID MUST ATTEND THIS WALK-THROUGH. ALL POTENTIAL BIDDERS MUST MEET AT THE PARKING STRUCTURE, CORNER OF 11TH & P STREETS, SACRAMENTO, CALIFORNIA. BALBIR JOHL AND MIKE FENTON WILL BE THE STATE REPRESENTATIVES CONDUCTING THE WALK-THROUGH. CONTACT MR. JOHL OR MR. FENTON BY 4:30 P.M. THE DAY BEFORE THE WALK-THROUGH TO CONFIRM YOUR ATTENDANCE. EACH BIDDER WILL BE REQUIRED TO SIGN A "VERIFICATION OF FACILITY INSPECTION BY POTENTIAL CONTRACTOR" FORM STATING THAT THE FACILITY(S) WERE INSPECTED IN THE PRESENCE OF AN AUTHORIZED STATE REPRESENTATIVE. FAILURE TO DO SO WILL RESULT IN YOUR BID BEING REJECTED!!!!!</u>

Figure 9.2 Bid specifications—California Telecommunications Division.

To All Interested Bidders   -2-   April 28, 1992

The bid deadline date is May 21, 1992 by 2:00 p.m. All bids must be received at the Department of General Services, Telecommunications Division, 601 Sequoia Pacific Boulevard, Sacramento, California, by this date/time. Bids received after this date/time will be returned unopened to the sender. The bid opening date is May 21, 1992 at 2:00 p.m. and will be held at the State facility mentioned above.

For information regarding technical requirements, call Mr. Johl at (916) 657-6131; for information regarding administrative requirements in this bid package, call Cherlyn Thomas at (916) 657-9368 or Mark MacRae at (916) 657-9269.

Sincerely,

DAPHNE RHOE
Telecommunications Manager

Enclosure

cc: Balbir Johl
     Mike Fenton
     Cherlyn Thomas
     Mark MacRae

Figure 9.2 (*Continued*) Bid specifications—California Telecommunications Division.

R-92-015

BID SPECIFICATIONS

1. GENERAL INFORMATION

The following is the specification for the maintenance of a lightwave communication loop located in Sacramento, California. The successful bidder must meet the requirements of this specification. Failure to perform in accordance with these specifications may result in substantial penalties. Site locations are shown at the end of this specification. The fiber optic cable plant consists of a main loop going to the major State office buildings with spurs extending out to serve local areas.

2. BACKGROUND

The system provides lightwave facilities for the State office buildings in downtown Sacramento. The State's needs can be best described in the following:

a. The State uses the facility to provide a high speed transport media. It is primarily to provide a media for high speed data networks. It is also used as a means to provide dedicated facilities for data transport and video distribution. Consequently the reliability of the lightwave network is paramount to the users.

b. The facility is nearly 5 years old and still growing. As a result, the system undergoes changes continually. The ability to respond to the dynamic nature of the system is essential.

c. The system was initially installed prior to the availability of effective fiber management hardware or systems. Consequently, the current facilities require updating in becoming more representative of current industry standards.

3. BID TECHNICAL REQUIREMENTS

a. Bidder must propose the necessary hardware to provide cable and jumper management within the fiber optic patch panels listed in Rider B. The Main objective of this requirement is the development of an orderly and systematic management scheme that will assist in developing system documentation and in readily identifying specific circuits. Vendor shall also periodically update current documentation to accurately reflect system configuration.

b. Bidder must maintain the lightwave terminal equipment and the fiber optic cable plant listed in Rider B. The main objective of system maintenance is to strive for system performance equal to the original system specifications. System maintenance will be considered acceptable if the following objectives are met:

-1-

Figure 9.2 (Continued) Bid specifications—California Telecommunications Division.

R-92-015

(1) The transmission performance objective is error-free transmission in at least 99.999 percent of all one second intervals. If there are more than 30 error seconds per hour, the system service will be considered impaired. System performance is measured in "error seconds" , where error second is any 1 second interval during which one or more bit errors occurs.

(2) The reliability objective is that circuit integrity between stations remain intact and error free 99.9 percent of the time, based on a long term average. In terms of cumulative outage time, this objective equates to an average downtime of less than 8.75 hours per circuit per year.

(3) The availability objective is that no single outage exceed four hours. The objective recognizes the impact of long outages on the States's data networks and the perishable nature of some forms of data.

c. Bidder must have proven experience in work of this type. Bidder shall provide three (3) references in their response representing paying customers that typify their experience with this type of work. Each reference must include the name of the contact person and phone number. Bidder must also list in the response the experience of person (s) assigned to perform the maintenance. The Bidders maintenance personnel must be certified by the manufacture to maintain the equipment, and have attended applicable schools and training. Copies of the certification or letter of credentials from the manufacture shall be included in the bid. See Bid Proposal Form.

d. Bidder must take responsibility for system maintenance by being able to respond to emergency requests for maintenance 24 hours a day, seven days a week. Bidder must have a 24 hour answering service to receive calls.

Emergency on-site maintenance must begin within two (2) hours after notification. Non-emergency requests for on-site maintenance must begin by the next business day. An emergency condition exists when the bit error rate (BERT) of any in-service T-1, or other high capacity circuit, exceeds 10^{-6} over a period of one half hour or greater. Bidder must explain in the response how this requirement will be met. See Bid Proposal Form.

e. Bidder must provide itemized prices as requested. This will allow the State to have the option of deleting or augmenting quantities. Price quotes within the bid will remain in effect for a two/three year period.

-2-

Figure 9.2 (Continued)   Bid specifications—California Telecommunications Division.

R-92-015

f. Bidder shall list all test equipment and tools to be utilized for maintaining the system.

  (1) The test equipment shall include as a minimum:

    (a) Optic power meter/source - ExFo FOT90/FLS210 or equivalent.
      (Quantity of two required)

    (b) OTDR - Tektronix FiberMaster or unit with similar resolution, accuracy and documenting capabilities.

    (c) Fiber Probe - Photodyne 8000 or unit with equivalent capabilities for identifying active optical circuits.

    (d) T1 BERT Tester- TTC T-Bird 220 or equivalent.
      (Quantity of two preferred).

    (e) DS-3 Tester- TTC T-Bird 430 or equivalent.
      (Required only upon advance notice from the State.)

  (2) The tools shall include as a minimum:

    (a) Fusion Splicer- Fujikura model FSM-20 ARC or equivalent

    (b) Microscope- Buehler Fibrskope or equivalent.
      (portable 100x w/Biconic stage)

    (c) Termination Kits - Dorran/3M Biconic Termination Kit, ATT ST Termination Kit, AMP SMA Termination Kit, or their equivalents.

    This equipment will also be utilized to verify that the system performance meets the requirements of paragraph [3.b].

4. EQUIPMENT

The Bidder must:

a. Provide all labor, materials, transportation and equipment required to maintain the lightwave terminals and fiber optic cable plant at all locations specified. The list of equipment and facilities is given in Rider B.

-3-

Figure 9.2 (*Continued*)   Bid specifications—California Telecommunications Division.

R-92-015

b. Conduct end-to-end proof-of-performance tests once a year, and on demand, to ensure the system meets the performance requirements of paragraph [3.b]. The proof-of-performance test will be witnessed by the State. A proof-of-performance test plan must be submitted and approved by the State, 30 calendar days after award. The measurement time for testing to determine compliance with the 'error seconds' criteria shall occur over a period of at least 360 hours.

5. SPARE PARTS

a. Bidders must keep within 90 miles of Sacramento a stock of spare parts and modules necessary for maintaining the system. Bidder shall list these spare parts in the bid and provide an address for the local stocking point. The address may be a State facility if agreed to by the State. See the Bid Proposal Form.

b. Bidder shall maintain a stock of the spare parts necessary for maintaining the system. Purchase of the existing parts inventory is possible through the current contract. (See Rider B) These shall remain the property of the bidder during the maintenance period. At the end of the maintenance period the bidder shall provide the State with an opportunity to purchase the spare parts for continued maintenance by others. The bidder shall provide in the response a cost to the State for purchasing these spares. The required spares for the lightwave terminals are listed in Rider B.

c. For the purposes of evaluation, the initial cost of the spare parts shall be considered a common cost among all bidders, and shall therefore not be considered part of the responses or quotations.

6. STANDARDS

The Bidder must meet the following:

a. Technical Standards:

All materials and craftsmanship furnished by the Bidder under this specification shall conform to the latest applicable standards of the Electronics Industry Association (EIA), The National Electrical Manufactures Association (NEMA), and the Institute of Electrical and Electronic Engineers (IEEE). Additionally all work must meet the applicable Rules and Regulations of the Federal Communications Commission, and the California Occupational Safety Administration, including EMI/RFI standards ensuring that equipment and installation will not cause, or be subject to electrical interference.

-4-

Figure 9.2 (Continued) Bid specifications—California Telecommunications Division.

R-92-015

Grounding of all equipment shall be in accordance with Article 250 of the National Electric Code, and shall conform to DGS/TD standard DM 220.2.1, <u>Grounding Procedures for Communications Facilities</u> including, but not limited to paragraph 6.2.3.b.(3). Copies will be available on the tour.

Design and installation of DC power systems and equipment shall be in accordance with Article 110 of the National Electric Code, and shall conform to DGS/TD standard DM 220.1.1, <u>DC Power Distribution Systems Installation Procedures</u>, and DM 550.1.1 <u>DC Power Distribution Systems Engineering Procedures.</u>

All new fiber optic cable installation shall be in accordance with Article 770 (or Article 800 where applicable) of the National Electric Code.

b. Maintenance Standards

The Bidder shall provide a copy of his standard maintenance procedures as part of the proposal. Bids submitted without Bidder's Maintenance Procedure Manual shall not be considered compliant.

The procedures should include the following topics:

(1) Safety Procedures

   (a) Personnel Safety
   (b) Hazard Protection
   (c) Equipment Protection
   (d) Fire Protection
   (e) Accident Reporting

(2) General Procedures

   (a) Remedial Assistance Support (See Rider A, 5.d)
   (b) Warehousing
   (c) Tool and equipment outfitting
   (d) Safety Meetings
   (e) Progress Reports
   (f) Quality Assurance Program

(3) Maintenance Practices

   (a) Equipment troubleshooting
   (b) Fiber Optic Cable splicing and maintenance
   (c) Conduit and raceway maintenance
   (d) Battery and charger maintenance
   (e) Cable routing and termination
   (f) Grounding, Bonding and Shielding

-5-

Figure 9.2 (Continued)   Bid specifications—California Telecommunications Division.

R-92-015

   c. Secure Anchoring Standards

      All equipment shall be mounted and braced to withstand, without
      damage, seismic acceleration forces in both horizontal and
      vertical directions of 0.33g, for installation in Seismic Zone
      3, in accordance with Uniform Building Code, Section 2312.
      Bidder shall submit proposed methods of securing equipment. The
      methods are to be submitted for State review after contract
      award. The installation shall meet the requirements of the
      Essential Services Building Seismic Safety Act, Senate Bill 239
      -1986.

7. **MANDATORY WALK-THROUGH OF FACILITIES/EQUIPMENT**

    A MANDATORY WALK-THROUGH OF THE FACILITIES/EQUIPMENT WILL BE HELD
    ON MAY 6, 1992 AT 8:30 A.M. EACH BIDDER WHO WISHES TO SUBMIT A BID
    MUST ATTEND THIS WALK-THROUGH. ALL POTENTIAL BIDDERS MUST MEET AT
    THE PARKING STRUCTURE, CORNER OF 11TH & P STREETS, SACRAMENTO,
    CALIFORNIA.

    BALBIR JOHL AND MIKE FENTON WILL BE THE STATE REPRESENTATIVES
    CONDUCTING THE WALK-THROUGH. CONTACT MR. JOHL OR MR. FENTON BY
    4:30 P.M. THE DAY BEFORE THE WALK-THROUGH TO CONFIRM YOUR
    ATTENDANCE.

    EACH BIDDER WILL BE REQUIRED TO SIGN A "VERIFICATION OF FACILITY
    INSPECTION BY POTENTIAL CONTRACTOR" FORM STATING THAT THE
    FACILITY(S) WERE INSPECTED IN THE PRESENCE OF AN AUTHORIZED STATE
    REPRESENTATIVE.

    **FAILURE TO ATTEND THE MANDATORY WALK-THROUGH WILL RESULT IN YOUR
    BID BEING REJECTED!!!!!**

-6-

Figure 9.2 (*Continued*) Bid specifications—California Telecommunications Division.

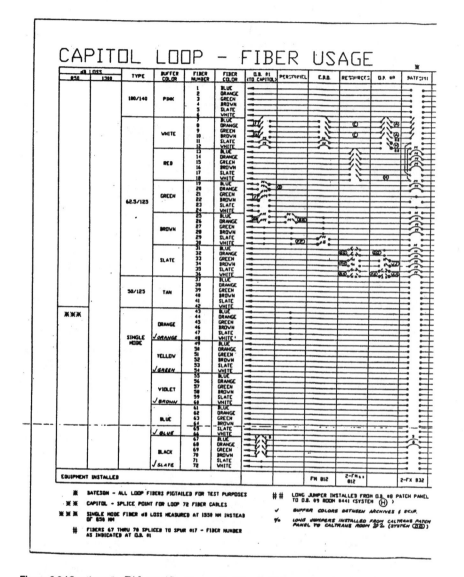

Figure 9.2 (Continued) Bid specifications—California Telecommunications Division.

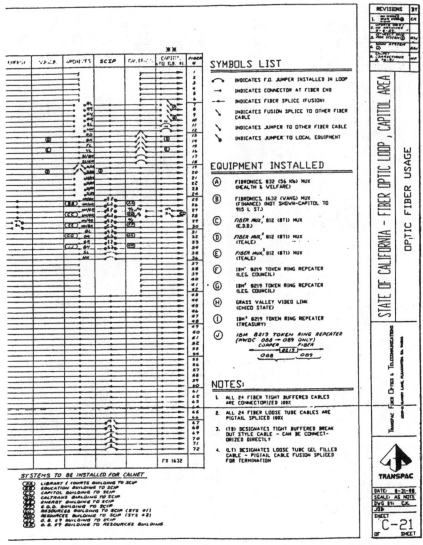

Figure 9.2 (Continued) Bid specifications—California Telecommunications Division.

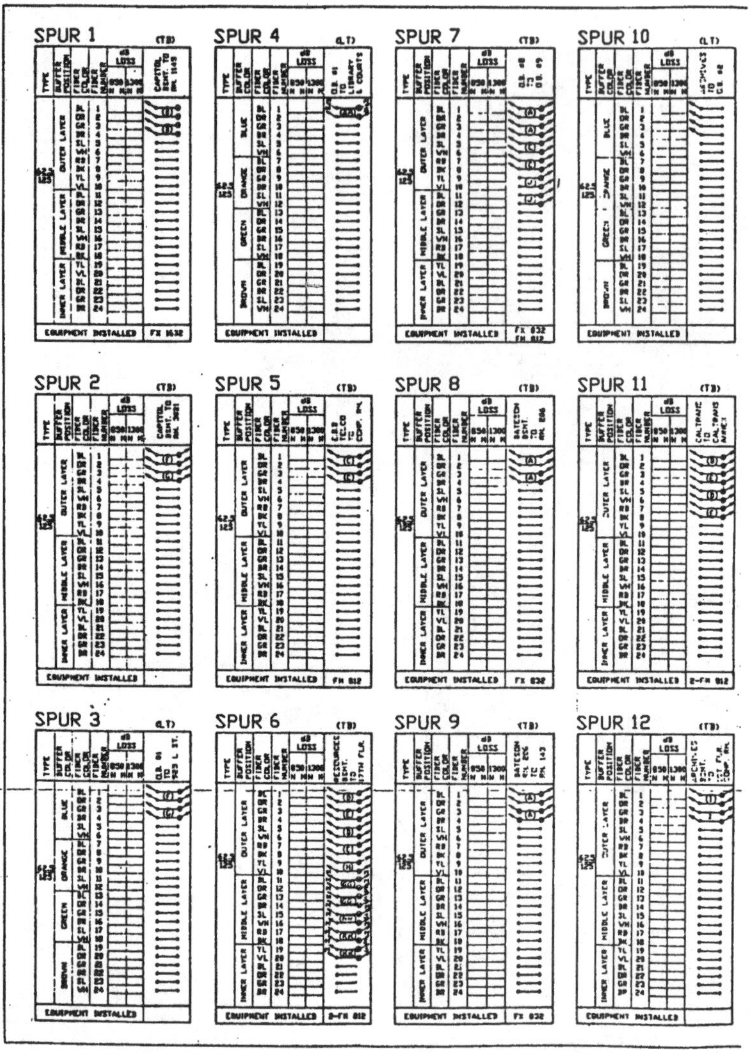

Figure 9.2 (Continued) Bid specifications—California Telecommunications Division.

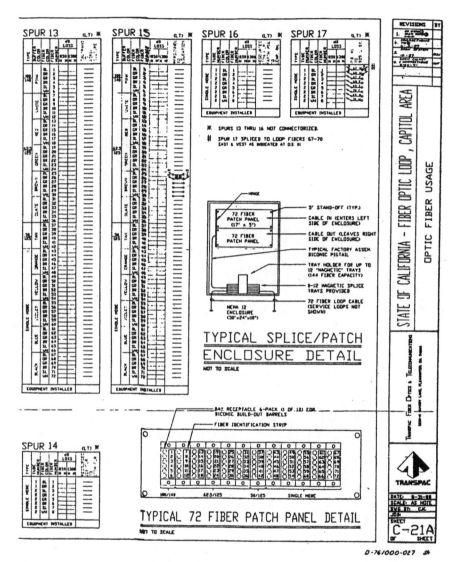

Figure 9.2 (*Continued*)   Bid specifications—California Telecommunications Division.

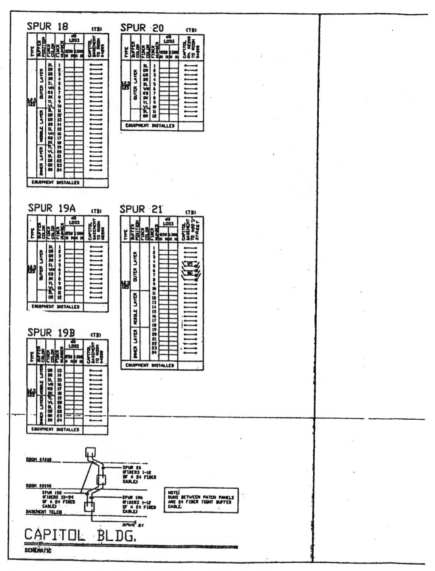

Figure 9.2 (*Continued*) Bid specifications—California Telecommunications Division.

Figure 9.2 (Continued) Bid specifications—California Telecommunications Division.

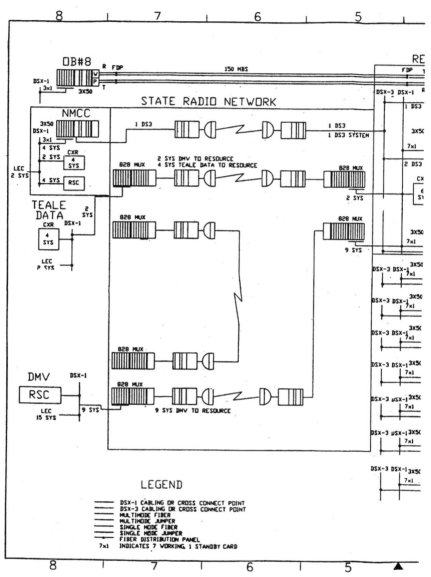

Figure 9.2 (Continued) Bid specifications—California Telecommunications Division.

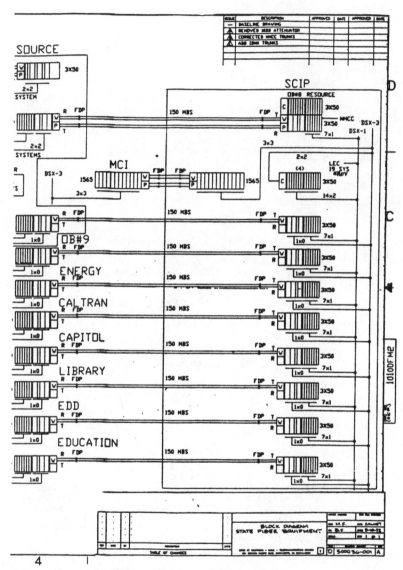

Figure 9.2 (Continued) Bid specifications—California Telecommunications Division.

R-92-015

RIDER A

*TERMS AND CONDITIONS APPLICABLE TO MAINTENANCE
OF FIBER OPTIC TELECOMMUNICATIONS EQUIPMENT*

1. *Purpose*

The purpose of this contract is to define the terms and conditions for maintenance service on the State's fiber optic telecommunications system in Sacramento hereafter called the Capitol Loop System. The maintenance is to be provided by the Contractor to the State for State-owned equipment listed in Rider B. Contractor must take system maintenance responsibility and be able to respond in 2 hours to requests for maintenance 24 hours a day seven days a week.

2. *Definition of Terms*

    a. *Capitol Loop System* -- The total complement of State furnished cable plant and telecommunications equipments or components which are acquired to operate as an integrated group.

    b. *Equipment* -- An all-inclusive term which refers either to individual apparatus or components or to a complete telecommunications system.

    c. *Equipment Failure* -- A malfunction in the equipment, excluding all external factors, which prevents the accomplishment of a job.

    d. *Apparatus* -- An individual unit, including special features installed thereon, of a telecommunications system or subsystem, separately identified by a type and/or model number, such as a fiber optic terminal, M13 multiplexer, etc.

    e. *Operational Use Time* -- That time during which equipment is in actual operation and is not synonymous with power-on time.

    f. *Preventive Maintenance* -- That maintenance performed by the Contractor which is designed to keep the equipment in proper condition.

    g. *Principal Period of Maintenance* -- 24 Hours per day, 7 days a week including all holidays and weekends.

    h. *Remedial Maintenance* -- That maintenance performed by the Contractor which results from equipment failure and which is performed as required; i.e., on an unscheduled basis. Remedial service is also known as *Emergency Service* or *Routine Service* depending on the nature of the problem. *Emergency Service* will be performed by the Contractor without regard to the time of day or the day of the week. *Routine Service* will be performed by the Contractor at a time agreed to by the State and the Contractor. This time may be on a weekend or at a time other than 8 am to 5 pm.

Figure 9.2 (Continued)   Bid specifications—California Telecommunications Division.

R-92-015

i. *Period of Maintenance Coverage* -- The period of time, as selected by the State, during which maintenance services are provided by the Contractor for a fixed monthly charge, as opposed to an hourly charge for services rendered. The Period of Maintenance Coverage consists of the Principal Period of Maintenance.

j. *Maintenance Diagnostic Routines* -- The system diagnostic programs used to test the system for proper functioning and reliability.

k. *2nd Level Technical Support* - A particular equipment specialist with unique training and/or experience who specializes in providing diagnostic assistance and/or repair expertise when a service call is particularly difficult.

l. *3rd Level Technical Support* - A particular equipment specialist whose geographic responsibilities normally include multiple Field Engineering Branch Offices and who has received in-depth specialized training and experience and possesses extensive diagnostic ability specifically designed to assist on unusually complex problems.

m. *Notification - - * The act by the State of placing a call to the Contractor's 24 hour per day telephone number requesting that remedial service be performed. The Contractor shall provide to the name of the person being dispatched at that time and an estimate of his arrival time.

3. *Term of Contract and Contract Termination*

a. This contract is effective after signing by the Contractor and on the date of its last approval or certification of exemption from approval by or on behalf of the Department of General Services, Telecommunications Division.

b. The term of this contract shall be as stated on the face of the State's Standard Agreement Form, subject to the availability of funds.

c. The State or Contractor may terminate this Contract upon one month's prior written notice.

d. The State may withdraw any equipment or fiber optic cable from this contract at any time by giving one month's prior written notice.

e. The maximum amount of this contract shall not exceed that amount stated on the face of the State's Standard Agreement Form. The State's obligation is payable only and solely from funds appropriated for the purpose of this agreement. The States monetary obligation under this agreement in subsequent fiscal years is subject to and contingent upon availability of funds appropriated for the purpose of this agreement.

RIDER A - PAGE 2

Figure 9.2 (*Continued*)   Bid specifications—California Telecommunications Division.

R-92-015

4. **Maintenance Service, Parts, and Documentation**

    a. Contractor agrees to provide maintenance coverage during periods selected by the State as shown in Rider B to keep the cable plant and listed terminal equipment operating in accordance with the bid specifications. This maintenance service includes:

        (1) Scheduled preventive maintenance based upon the specific needs of the cable equipment and individual equipments as determined by Contractor and approved by the State.

        (2) Unscheduled, on-call remedial maintenance. Such maintenance will include repair, adjustments, and replacement of maintenance parts deemed necessary by Contractor and approved by the State.

    b. Contractor must maintain a local stock of spare parts and modules necessary to maintain the system. Maintenance parts will be furnished by Contractor and will be new or equivalent to new in performance when used in these equipments. Replaced maintenance parts become the property of Contractor.

    c. Maintenance service does not include:

        (1) Electrical work external to the cable plant or maintenance of accessories, alterations, attachments, or other devices not furnished by Contractor hereunder.

        (2) Repair of damage or increase in service time caused by: accident, transportation, neglect, or misuse; alterations, which shall include but are not limited to, any deviation from Contractor's physical, mechanical, or electrical equipment design; and attachments, which are defined as the mechanical, electrical or electronic interconnection to a Contractor equipment of non-contractor equipment and devices not supplied by Contractor.

        (3) Repair of damage or increase in service time resulting from failure to provide a suitable installation environment with all facilities prescribed by the appropriate Contractor Installation Manual--Physical Planning (including, but not limited to, failure of, or failure to provide adequate electrical power, air conditioning, or humidity control) or from use of supplies or materials not meeting Contractor specifications for such equipment.

        (4) Repair of damage or increase in service time attributable to the use of the cable plant from other than telecommunications purposes for which acquired.

RIDER A - PAGE 3

Figure 9.2 (Continued) Bid specifications—California Telecommunications Division.

R-92-015

(5)   *Furnishing supplies, or accessories; painting or refinishing the cable plant or furnishing material therefor; making specification changes or performing services connected with relocation of cable plant; or adding or removing accessories, attachments, or other devices.*

(6)   *Such service which is impractical for Contractor to render because of alterations in the cable plant or their connection by mechanical or electrical means to another equipment or device.*

(7)   *Repair of damage or increase in service time caused by catastrophe, by fault or negligence of the State, or by causes external to the equipment.*

(8)   *Repair or maintenance by Contractor that is required, in Contractor's sole opinion, to restore such equipment to proper operating conditions after any person other than Contractor's employee had performed maintenance or otherwise repaired an item of equipment. An additional charge for such repair or maintenance shall be at the Contractor's ates listed in Rider C.*

(9)   *Repair of damage, replacement of parts (due to other than normal wear) or repetitive service calls caused by the use of supplies or materials not meeting State's specifications for such supplies or materials before the commencement of the maintenance agreement.*

(10)   *Repair of damage or increase in service time caused by conversion from one model to another, or the installation or removal of a feature whenever any of the foregoing was caused by other than the authorized Contractor personnel.*

d.   *The Contractor will be temporarily excused from its performance when, in the opinion of its service representative, dangerous or hazardous conditions exist at the Service Site listed in Rider B that make performance under this Agreement temporarily impractical.*

RIDER A - PAGE 4

Figure 9.2 (Continued)   Bid specifications—California Telecommunications Division.

R-92-015

e.   Documentation

The Contractor will be provided with original drawings of the facilities. The State shall not be responsible for any omission or inaccuracy in drawings or other information it supplies to the Contractor. The Contractor shall verify all dimensions, equipment configurations and equipment locations before commencing work. Any subsequent changes to the facilities by the Contractor shall be recorded on the drawings by the Contractor. The Contractor will then send to the State one mylar copy and four blueline copy's of the changed drawing.

5.   Responsibilities of the Contractor

a.   The Contractor shall provide maintenance (labor and parts) and keep the cable plant and equipment in good operating condition.

b.   The Contractor shall specify the frequency and duration of preventive maintenance for the equipment listed in Rider B.   Preventive maintenance shall be performed on a schedule mutually agreed to by the State and Contractor.   Preventive maintenance schedule periods may be modified by mutual agreement.

c.   Contractor shall provide remedial maintenance promptly after notification. The Contractor's maintenance personnel will normally arrive at the service site within two (2) hours after notification by the State that remedial maintenance is required.   For this purpose, Contractor shall have full and free access to the cable plant consistent with the State's operating requirements and security regulations.

d.   When a Contractor's technician responds to a remedial maintenance call and the equipment malfunction has not been diagnosed and repair begun within two (2) hours from the time of arrival of the technician, the Contractor will utilize 2nd Level Technical Support. In the event that four (4) additional hours elapse from the time of response of the 2nd level of technical support and the equipment's malfunction has not been diagnosed and repair begun, the Contractor will utilize 3rd Level Technical Support.   In any event, the Contractor will assign one or more levels of support for analysis and repair of the problem until the equipment has been returned to good operating condition.

e.   If after the Contractor's third level technical support, the problem has not been diagnosed and repair begun within two hours, then management escalation will begin.   In any case, State reserves the right to escalate to Contractor management any problems the State deems appropriate.

RIDER A - PAGE 5

Figure 9.2 (Continued)   Bid specifications—California Telecommunications Division.

R-92-015

    f.   Contractor must provide a list of Contractor's management, including names, titles, and work phone numbers.

6.   _Responsibilities of the State_

    a.   The State shall provide adequate storage space for spare parts, and adequate working space, including heat, light, ventilation, an electric current and outlets, for the use of the Contractor's maintenance personnel. These facilities shall be within a reasonable distance of the equipment to be serviced and shall be provided at no charge to the Contractor.

    b.   Unless mutually agreed to by Contractor and the State, State personnel shall not perform maintenance or attempt repairs to equipment while such equipment is under this contract.

    c.   Subject to the State's security regulations, the State shall permit access to the equipment which is to be maintained.

    d.   The State will be responsible for the additional cost of remedial maintenance of a system which has been modified by substitution and/or additions provided by a third party, and the system or equipment failure was caused by the system modification.

    e.   The State shall provide an appropriate operating environment, including temperature, humidity, and electrical power, in accordance with the environmental requirements contained in the manufacture's published specifications for the equipment listed in Rider B.

7.   _Maintenance_

    a.   Period of maintenance coverage:

        (1)   A basic monthly maintenance charge entitles the State to maintenance coverage during the Principal Period of Maintenance.

        (2)   The State may change its selected period of maintenance coverage by giving Contractor fifteen (15) days prior written notice.

        (3)   The Contractor may not charge for travel expenses.

        (4)   Maintenance charges as specified in Rider B for a equipment will commence on the first day of the contract period or the day (Monday through Friday) following the last day of any guarantee period.

RIDER A - PAGE 6

Figure 9.2 (*Continued*) Bid specifications—California Telecommunications Division.

R-92-015

b. Preventive Maintenance (scheduled)

   (1) Preventive maintenance shall be performed consistent with the
        State's operating requirements and security regulations
        within the Principal Period of Maintenance. The Contractor
        shall schedule these times with the State's concurrence.

c. Remedial Maintenance (scheduled)

   (1) Remedial maintenance shall be performed after notification by
        authorized State personnel at a time agreed to by the State
        and the Contractor. This time may be on a weekend or at a
        time other than 8 am to 5 pm.

   (2) The Contractor shall provide the State with a designated
        point of contact and will make arrangements to enable its
        maintenance representative to receive such notification.

   (3) There shall be no additional maintenance charges for:

        (a) Remedial maintenance performed during the period of
             maintenance coverage unless the remedial maintenance
             is due to the fault or negligence of the State.

        (b) Time spent by maintenance personnel after arrival at
             the site awaiting the arrival of additional
             maintenance personnel and/or delivery of parts, etc.,
             after a service call has been commenced.

        (c) Remedial maintenance required because the scheduled
             preventive maintenance preceding the malfunction had
             not been performed, unless the State had failed to
             provide access to the equipment.

d. Maintenance Charges

   (1) The monthly maintenance charges described includes 24-hour
        service, 7 days per week, year-round, and includes holidays,
        Saturdays, and Sundays.

   (2) Maintenance charges for fractions of a calendar month shall
        be computed at the rate of 1/30 of the applicable Total
        Monthly Maintenance Charge, as shown in the Bid, for each day
        maintenance was provided.

   (3) There will be no charge for travel expense associated with
        maintenance service or programming service under this
        Agreement.

RIDER A - PAGE 7

Figure 9.2 (Continued) Bid specifications—California Telecommunications Division.

*(4)   Maintenance rates shall be firm for the contract period
subject to paragraph 3, Termination.*

*(5)   The Contractor shall not be responsible for failure to render
service due to causes beyond its control. When a non-
Contractor system fails and as a result the Contractor's
Maintenance Diagnostic procedures do not pinpoint the
failure, the State shall pay for the time spent by the
Contractor in diagnosing the failure at the hourly
maintenance rate listed in Rider C.*

*(6)   When Contractor is called to perform remedial maintenance
service on the equipment and by mutual agreement it is
determined that either no failure existed or that the service
was outside the scope of the Contract, the State shall pay
for the time spent by the Contractor at the hourly
maintenance rate as listed in Rider C. If mutual agreement
cannot be reached the dispute shall be resolved in accordancw
with paragraph 12, Disputes.*

*(7)   Unless otherwise mutually agreed, the Contractor shall not be
required to adjust or repair any machine or part if it would
be impractical for Contractor personnel to do so because of
alterations made by or on behalf of the State. Increased
service pursuant to this Bid caused by alteration or
attachment shall be paid for by the State unless the
Contractor elects not to apply such charge on an individual
instance basis.*

8. *Maintenance Credit for Inoperative Cable Plant*

a. *The Contractor shall grant a proportionate maintenance credit on a
particular equipment shown in Rider B when the equipment is
inoperative for consecutive scheduled work periods totaling 24 hours
from the time the State notifies the Contractor the equipment was
inoperative, provided (1) the equipment became inoperative through
no fault of the State and not caused by the items listed in paragraph
4.c., and (2) the breakdown was attributable to equipment failure.*

b. *Contractor shall grant a credit to the State for each such hour in
the amount of 1/168 of the total monthly maintenance charges for the
inoperative equipment plus 1/168 of the total monthly maintenance
charges for any Contractor-supplied interconnected equipment which
became unusable as a result of a breakdown. Credit will be from the
first inoperative hour when equipment has been inoperative for more
than eight hours. For disputes arising from the State's request for
credit see paragraph 12, Disputes.*

RIDER A - PAGE 8

Figure 9.2 (Continued) Bid specifications—California Telecommunications Division.

R-92-015

    c.   The amount of credit granted for each equipment during a calendar month shall not exceed the total monthly maintenance charges.

9.   <u>Engineering Changes</u>

    a.   CONTRACTOR INITIATED ENGINEERING CHANGES

Engineering changes, determined applicable by Contractor, will be controlled and installed by the Contractor on equipment covered by this contract. The State may elect to have only mandatory changes, as determined by Contractor, installed on the cable plant so designated. A written notice of this election must be provided to the Contractor for written confirmation. There shall be no charge for engineering-changes made. Any Contractor-initiated change shall be installed at a time mutually agreeable to the State and the Contractor. Contractor reserves the right to charge, at the rates listed in Rider C, for additional service time and materials required due to non-installation of applicable engineering changes after Contractor has made a reasonable effort to secure time to install such changes.

    b.   STATE INITIATED ENGINEERING CHANGES

Engineering changes, determined applicable by the State, will be controlled by the State and installed by the Contractor on equipment covered by this contract. The Contractor shall:

    (1)   Install and remove State furnished terminal equipment as needed at a time mutually agreed to by the State and Contractor.

    (2)   Furnish matching and inter-membering modules and support equipment for existing lightwave terminal equipment as needed and as available.

    (3)   Make additions, deletions and modifications to the fiber optic cable plant as needed.

The labor charges for the Contractor to make the changes shall be at the rate set forth in Rider C. Cost for the fiber optic cable, modules and support Charges for the equipment is to be mutually agreed to. The State reserves the right to supply any and all of the above labor and material.

RIDER A - PAGE 9

Figure 9.2 (*Continued*)   Bid specifications—California Telecommunications Division.

R-92-015

c. In the event the equipment maintained under the terms and conditions of this contract is the subject of the engineering change(s) and the changes are made by the Contractor, the contractor shall continue to maintain the equipment as modified. If the equipment is changed, Rider B of this contract shall be amended to recognize engineering change.

10. Relocation of Equipment

a. In the event the equipment being maintained under the terms and conditions of this contract is moved to another location within metropolitan Sacramento, the Contractor shall continue to maintain the equipment at the new location.

b. The charges of the Contractor to dismantle and pack the equipment and installation at the new location shall be at the rates mutually agreed to at the time of relocation.

c. The State agrees to pay all costs incidental to any inter-building move, including costs for packing, crating, rigging, transportation, unpacking, uncrating, insurance, installation, and State and local tax, if any.

d. Any physical planning assistance requested by the State and provided by the Contractor after the initial installation on the same site shall be at the State's expense.

f. To the extent practicable, the State shall give at least sixty (60) days written notice of the movement of the equipment unless a shorter time period is agreed to by the Contractor.

11. Invoices and Payments

a. Payments will be allowed monthly in arrears upon satisfactory performance of service, acceptance of performance by State personnel, and the State's receipt of an itemized invoice in triplicate. Invoices submitted for payment must include:

     o   Invoice number
     o   Invoice Date
     o   Date service performed
     o   Description of service performed
     o   Dollar amount (in accordance with the contract)
     o   Contract number
     o   Contractor personnel (names/titles) performing service

RIDER A - PAGE 10

Figure 9.2 (Continued) Bid specifications—California Telecommunications Division.

R-92-015

The State shall pay all such invoices in accordance with applicable State policy. The Contractor shall render invoices for total monthly charges in the month following the month for which the charges accrue. Unless notified otherwise by the State, invoices will be submitted in triplicate to:

Attention: Mark MacRae, M.S. 09E
Department of General Services
Telecommunications Division
601 Sequoia Pacific Blvd.
Sacramento, CA 95814-0282

b. Charges for fractions of a calendar month shall be computed at the rate of 1/30 of the applicable Total Monthly Charge, as shown in Rider B., for each day maintenance was provided.

c. The State of California is exempt from Federal excise taxes, and no payment will be made for any taxes levied on employee's wages. The State will pay only for any State of California or local sales or use taxes on the services rendered or equipment or parts supplied pursuant to this Contract.

d. The monthly charges, equipment groups, period of maintenance coverage offered, and optional period percentages are subject to change by Contractor, only annually, three months following written notice to the State. Such changes will be made by amendment to this Contract in accordance with Paragraph 13c. In no case will an increase in monthly charges exceed 10 percent of the preceding year's rate.

12. <u>Disputes</u>

a. Any dispute concerning a question of fact arising under the terms of this agreement which is not disposed of within a reasonable period of time by the Contractor and State employees normally responsible for the administration of this contract shall be brought to the attention of the Chief Executive Officer (or designated representative) of each organization for joint resolution. At the request of either party, the State shall provide a forum for discussion of the disputed item(s), at which time the State Chief of Procurement, or his representative, shall be available to assist in resolution by providing advice to both parties as to the State of California policies and procedures. If agreement cannot be reached through the application of high level management attention, either party may assert its other rights and remedies within this contract or within a court of competent jurisdiction.

b. The rights and remedies of the State provided above shall not be exclusive and are in addition to any other rights and remedies provided by law or under the contract.

RIDER A - PAGE 11

Figure 9.2 (*Continued*) Bid specifications—California Telecommunications Division.

R-92-015

13. General

    a.   Contractor is not responsible for delay or failure to render service due to causes beyond the Contractor's control.

    b.   This contract shall be governed by the laws of the State of California. The term "this contract" as used herein include any future written amendments made in accordance herewith.

    c.   This contract may be amended or modified only by mutual agreement of the parties in writing.

14. Confidentiality of Data

    a.   All financial, statistical, personal, technical and other data and information relating to the State's operation which are designated confidential by the State and made available to the Contractor in order to carry out this agreement, shall be protected by the Contractor from unauthorized use and disclosure through the observance of the same or more effective procedural requirements as are applicable to the State. The identification of all such confidential data and information as well as the State's procedural requirements for protection of such data and information from unauthorized use and disclosure shall be provided by the State in writing to the Contractor.

    b.   If the methods and procedures employed by the Contractor for the protection of the Contractor's data and information are deemed by the State to be adequate for the protection of the State's confidential information, such methods and procedures may be used, with the written consent of the State, to carry out the intent of this paragraph. The Contractor shall not be required under the provisions of this paragraph to keep confidential any data or information which is or becomes publicly available, is already rightfully in the Contractor's possession is independently developed by the Contractor outside the scope of this agreement, or is rightfully obtained from third parties.

15. Contractor Evaluation

The Contractor's performance will be evaluated by the State within thirty days of completion of the contract. Contractor will be subject to examination and audit of the Auditor General for a period of three years after final payment under the contract.

RIDER A - PAGE 12

Figure 9.2 (Continued) Bid specifications—California Telecommunications Division.

R-92-015

16. <u>Assignment of Antitrust Actions</u>

The following provisions of Government Code Section 4552, 4553, and 4554 (Statutes of 1978, Ch. 414) shall be applicable to the Contractor.

"In submitting a bid to a public purchasing body, the bidder offers and agrees that if the bid is accepted, it will assign to the purchasing body all rights, title, and interest in and to all causes of action it may have under Section 4 of the Clayton Act (15 U.S.C. Sec. 15) or under the Cartwright Act [Chapter 2 (commencing with Section 16700) of part 2 of division 7 of the Business and Professions Code], arising from purchases of goods, materials, or services by the bidder for sale to the purchasing body pursuant to the bid. Such assignment shall be made and become effective at the time the purchasing body tenders final payment to the bidder."

"If an awarding body or public purchasing body receives, either through judgment or a settlement, a monetary recovery for a cause of action assigned under this chapter, the assignor shall be entitled to receive reimbursement for actual legal costs incurred and may, upon demand, recover from the public body and any portion of the recovery, including treble damages, attributable to overcharges that were paid by the assignor but were not paid by the public body as part of the bid price, less the expenses incurred in obtaining that portion of the recovery."

"Upon demand in writing by the assignor, the assignee shall, within one year from such demand, reassign the cause of action assigned under this part if the assignor has been or may have been injured by the violation of law for which the cause of action arose and (a) the assignee has not been injured thereby, or (b) the assignee declines to file a court action for the cause of action."

17. <u>National Labor Relations Board Certification</u>

By signing hereon the Contractor swears under penalty of perjury that no more than one final, unappealable finding of contempt of court by a federal court has been issued against the Contractor within the immediately preceding two-year period because of the Contractor's failure to comply with an order of the National Labor Relations Board. This provision is required by, and shall be construed in accordance with Public Contract Code Section 10296.

RIDER A - PAGE 13

Figure 9.2 (Continued) Bid specifications—California Telecommunications Division.

R-92-015

18. *Statement of Compliance*

   The Contractor's signature affixed hereon and dated shall constitute a certification under the penalty of perjury under the laws of the State of California that the Contractor has, unless exempted, complied with the nondiscrimination program requirements of Government Code Section 12990 and Title 2, California Administrative Code, Section 8103.

19. *Examination and Audit*

   The contracting parties shall be subject to the examination and audit of the State Auditor General or State representative for a period of three (3) years after final payment under the contract in accordance with Government Code Section 10532. The examination and audit shall be confined to those matters connected with the performance of the contract, including, but not limited to, the costs of administering the contract.

20. *WAGES:*

   The Contractor shall pay his/her employees wages not less than those required by any applicable law.

21. *LICENSES/PERMITS:*

   The Contractor shall obtain and at his/her expense pay for any/all licenses/permits required by law for accomplishing any work required in connection with this contract.

22. *FAIR EMPLOYMENT PRACTICES:*

   Standard Form 17A, Nondiscrimination Clause, may be attached and made a part of the contract.

23. *STATE CONTRACTORS LICENSE*

   The Contractor must hold a license issued by the State Contractors Board which authorizes the Contractor to perform the work described in this contract.

24. *BOND*

   A Performance bond (or acceptable substitutes) is required in an amount equal to 100 percent of the contract price. In the bid the Bidder shall provide a letter from a Surety authorized to do business in the State of California stating that the Surety will issue the bond (or equal) as required. At the time of contract award the Bidder shall deliver to the State the required bond or equal for the faithful performance of the contract. The bond, or equal, are subject to approval by the State.

RIDER A - PAGE 14

Figure 9.2 (Continued)   Bid specifications—California Telecommunications Division.

R-92-015

25. _INSURANCE_

The Contractor shall, at their own expense, procure, carry, and maintain insurance on all of its operations as follows:

a. Workmen's Compensation and Employers Liability Insurance for the protection of Contractor's employees as required by law._ The Employer's Liability Insurance shall be in an amount not less than Five Hundred Thousand Dollars ($500,000) per claim.

b. Insurance for liability because of personal injury, bodily injury and/or property damage sustained or alleged to have been sustained by any person._ The insurance shall name the State as an insured and shall cover all operations of the Contractor including, but not limited to, the following:

  *   Premises;

  *   Operations and Mobile Equipment Liability;

  *   Completed operations and products liability;

  *   Contractual Liability insuring the obligation assumed by the Contractor in the Contract;

  *   Liability which the Contractor may incur and a result of the operations, acts, of omissions of its subcontractors, suppliers of material men and their agents of employees; and

  *   Automobile liability including owned, non-owned and hired automobiles.

Such policy or policies shall be endorsed to include the State, its officers, directors and employees as additionally insured and shall stipulate that the insurance afforded for the State, its officers, directors, and employees, shall be primary insurance and that any insurance carried by the State, its officers, directors of employees shall be excess and not contributory insurance.

Insurance for liability shall be on an occurrence basis and shall provide coverage with limits not less than:

  (i)   Personal injury and bodily injury: One Million dollars ($1,000,000) each person, One Million dollars ($1,000,000) each occurrence.

  (ii)   Property Damage: One Million dollars ($1,000,000) each occurrence.

RIDER A - PAGE 15

Figure 9.2 (Continued) Bid specifications—California Telecommunications Division.

R-92-015

c. *At the time of contract award the Contractor shall, at its own expense, furnish to the State for inclusion in the final contract Certificates of Insurance from companies acceptable to the State. The Certificates of Insurance shall provide that there shall be no cancellation, reduction or modification of coverage without thirty (30) days written notice to the State.*

26. ADMINISTRATIVE REQUIREMENTS

a) *Bidder is required to complete, sign, and return the enclosed forms with his/her bid:*

   Rider D, Statement of Compliance
   Rider E, Confidentiality Statement
   *Rider F, Minority/Women/Disabled Veterans Business
       Participation/Good Faith Effort Requirements*
   Rider G, Drug-Free Work-place Certification.

b) *If applicable, bidder is required to complete, sign, and return the enclosed forms with his/her bid:*

   Rider H, Small Business Preference

27. BID EVALUATION AND AWARD

a) *Bids are required for all three years on the Bid Proposal Cost Sheet. The bidder shall set forth for each year, in clearly legible figures, a Monthly Cost and a Yearly Cost. The Yearly Cost should be calculated by multiplying each Monthly Cost by 12 months. The amount of the bid, for comparison purposes, shall be the Grand Total [d] of all three (3) years ([a] + [b] + [c] = [d]), which should be entered onto the Bid Proposal Summary.*

b) *In the case of a discrepancy between the "Monthly Total" and the "Yearly Total", the Monthly Total shall prevail. However, if the amount set forth as a Monthly Total is ambiguous, unintelligible or uncertain for any cause, or it is omitted, the amount set forth in the Yearly Total shall prevail and be recalculated by dividing the total amount by 12 months to arrive at an accurate Monthly Total. The same will apply for Year Two and Year Three by applying the percentage quoted by the bidder if/as applicable.*

RIDER A – PAGE 16

Figure 9.2 (*Continued*)   Bid specifications—California Telecommunications Division.

R-92-015

c)   *A bidder will be allowed to increase rates for Year Two and Year Three of the contract term. A price escalation shall include a percentage NOT TO EXCEED Ten-percent. The bidder shall calculate the Year Two and Year Three prices as follows using the percentage of increase the bidder has established:*

(% of increase) x (previous rates)

where "previous rates" are the bid rates for the Year One.

(% of increase) x (Year Two rates)

where "Year Two rates" are the bid rates for the second year.

d)   *A bidder is required to complete the Bid Proposal Summary by adding [a] + [b] + [c] from the Bid Proposal Cost Sheet and transferring the figure to Grand Total [d] on the Bid Proposal Summary.*

e)   *The contract term is estimated to be June 1, 1992 through May 31, 1995. If awarded, the contract duration is effective upon final approval by the Department of General Services, Telecommunications Division.*

f)   *Intent to award, if made, will be to the lowest responsible bidder of all the conditions required in this bid.*

28.   *CONTRACT AUDITS FOR MINORITY/WOMEN/DISABLED VETERAN BUSINESS ENTERPRISE REQUIREMENTS*

Contractor agrees that the State or its delegatee will have the right to review, obtain, and copy all records pertaining to performance of the contract. Contractor agrees to provide the State or its delegatee with any relevant information requested and shall permit the State or its delegatee access to its premises, upon reasonable notice, during normal business hours for the purpose of interviewing employees, and inspecting and copying such books, records, accounts, and other material that may be relevant to a matter under investigation for the purpose of determining compliance with this requirement. Contractor further agrees to maintain such records for a period of three years after final payment under the contract.

RIDER A - PAGE 17

Figure 9.2 (*Continued*)   Bid specifications—California Telecommunications Division.

CAPITAL FIBER LOOP MAINTENANCE

RIDER B

EQUIPMENT LIST FOR FIBER MANAGEMENT

COST FOR UPGRADING CURRENT FIBER PATCH PANELS TO PROVIDE	*$* _____
ORDERLY, SYSTEMATIC CABLE AND JUMPER MANAGEMENT, AND FOR	
UPDATING CURRENT DOCUMENTATION	

EQUIPMENT LIST

ITEM	*DESCRIPTION*	*QUANTITY*	*COST*

RIDER B - PAGE 1

Figure 9.2 (*Continued*)   Bid specifications—California Telecommunications Division.

CAPITAL FIBER LOOP MAINTENANCE

RIDER B

EQUIPMENT LIST AND MAINTENANCE CHARGES

BASIC MONTHLY MAINTENANCE CHARGE FOR 24 HOUR PER DAY, $_____ /MO
7 DAY A WEEK, 2 HOUR RESPONSE, MAINTENANCE SERVICES ON
EQUIPMENT AND CABLE LISTED ON RIDER B IN ACCORDANCE WITH
THE TERMS AND CONDITIONS LISTED HEREIN.

MAXIMUM YEARLY PERCENTAGE INCREASE IN THE MONTHLY   _____ %
MAINTENANCE CHARGE.

EQUIPMENT LIST		
LOCATION		
MANUFACTURER/MODEL/DESCRIPTION	*QUANTITY*	*RESPONSIBILITY*

OB-1
IBM 8219 TOKEN RING REPEATER	*2*	*ASSIST AS REQ'D*

LIBRARY AND COURTS
ROCKWELL 3X50 MULTIPLEXER	*1*	*ASSIST AS REQ'D*

EDUCATION
ROCKWELL 3X50 MULTIPLEXER	*1*	*ASSIST AS REQ'D*

EDD
FIBERMUX 812 (8 T1) TERMINAL	*2*	*FULL*
ROCKWELL 3X50 MULTIPLEXER	*1*	*ASSIST AS REQ'D*

RESOURCES
FIBERMUX 812 (8 T1) TERMINAL	*4*	*FULL*
GRASS VALLEY VIDEO LINK	*1*	*NONE*
ROCKWELL 3X50 MULTIPLEXER	*3*	*ASSIST AS REQ'D*

OB-8
FIBERMUX 812 (8 T1) TERMINAL	*2*	*FULL*
FIBRONICS 832 (56 Kb) TERMINAL	*2*	*FULL*
GRASS VALLEY VIDEO LINK	*1*	*NONE*
ROCKWELL 3X50 MULTIPLEXER	*1*	*ASSIST AS REQ'D*

BATESON
FIBRONICS 832 (56 Kb) TERMINAL	*2*	*FULL*

ENERGY
ROCKWELL 3X50 MULTIPLEXER	*1*	*ASSIST AS REQ'D*

ARCHIVES
IBM 8219 TOKEN RING REPEATER	*2*	*ASSIST AS REQ'D*

RIDER B - PAGE 2

Figure 9.2 (*Continued*) Bid specifications—California Telecommunications Division.

CALTRANS
FIBERMUX 812 (8 T1)   4   FULL
ROCKWELL 3X50 MULTIPLEXER   1   ASSIST AS REQ'D

THE CAPITOL
IBM 8219 TOKEN RING REPEATER   2   ASSIST AS REQ'D
ROCKWELL 3X50 MULTIPLEXER   1   ASSIST AS REQ'D

SACRAMENTO SCIP
ROCKWELL 3X50 MULTIPLEXER   11   ASSIST AS REQ'D
ROCKWELL 1565 MULTIPLEXER   1   ASSIST AS REQ'D

SACRAMENTO NMCC
ROCKWELL 3X50 MULTIPLEXER   1   ASSIST AS REQ'D

MCI - 925 L STREET
ROCKWELL 1565 MULTIPLEXER   1   ASSIST AS REQ'D

RIDER B - PAGE 3

Figure 9.2 (_Continued_)   Bid specifications—California Telecommunications Division.

CAPITAL FIBER LOOP MAINTENANCE

RIDER B

EQUIPMENT LIST - SPARE PARTS FOR MAINTENANCE

VALUE OF RIDER B INVENTORY 5/15/92 $_____15,100_
IF PURCHASED THROUGH THE CURRENT CONTRACT

COST TO PURCHASE SPARES LISTED IN RIDER B AT THE END OF $_____
THE MAINTENANCE PERIOD. (GUARANTEED 100% FUNCTIONAL)

LIST OF REQUIRED SPARE PARTS

FIBERMUX 812 LIGHTWAVE TERMINAL

QUANTITY	DESCRIPTION	PART NUMBER
2 EACH	Common Logic Module	CC671
4 EACH	I/O Module	CC661E
2 EACH	Power Supply	CC127
2 EACH	Switch Logic	CC130/V

FIBRONICS 832 LIGHTWAVE TERMINAL

1 EACH	RS 232 Channel Card	CC83210
1 EACH	Logic Unit Tranciever Model	LU83208
1 EACH	Power Supply	PS83297

FIBRONICS 1632-10 LIGHTWAVE TERMINAL

(Not Applicable at this time)
2 EACH   Fiber Tranciever Card   FO 30

RIDER B - PAGE 4

Figure 9.2 (Continued) Bid specifications—California Telecommunications Division.

CAPITAL FIBER LOOP MAINTENANCE

RIDER B

FIBER OPTIC CABLE PLANT LIST

DESCRIPTION LOCATION OR NAME	CABLE	QUANTITY
72 FIBER INTER-BUILDING CABLE		
CAPITOL TO OB-1		*1*
OB-1 TO PERSONNEL		*1*
PERSONNEL TO EDD		*1*
EDD TO RESOURCES		*1*
RESOURCES TO OB-8		*1*
OB-8 TO BATESON		*1*
BATESON TO ENERGY		*1*
ENERGY TO WRCB		*1*
WRCB TO ARCHIVES		*1*
ARCHIVES TO THE SCIP		*1*
THE SCIP TO CALTRANS		*1*
CALTRANS TO THE CAPITOL		*1*
SPUR CABLES		
#1 24 FIBER, TB, CAPITOL		*1*
#2 24 FIBER, TB, CAPITOL		*1*
#3 24 FIBER, LT, OB-1		*1*
#4 24 FIBER LT, OB-1		*1*
#5 24 FIBER TB, EDD		*1*
#6 24 FIBER TB, RESOURCES		*1*
#7 24 FIBER TB, OB-8 TO OB-9		*1*
#8 24 FIBER TB, BATESON		*1*
#9 24 FIBER TB, BATESON		*1*
#10 24 FIBER LT, ARCHIVES TO OB-2		*INACTIVE*
#11 24 FIBER TB, CALTRANS TO CT ANNEX		*1*
#12 24 FIBER TB, CAPITOL		*1*
#13 72 FIBER LT, CALTRANS TO AGRICULTURE		*1*
#14 8 FIBER LT, RESOURCES		*1*
#15 72 FIBER LT, PERSONNEL TO EDUCATION		*1*
#16 8 FIBER LT, RESOURCES		*1*
#17 8 FIBER LT, OB-1 TO 925 L STREET		*1*
#18 24 FIBER TB, CAPITOL		*1*
#19A 12 FIBER TB, CAPITOL		*1*
#19B 12 FIBER TB, CAPITOL		*1*
#20 24 FIBER TB, CAPITOL		*1*

RIDER B - PAGE 5

Figure 9.2 *(Continued)* Bid specifications—California Telecommunications Division.

RIDER C
EQUIPMENT LIST AND MAINTENANCE CHARGES

BASIC HOURLY MAINTENANCE CHARGE FOR 24 HOUR PER DAY, $_____ /HR
7 DAY A WEEK, 2 HOUR RESPONSE, MAINTENANCE SERVICES ON
EQUIPMENT AND CABLE LISTED ON RIDER B IN ACCORDANCE WITH
THE TERMS AND CONDITIONS LISTED HEREIN.

MAXIMUM YEARLY PERCENTAGE INCREASE IN THE HOURLY   _____ %
MAINTENANCE CHARGE.

EQUIPMENT LIST

LOCATION
MANUFACTURER/MODEL/DESCRIPTION   QUANTITY   RESPONSIBILITY

OB-1
IBM 8219 TOKEN RING REPEATER   2   ASSIST AS REQ'D

LIBRARY AND COURTS
ROCKWELL 3X50 MULTIPLEXER   1   ASSIST AS REQ'D

EDUCATION
ROCKWELL 3X50 MULTIPLEXER   1   ASSIST AS REQ'D

EDD
FIBERMUX 812 (8 T1) TERMINAL   2   FULL
ROCKWELL 3X50 MULTIPLEXER   1   ASSIST AS REQ'D

RESOURCES
FIBERMUX 812 (8 T1) TERMINAL   4   FULL
GRASS VALLEY VIDEO LINK   1   NONE
ROCKWELL 3X50 MULTIPLEXER   3   ASSIST AS REQ'D

OB-8
FIBERMUX 812 (8 T1) TERMINAL   2   FULL
FIBRONICS 832 (56 Kb) TERMINAL   2   FULL
GRASS VALLEY VIDEO LINK   1   NONE
ROCKWELL 3X50 MULTIPLEXER   1   ASSIST AS REQ'D

BATESON
FIBRONICS 832 (56 Kb) TERMINAL   2   FULL

ENERGY
ROCKWELL 3X50 MULTIPLEXER   1   ASSIST AS REQ'D

ARCHIVES
IBM 8219 TOKEN RING REPEATER   2   ASSIST AS REQ'D

CALTRANS
FIBERMUX 812 (8 T1)   4   FULL
ROCKWELL 3X50 MULTIPLEXER   1   ASSIST AS REQ'D

RIDER C - PAGE 1

Figure 9.2 (Continued) Bid specifications—California Telecommunications Division.

THE CAPITOL
IBM 8219 TOKEN RING REPEATER   2   ASSIST AS REQ'D
ROCKWELL 3X50 MULTIPLEXER   1   ASSIST AS REQ'D

SACRAMENTO SCIP
ROCKWELL 3X50 MULTIPLEXER   11   ASSIST AS REQ'D
ROCKWELL 1565 MULTIPLEXER   1   ASSIST AS REQ'D

SACRAMENTO NMCC
ROCKWELL 3X50 MULTIPLEXER   1   ASSIST AS REQ'D

MCI - 925 L STREET
ROCKWELL 1565 MULTIPLEXER   1   ASSIST AS REQ'D

RIDER C - PAGE 2

Figure 9.2 (Continued) Bid specifications—California Telecommunications Division.

CAPITAL LOOP FIBER MAINTENANCE

RIDER C
CONTRACTOR'S RATES

CONTRACTORS HOURLY RATE (24 HOURS PER DAY, 7 DAYS A $_____/HR
WEEK), FOR ADDITIONS, DELETIONS, MOVES AND CHANGES TO
THE EXISTING SYSTEM.

MAXIMUM PERCENTAGE YEARLY INCREASE IN THE CONTRACTOR'S   _____%
HOURLY RATE CHARGE.

RIDER C - PAGE 3

Figure 9.2 (Continued) Bid specifications—California Telecommunications Division.

RIDER D

STATEMENT OF COMPLIANCE

_____ (hereinafter referred to as
(Company Name)

"prospective contractor") hereby certifies, unless specifically

exempted, compliance with Government Code Section 12990 and

California Administrative Code, Title II, Division 4, Chapter 5

in matters relating to the development, implementation and main-

tenance of a nondiscrimination program. Prospective contractor

agrees not to unlawfully discriminate against any employee or

applicants for employment because of race, religion, color,

national origin, ancestry, physical handicap, medical condition,

marital status, sex or age (over forty).

I _____ hereby swear that I am
  (Name of Official)

duly authorized to legally bind the prospective contractor to

the above described certification. I am fully aware that this

certification executed on _____ in the county
                                  (Date)

of _____ is made under the penalty of perjury
       (County)

under the laws of the State of California.

_____
     Signature

_____
     Title

Figure 9.2 (_Continued_)   Bid specifications—California Telecommunications Division.

RIDER E

CONFIDENTIALITY STATEMENT

As an authorized representative and/or corporate officer of the company named below, I warrant my company and its employees will not disclose any financial, statistical, personal, technical or other data and information relating to the State's operation which are designated confidential by the State and made available to us by the State for the purpose of responding to this bid or in conjunction with any contract arising therefrom. I warrant that only those employees who are authorized and required to use such materials will have access to them.

I further warrant that all materials if/as provided by the State will be returned promptly after use and that all copies or derivations of the materials will be physically and/or electronically destroyed. I will include with the returned materials, a letter attesting to the complete return of materials and documenting the destruction of copies and derivations. Failure to comply will subject this company to liability, both criminal and civil, including all damages to the State and third parties. I authorize the State to inspect and verify the above.

I warrant that if my company is awarded the contract, it will not enter into any agreements or discussions with a third party concerning such materials prior to receiving written confirmation from the State that such third party has a confidentiality agreement with the State similar in nature to this one.

_____

(Signature of Represenative)   (Date)

_____

(Typed Name of Representative)

_____

(Typed Title of Representative)

.-
_____

(Typed Name of Company)

-26-

Figure 9.2 (Continued) Bid specifications—California Telecommunications Division.

RIDER F

MINORITY, WOMEN, AND DISABLED VETERAN BUSINESS ENTERPRISE PARTICIPATION (M/W/DVBE) REQUIREMENT

State law requires that state contracts have participation goals of 15 percent for Minority Business Enterprises (MBEs), 5 percent for Women Business Enterprises (WBEs), and 3 percent for Disabled Veteran Business Enterprises (DVBEs) for contracts over the amount of $10,000.00.

PLEASE READ THESE REQUIREMENTS CAREFULLY. FAILURE TO COMPLY WITH THE M/W/DVBE REQUIREMENT MAY CAUSE YOUR BID TO BE DEEMED NON-RESPONSIVE AND YOU TO BE INELIGIBLE FOR AWARD OF THIS CONTRACT.

I. **CONTRACT GOALS/"GOOD FAITH EFFORT"**

In order to be "responsive" to this requirement, the bidder must do either of the two following alternatives:

A. Meet or exceed the goals of 15 percent MBE, five percent WBE, and three percent DVBE participation for the proposed contract by one of the following four ways:

1. Commit to use MBEs for not less than 15 percent, WBEs for not less than five percent and DVBEs for not less than three percent of the contract amount; or

2. If the bidder is an MBE, commit to performing not less than 15 percent of the contract amount with its own forces, commit to use WBEs for not less than five percent and DVBEs for not less than three percent of the contract amount; or

3. If the bidder is a WBE, commit to performing not less than five percent of the contract amount with its own forces, commit to use MBEs for not less than 15 percent and DVBEs for not less than three percent of the contract amount; or

4. If the bidder is a DVBE, commit to performing not less than three percent of the contract amount with its own forces, commit to use MBEs for not less than 15 percent and WBEs for not less than five percent of the contract amount.

NOTE: If the DVBE is an MBE or WBE, they can meet either specification as either an MBE/DVBE or WBE/DVBE.

OR

B. Make a "good faith effort" to meet the goals by doing all of the following by the final bid/proposal date:

1. Contact either Mark MacRae at (916) 657-9269 or Cherlyn Thomas at (916) 657-9368 at the Department of General Services, Telecommunications Division, to identify M/W/DVBEs;

M/W/DVBE REQUIREMENTS   Page 1 of 5

Figure 9.2 (Continued) Bid specifications—California Telecommunications Division.

2. Contact other state **AND** federal agencies **AND** local M/W/DVBE organizations to identify potential M/W/DVBEs for this contract (see Exhibit 1 for information);

3. Advertise in trade papers and papers focusing on M/W/DVBEs; (BEGIN THIS PROCESS IMMEDIATELY UPON RECEIPT OF YOUR BID PACKAGE TO ALLOW SUFFICIENT TIME FOR RESPONSES, ETC.).

4. Send solicitations to potential M/W/DVBE subcontractor/suppliers for this contract with sufficient lead time to fully entertain and consider responding bids; and,

5. Consider responding M/W/DVBEs for participation in this contract.

II. DOCUMENTATION REQUIREMENTS

Exhibits 2 and 2A
Whether the contract goal or the "good faith effort" alternative is chosen, Exhibit 2 (to be used for MBEs and WBEs) AND Exhibit 2A (to be used for DVBEs) must be completed and included in the final Bid/Proposal. These exhibits show the type of work and company proposed for M/W/DVBE participation, the subcontractors (if any), and other related information. If no participation is obtained, state "None" or "N/A" on the first line of both Exhibit 2 AND Exhibit 2A. The contracting tier should be indicated with the following level designations:

0 = Prime or Joint Contractor;
1 = Primary Subcontractor/Supplier;
2 = Subcontractor/Supplier of Level 1 Subcontractor/Supplier;
3 = Subcontractor/Supplier of Level 2 Subcontractor/Supplier; etc.

NOTE: Exhibit 2 AND 2A MUST be included with your bid response.

M/WBE CERTIFICATION

Exhibit 3
The bidder must include a signed certification (See Exhibit 3) that each firm listed on Exhibit 2 as a M/WBE complies with the legal definition of M/WBE. (Exhibit 3 MUST be included with your bid response).

Exhibit 4
Each proposed M/WBE listed on Exhibit 2 must also sign and complete a certification (See Exhibit 4) that it complies with the legal definition of M/WBE. (Exhibit 4 MUST be included with your bid response).

DVBE CERTIFICATION

The bidder must include a signed certification letter from the Office of Small and Minority Business (OSMB) for each firm listed on Exhibit 2A.

M/W/DVBE REQUIREMENTS   Page 2 of 5

Figure 9.2 (*Continued*) Bid specifications—California Telecommunications Division.

GOOD FAITH EFFORT

In addition to the above, for those bidders opting to document a "good faith effort", the bidder MUST document contacts with other state and federal agencies, and other organizations that helped identify or provided a list of interested M/W/DVBEs for this procurement. Dates, times, organizations contacted, and contact names and telephone numbers must be provided. Exhibit 5 should be utilized to document this information.

Bidders who propose goal attainment are permitted to submit documentation for making a "good faith effort" to ensure against the possibility that the State will not agree that goal attainment has, in fact, been met.

The bidder's efforts to meet the contract goal and/or make a "good faith effort" to meet the goal must be sincere and the documentation must be sufficient to reasonably demonstrate that sincerity to the State. FINAL DETERMINATION OF GOAL ATTAINMENT OR "GOOD FAITH EFFORT" BY THE BIDDER SHALL BE AT THE STATE'S SOLE DISCRETION.

III. **USE OF PROPOSED M/W/DVBE**

If awarded the contract(s), the successful bidder must use the M/W/DVBE subcontractor/supplier proposed in its final Bid/Proposal unless the Contractor requested substitution via prior written notice to the State and the State has approved such substitution. At a minimum, the request must include:

A. A written explanation of the reason for the substitution; and

B. The identity of the person or firm substituted.

THE REQUEST AND THE STATE'S APPROVAL OR DISAPPROVAL IS NOT TO BE CONSTRUED AS AN EXCUSE FOR NONCOMPLIANCE WITH ANY OTHER PROVISION OF LAW, INCLUDING, BUT NOT LIMITED TO THE SUBLETTING AND SUBCONTRACTING FAIR PRACTICES ACT OR ANY OTHER CONTRACT REQUIREMENTS RELATING TO SUBSTITUTION OF SUBCONTRACTORS.

FAILURE TO ADHERE TO AT LEAST THE M/W/DVBE PARTICIPATION PROPOSED BY THE SUCCESSFUL BIDDER MAY BE CAUSE FOR CONTRACT TERMINATION AND RECOVERY OF DAMAGES UNDER THE RIGHTS AND REMEDIES DUE THE STATE UNDER THE DEFAULT SECTION OF THE CONTRACT(S).

M/W/DVBE REQUIREMENTS   Page 3 of 5

Figure 9.2 (Continued)   Bid specifications—California Telecommunications Division.

IV. __MBE/WBE/DVBE PARTICIPATION REQUIREMENT__

For the purposes of this M/WBE participation requirement, the following apply:

A. Any bid amount proposed for MBE or WBE participation can only be counted once. That is, any further subcontracting or spending of MBE or WBE designated bid amounts to another subcontractor/supplier will not count toward meeting the contract goal. Moreover, any part of a MBE or WBE designated bid amount that is further designated for any other subcontractor involved in the same final Bid/Proposal (Suppliers are acceptable) will not count toward meeting the contract goal.

B. If the State reserves the right to make multiple awards or a single contract award as a result of this solicitation, the bidder is deemed responsive to this requirement if there would be compliance based on award of a single contract, notwithstanding that the bidder may be unable to achieve compliance to meet the established goal if the State exercises its right to make multiple awards.

V. __DVBE PARTICIPATION__

For the purposes of this DVBE participation requirement, the following definitions apply:

A. Disabled Veteran means a veteran of the military, naval or air services of the United States with at least a 10 percent service-connected disability who is a resident of the State of California.

B. DVBE means a business concern certified by the Office of Small and Minority Business as meeting all of the following:

1. A sole proprietorship owned by a disabled veteran; or a firm or partnership, 100 percent of the stock or partnership interests of which are owned by one or more disabled veterans.

2. Managed by, and the daily business operations are controlled by, one or more disabled veterans.

3. A sole proprietorship, corporation, or partnership with its home office located in the United States, which is not a branch or subsidiary of a foreign corporation, firm, or other business.

M/W/DVBE REQUIREMENTS   Page 4 of 5

Figure 9.2 (*Continued*)   Bid specifications—California Telecommunications Division.

VI. CONTRACT AUDITS

Contractor agrees that the State or its delegate will have the right to review, obtain, and copy all records pertaining to performance of the contract. Contractor agrees to provide the State or its delegate with any relevant information requested and shall permit the State or it delegate access to its premises, upon reasonable notice, during normal business hours for the purpose of interviewing employees and inspecting and copying such books, records, accounts, and other material that may be relevant to a matter under investigation for the purpose of determining compliance with this requirement. Contractor further agrees to maintain such records for a period of three (3) years after final payment under the contract.

M/W/DVBE REQUIREMENTS   Page 5 of 5

Figure 9.2 (*Continued*)   Bid specifications—California Telecommunications Division.

STATE OF CALIFORNIA—STATE AND CONSUMER SERVICES AGENCY   PETE WILSON, *Governor*

DEPARTMENT OF GENERAL SERVICES
Office of Small and Minority Business
1808 14th Street, Suite 100
Sacramento, CA 95814-7120

EXHIBIT 1

Thank you for contacting our office regarding the Minority and Women Business Enterprise Participation Program as required in your bid package.

Enclosed you will find resources which will assist you in identifying M/WBE firms interested in participating in state contracts.

The resources are not an exclusive list, but rather a few of the organizations available throughout the State of California.

If you have any further questions, please do not hesitate to contact an outreach specialist at (916) 452-6494.

Sincerely,

Alice M. Flissinger, Chief
Office of Small and Minority Business

Figure 9.2 (*Continued*) Bid specifications—California Telecommunications Division.

RESOURCES FOR SOLICITING
MINORITY, WOMEN, AND DISABLED VETERAN BUSINESS ENTERPRISES

ADVERTISING SOURCES

Sacramento Area

Construction Data and News
1791 Tribute Road, Ste. D
Sacramento, CA 95815
(916) 920-2240

Sacramento Builders Exchange
1331 T Street
Sacramento, CA 95814
(916) 442-8991

NEDA San Joaquin Valley, Inc.
2010 N. Fine Street, Ste. 103
Fresno, CA 93727
(209) 252-7551

El Tiempo
P. O. Box 2504
Sacramento, CA 95814
(916) 972-8402

El Hispano
P. O. Box 2856
Sacramento, CA 95812
(916) 442-0267

Sacramento Observer
3540 4th Ave.
Sacramento, CA 95817
(916) 452-4781

Sacramento Builders Exchange North
1 Sierragate Plaza, Ste. 120-A
Roseville, CA 95689
(916) 969-5315

Small Business Exchange
1400 S Street, Ste. 208
Sacramento, CA 95814
(916) 442-4942

San Francisco/Bay Area

Coalition of Woman-Owned
  Businesses News
4650 Scotia Ave.
Oakland, CA 94605
(415) 635-8898

Minority Business & Professional
  Directory
P. O. Box 77226
San Francisco, CA 94107
(415) 836-3473

Daily Pacific Builder
Public Notice Section
221 Main Street, 8th Floor
San Francisco, CA 94105
(415) 495-4200

Small Business Exchange, Inc.
926 Natoma Street
San Francisco, CA 94103
(415) 255-6411

United Minority Business
  Entrepreneurs (UMBE)
413 Josefa Street
San Jose, CA 95126
(408) 995-5000

Southern California Area

Hispanic Business
360 S. Hope Ave., Ste. 300C
Santa Barbara, CA 93105
(805) 682-5843

Minority Business Entrepreneur
924 N. Market Street
Inglewood, CA 90302
(213) 673-9398

June 18, 1991   2

Figure 9.2 (Continued)   Bid specifications—California Telecommunications Division.

The Forum
1446 Front Street, Ste. 203
San Diego, CA 92101
(619) 232-2244

La Prensa San Diego
1950 5th Ave., Ste. 1,2,3
San Diego, CA 92101
(619) 231-2873

Small Business Exchange
6033 W. Century Blvd., Ste 1200
Los Angeles, CA 90045
(213) 412-8425

Indian Business & Management
  National Headquarters
9650 Flair Dr., Ste. 303
El Monte, CA 91731
(818) 442-3701

La Opinion
411 W. 5th Street
Los Angeles, CA 90013
(213) 891-9191

AGENCIES AND ORGANIZATIONS

State and Federal Agencies

State Department of Transportation
Office of Civil Rights
1120 N Street, Room 3400
Sacramento, CA 95814
(916) 445-5503
FAX (916) 323-6090

State Department of Transportation
Office of Civil Rights
120 S. Spring Street, Room 3-5F
Los Angeles, CA 90012
(213) 620-2325
FAX (213) 620-5805

U.S. Small Business
  Administration
71 Stevenson Street, 20th Floor
San Francisco, CA 94105
(415) 744-6423
FAX (415) 744-6435

Department of General Services
Office of Small and Minority
  Business
1808 14th Street, Ste. 100
Sacramento, CA 95814-7120
(916) 452-6494
FAX (916) 442-7855

Local Agencies
Northern California

Minority Business Development
  Center (NEDA)
1779 Tribute Road, Ste. J
Sacramento, CA 95815
(916) 649-2551

Stockton Minority Business
  Development Center
5361 N. Pershing Ave., Ste. A-1
Stockton, CA 95207
(209) 477-2098

North Coast Small Business
  Resource Center
P. O. Box 728
Crescent City, CA 95531
(707) 464-2169

Sacramento Hispanic Chamber
  of Commerce
530 Bercut Dr., Ste. C
Sacramento, CA 95814
(916) 441-6244

Mexican American Chamber of
  Commerce
246 N. Sutter Street
Stockton, CA 95202
(209) 943-6117 Business

June 21, 1991   3

Figure 9.2 (*Continued*)   Bid specifications—California Telecommunications Division.

San Francisco/Bay Area

American Society for Training
& Development
100 Marin Center Dr., Ste. 6
San Rafael, CA 94903
(415) 479-6631

California Business Fund
1050 Northgate Dr., Ste 358
San Rafael, CA 94903
(415) 472-0900

Elinor Sue Coates, CPCM
1705 11th Ave.
San Francisco, CA 94122
(415) 665-0612

Bay Area Women Entrepreneurs
Office of Community Relations
Laney College
900 Fallon Street
Oakland, CA 94607
(415) 834-5740

Asian Business League of
San Francisco
166 Geary Blvd., Ste. 405
San Francisco, CA 94108
(415) 788-4664

Chicana Foundation of
N. California
P. O. Box 27083
Oakland, CA 94602
(415) 525-0364

Conference of Minority
Transportation Officials
1600 Franklin Street
Oakland, CA 94612
(415) 891-4882

Contra Costa Private Industry
Council (PIC)
Business Resource Center
2425 Bisso Lane, Ste. 100
Concord, CA 94520
(415) 646-5377

Small Business Institute
P. O. Box 65
Kentfield, CA 94904
(415) 479-8208

Asian American Architects
Engineers (AAAE)
1670 Pine Street
San Francisco, CA 94109
(415) 928-5910

Bascom Enterprises
329 S. Mayfair, Ste. 202
Daly City, CA 94105
(415) 333-7887

Bay Area Purchasing Council
1970 Broadway, Ste. 710
Oakland, CA 94612
(415) 763-8162

Minority Business Development
Agency (MBDA)
U. S. Department of Commerce
221 Main Street, Room 1280
San Francisco, CA 94105
(415) 974-9597

California Regional Urban
Development Corporation
3932 Harrison Street
Oakland, CA 94611
(415) 652-5262

Urban Economic Development
Corporation
1426 Filmore Street, Ste. 205
San Francisco, CA 94117
(415) 923-0105

Contra Costa Black Chamber
of Commerce
3101 MacDonald Avenue
Richmond, CA 94804
(415) 235-9350

Martin L. Dinkins
Construction Industry
Consultant
855 La Playa, Ste. 259
San Francisco, CA 94121
(415) 668-8830

Spanish Speaking Unity Council
1900 Fruitvale Ave., Ste. 2A
Oakland, CA 94601
(415) 534-7764

Figure 9.2 (Continued) Bid specifications—California Telecommunications Division.

Entreprenews/Senate Select
Committee on Small Business
Enterprises
1100 J Street, Room 516
Sacramento, CA 95814
(916) 322-3960

California Department of
Commerce
Office of Business Development
111 N. Market Street, Ste. 815
San Jose, CA 95113
(418) 277-9799

Family Rights of Marin, Inc.
P. O. Box 3614
San Rafael, CA 94912
(415) 897-8100

Horizons Unlimited
440 Potrero Ave.
San Francisco, CA 94110
(415) 864-3366

Hispanic Chamber of Commerce
of Alameda County
22412 Main Street
Hayward, CA 94541
(415) 643-6648

Minority Contractors Association
of N. California
1485 Bayshore Blvd., Ste. 455
San Francisco, CA 94124
(415) 467-3835

Marin Builders Exchange
110 Belvedere Street
San Rafael, CA 94901
(415) 456-3233

NAACP, Marin City Chapter
105 Drake Ave.
Marin City, CA 94965
(415) 332-3218

National Association for Female
Executives
470 N. Civic Dr., Ste. 303
Walnut Creek, CA 94596-3320
(415) 386-5764

National Black Minority Business
Association
P. O. Box 193683
San Francisco, CA 94119-3683
(415) 821-6227

Mission Economic Development
Association
2601 Mission Street, 9th Floor
San Francisco, CA 94110
(415) 282-3334

National Association of Women
Business Owners
2339 3rd Street, Ste. 33
San Francisco, CA 94107
(415) 978-0938

N. California Council of Black
Professional Engineers
P. O. Box 1686
Oakland, CA 94604
(415) 632-9736

San Francisco Bay Area Women's
Yellow Pages
5737 Thornhill Dr., Ste. 207
Oakland, CA 94611
(415) 468-7876

San Francisco Black Chamber of
Commerce
1426 Filmore Street, Ste. 205
San Francisco, CA 94117
(415) 922-8720

Community Entrepreneurs
Organization
P. O. Box 2781
San Rafael, CA 94912
(415) 435-4461

San Francisco Human Rights
Commission
1095 Market Street, Ste. 501
San Francisco, CA 94103
(415) 558-4901

San Francisco Women's
Foundation
3543 18th Street
San Francisco, CA 94110
(415) 431-1290

Figure 9.2 (Continued) Bid specifications—California Telecommunications Division.

San Francisco Minority Business
Opportunity Committee (MBOC)
U.S. Small Business Administration
211 Main Street, 4th Floor
San Francisco, CA 94105
(415) 974-0662

San Francisco Minority Business
Development Center
Grant Thornton Company
1 California Street, Ste. 2100
San Francisco, CA 94111
(415) 989-2920

Tradeswomen, Inc.
P. O. Box 40664
San Francisco, CA 94140
(415) 821-7334 (Tues/Thurs Only)

Solano County Private Industry
Council (PIC)
320 Campus Lane
Suisun, CA 94585
(707) 864-3370

Japanese Chamber of Commerce
of N. California
685 Market Street, Ste. 820
San Francisco, CA 94105
(415) 543-8522

Smart Enterprise
5741 Telegraph Ave., Ste. B
Oakland, CA 94609
(415) 547-5936

Port of Oakland
Contract Compliance Unit
530 Water Street
Oakland, CA 94604-2064
(415) 272-1316

Business Development, Inc. (BDI)
1485 Bayshore Blvd., Ste. 352
San Francisco, CA 94124
(415) 468-2200

United Minority Business
Entrepreneurs, Inc.
413 Josefa Street
San Jose, CA 95126
(408) 995-0500

Women Construction Owners &
Executives, USA
San Francisco Bay Area District
1533 Berger Dr.
San Jose, CA 95112
(408) 287-4303

Oakland Minority Business
Development Center
Grant Thornton Company
1000 Broadway, Ste. 270
Oakland, CA 94607
(415) 465-6756

San Francisco, City & County of
City Hall, Room 270
San Francisco, CA 94102
(415) 554-6749

U. S. Chinese Business Institute
San Francisco State University
1600 Holloway Ave.
San Francisco, CA 94132
(415) 338-2106

East Bay Small Business
Development Center
2201 Broadway, Ste. 814
Oakland, CA 94612
(415) 893-4114

City of Oakland
1333 Broadway
Oakland, CA 94162
(415) 273-3970

Asian, Inc. & Council of Asian
American Associations
1670 Pine Street
San Francisco, CA 94109
(415) 928-5910

Hispanic Chamber of Commerce
300 N. 1st Street, Ste. 201
San Jose, CA 95112
(408) 298-8472

Oakland/Alameda County Black
Chamber of Commerce
5741 Telegraph Ave., Ste. A
Oakland, CA 94608
(415) 601-5741

Figure 9.2 (*Continued*) Bid specifications—California Telecommunications Division.

Human Rights Commission
San Francisco International Airport
P. O. Box 8097
San Francisco, CA 94128
(415) 876-2255

Chinese Chamber of Commerce
730 Sacramento Street
San Francisco, CA 94108
(415) 982-3000

Central Valley Area

Fresno Minority Business
Development Center (NEDA)
2010 N. Fine Street, Ste. 103
Fresno, CA 93727
(209) 252-7551

Bakersfield Minority Business
Development Center (NEDA)
218 S. H Street, Ste. 103
Bakersfield, CA 93304
(805) 837-0291

City of Fresno
General Services/Purchasing
2326 Fresno Street
Fresno, CA 93721
(209) 498-1405

Tulare County Economic
Development Corp.
2380 W. Whitendale
Visalia, CA 93278
(209) 627-0766

Hispanic Chamber of Commerce
1900 Mariposa Mall, Ste. 100
Fresno, CA 93721
(209) 485-6640

Southern California

United Indian Development
Association Headquarters
9650 Flair Dr., Ste. 303
El Monte, CA 91731
(818) 442-3701

Los Angeles Minority
Business Development Center
601 Figeroa Street, Ste. 1450
Los Angeles, CA 90017
(213) 488-4949

Oxnard Minority Business
Development Center
451 W. 5th Street
Oxnard, CA 93030
(805) 483-1123

Anaheim/Orange County
Business Development Center
6 Hutton Center Dr., Ste. 10
Santa Ana, CA 92707
(714) 434-0444

Women Construction Owners
& Executives, USA
120 N. 2nd Ave.
Chula Vista, CA 91910
(619) 422-9204

Riverside Minority Business
Development Center
11016 Cooley Dr., Ste. F
Colton, CA 92324
(714) 824-9695

Minority Contractors
Association of Los Angeles
3707 West Jefferson
Los Angeles, CA 90016
(213) 293-3512

Mexican American Foundation
1446 Front Street, Ste. 203
San Diego, CA 92103
(619) 232-1010

June 18, 1991   7

Figure 9.2 (*Continued*) Bid specifications—California Telecommunications Division.

City of San Diego
Equal Opportunity Contracting
  Program
1010 Second Ave., Ste. 1515
San Diego, CA 92101
(619) 236-6945

San Diego Business Development
  Center
6363 Alvarado Court, Ste. 225
San Diego, CA 92120
(619) 265-3684

Professional American Indian
  Development (PAID)
7365 Carnelian Street, Ste. 225,226
Rancho Cucamonga, CA 91730
(714) 941-8180

Black Business Association
  of Los Angeles
5140 Crenshaw Blvd., Ste. B
Los Angeles, CA 90043
(213) 292-0271

San Diego County Certified
  Development Corporation
5353 Mission Center Road
San Diego, CA 92102
(619) 291-3594

Open Bid, Inc.
12750 Center Court Dr., Ste. 260
Cerritos, CA 90701
(213) 865-2002

Valley Economic Development
  Center
14540 Victory Blvd., Ste. 200
Van Nuys, CA 91411-1618
(818) 989-4377

M/WBE DIRECTORIES/GUIDES/DATABASES

M/WBE Directory
UC Berkeley Material Management
Office of Business Development
Berkeley, CA 94720
(415) 642-8600

M/WBE Directory
UCLA Material Management
Small Business Development Office
405 Holgard Ave., Broxton Plaza
Los Angeles, CA 90024-1482
(213) 206-2968

San Francisco Redevelopment
  Agency
770 Golden Gate Ave.
San Francisco, CA 94102
(415) 771-8800 (Ext. 240)

M/WBE Directory
Sacramento Housing & Redevelopment
Department of Community Development
630 I Street
Sacramento, CA 95812-1834
(916) 440-1325

Try Us, National Minority
  Business Directory
2105 Central Ave. N.E.
Minneapolis, MN 55418
(612) 781-6819

Human Rights Commission
San Francisco International
  Airport
P. O. Box 8097
San Francisco, CA 94128
(415) 876-2255

Asian Business Association
  of California
1871 Cloverdale Ave.
Los Angeles, CA 90019
(213) 933-1151

County of San Diego
Equal Opportunity Management
  Office
1600 Pacific Highway, Room 208
San Diego, CA 92101
(619) 531-5819

Figure 9.2 (*Continued*)   Bid specifications—California Telecommunications Division.

WMBE Clearing House
601 Montgomery Street, Ste. 1110
San Francisco, CA 94111
(415) 955-0888

Bay Area Rapid Transit District
800 Madison Street
P. O. Box 12688
Oakland, CA 94604
(415) 464-6110

Alameda County
399 Elmhurst Street
Hayward, CA 94544
(415) 881-6842

Southern California Rapid
 Transit District
425 S. Main Street
Los Angeles, CA 90013
(213) 972-4901

Port of Oakland
Contract Compliance
66 Jack London Square
Oakland, CA 94607
(415) 444-3188

Cordoba Corporation
5400 E. Olympic Blvd., Ste. 210
Los Angeles, CA 90022
(213) 724-6788

City of Berkeley
2180 Milvia Street
Berkeley, CA 94704
(415) 644-6630

Santa Clara County
 Transportation Agency
1555 Berger Dr., Room 207
San Jose, CA 95112
(408) 299-2884

City of Oakland
1 City Hall Plaza, 8th Floor
Oakland, CA 94162
(415) 273-3970

City of Fresno
Public Works Department
2326 Fresno Street
Fresno, CA 93721
(209) 488-1405

San Francisco Public Utilities
 Commission
949 Presidio Ave.
San Francisco, CA 94115
(415) 923-6139

Cordoba Corporation
1515 4th Ave., Ste. 503
San Diego, CA 92101
(619) 234-5848

June 18, 1991   9

Figure 9.2 (Continued) Bid specifications—California Telecommunications Division.

STATE OF CALIFORNIA

MINORITY/WOMEN-OWNED BUSINESS PARTICIPATION SUMMARY

EXHIBIT 2

STD. 840 (NEW 12-90)
(Formerly OSMB 630)

COMPANY NAME	NATURE OF WORK	CONTRACTING WITH	TIER	CLAIMED MBE VALUE	CLAIMED WBE VALUE	CERTIFICATION

Figure 9.2 (*Continued*) Bid specifications—California Telecommunications Division.

DISABLED VETERAN OWNED
BUSINESS PARTICIPATION SUMMARY

DEPARTMENT OF GENERAL SERVICES
OFFICE OF SMALL AND MINORITY BUSINESS

EXHIBIT 2–A

OSMB 830 (REV. 1-91)

COMPANY NAME	NATURE OF WORK	CONTRACTING WITH	TIER	CLAIMED DVBE VALUE	DVBE CERTIFICATION

Figure 9.2 (*Continued*) Bid specifications—California Telecommunications Division.

STATE OF CALIFORNIA

**BIDDER'S MINORITY/WOMEN BUSINESS ENTERPRISE (M/WBE) CERTIFICATION
(STATUS OF SUBCONTRACTORS AND SUPPLIERS)**

EXHIBIT 3

STD. 841 (NEW 12-90)
(Formerly DSMB 831)

I hereby certify that I have made a diligent effort to ascertain the facts with regard to the representations made herein and, to the best of my knowledge and belief, each firm set forth in this bid as a minority or women business enterprise complies with the relevant definition set forth in Section 1896.61 of Title 2, California Code of Regulations hereof. In making this certification, I am aware of Section 12650 et seq. of the Government Code providing for the imposition of treble damages for making false claims against the State and Section 10115.10 of the Public Contract Code making it a crime for intentionally making an untrue statement in this certificate.

CONTRACTOR/BIDDER

BY (Authorized Signature) | DATE SIGNED

PRINTED NAME AND TITLE OF PERSON SIGNING

Figure 9.2 (*Continued*)   Bid specifications—California Telecommunications Division.

STATE OF CALIFORNIA

MINORITY/WOMEN BUSINESS ENTERPRISE (M/WBE) CERTIFICATION

STD. 842 (NEW 12-80)
(Formerly OSMB 832)

I hereby certify that this firm is a M/WBE as defined in Title 2, California Code of Regulations, Section 1896.61. In making this certification, I am aware of Section 12650 et seq. of the Government Code providing for the imposition of treble damages for making false claims against the State and Section 10115.10 of the Public Contract Code making it a crime for intentionally making an untrue statement in this certificate.

CONTRACTOR/BIDDER

BY (Authorized Signature)		DATE SIGNED	TELEPHONE NUMBER (Include Are
PRINTED NAME AND TITLE OF PERSON SIGNING			

CONTRACTOR'S/BIDDER'S MAILING ADDRESS

Please check the box which best describes the ownership and control of your business.

I.   BUSINESS OWNERSHIP

(X)	ETHNIC/GENDER CLASSIFICATION (See ethnic definitions below.)	%	(X)	ETHNIC/GENDER CLASSIFICATION (See ethnic definitions below.)	
	Black			Woman-Owned	
	Hispanic				
	American Indian				
	Asian				
	Filipino				

Definitions:

"Minority" OR "women-owned business" (M/WBEs) is a business concern:

(1)   Which is at least 51% owned by one or more minorities or women or, in the case of a publicly owned business, at least 51% of the stock of which is owned by one or more minorities or women, and

(2)   Whose management and daily business operations are controlled by one or more such individuals, and

(3)   Domestic corporation - A corporation with its home office in the United States which is not a branch or subsidiary of a foreign corporation, firm or other business.

"Minority" means an ethnic person of color including American Indians, Asians (including, but not limited to, Chinese, Japanese, Koreans, Pacific Islanders, Samoans, and Southeast Asians), Blacks, Filipinos, and Hispanics.

Figure 9.2 (Continued)   Bid specifications—California Telecommunications Division.

DOCUMENTATION OF GOOD FAITH EFFORT

A.   Contact Documentation

Vendors are to list their contacts with state **AND** federal **AND** local organizations.

Date/Time	Agency/Organization	Contact Name	Phone Number
	Telecommunications Division		
___	___	___	___
___	___	___	___
	Other State Departments		
___	___	___	___
___	___	___	___
___	___	___	___
___	___	___	___
	Federal Organizations		
___	___	___	___
___	___	___	___
___	___	___	___
___	___	___	___
	Local Organizations		
___	___	___	___
___	___	___	___
___	___	___	___
___	___	___	___

Figure 9.2 (Continued)   Bid specifications—California Telecommunications Division.

B.   Advertisement Documentation

List publication(s) in which you advertised and include a copy of the advertisement.

Publication's Name   **Publication Date(s)**

_____   _____

_____   _____

                                       _____

If none, state reason:

_____

_____

_____

C.   Minority/Women/Disabled Veteran Solicitations

1. Solicitation Sample
    (Contractor/bidder should attach a sample of the solicitation sent to MBE and/or WBE and/or DVBE firms)

2. Bidders List
    (Contractor/bidder should attach the list of M/W/DVBE's to which the solicitation was mailed or FAXed)

3. Responses
    (Contractor/bidder should attach copies of the responses received by MBE and/or WBE and/or DVBE firms).

Firm Name(s)   **Selected/Reason For Non-Selection**

_____   _____

_____   _____

_____   _____

_____   _____

_____   _____

_____   _____

_____   _____

Figure 9.2 (Continued)   Bid specifications—California Telecommunications Division.

STATE OF CALIFORNIA
DRUG-FREE WORKPLACE CERTIFICATION
STD. 21 (NEW 11-90)

RIDER G

COMPANY/ORGANIZATION NAME

The contractor or grant recipient named above hereby certifies compliance with Government Code Section 8355 in matters relating to providing a drug-free workplace. The above named contractor or grant recipient will:

1. Publish a statement notifying employees that unlawful manufacture, distribution, dispensation, possession, or use of a controlled substance is prohibited and specifying actions to be taken against employees for violations, as required by Government Code Section 8355(a).

2. Establish a Drug-Free Awareness Program as required by Government Code Section 8355(b), to, inform employees about all of the following:

   (a) The dangers of drug abuse in the workplace,

   (b) The person's or organization's policy of maintaining a drug-free workplace,

   (c) Any available counseling, rehabilitation and employee assistance programs, and

   (d) Penalties that may be imposed upon employees for drug abuse violations.

3. Provide as required by Government Code Section 8355(c), that every employee who works on the proposed contract or grant:

   (a) Will receive a copy of the company's drug-free policy statement, and

   (b) Will agree to abide by the terms of the company's statement as a condition of employment on the contract or grant.

CERTIFICATION

I, the official named below, hereby swear that I am duly authorized legally to bind the contractor or grant recipient to the above described certification. I am fully aware that this certification, executed on the date and in the county below, is made under penalty of perjury under the laws of the State of California.

OFFICIAL'S NAME

DATE EXECUTED | EXECUTED IN THE COUNTY OF

CONTRACTOR or GRANT RECIPIENT SIGNATURE

TITLE

Figure 9.2 (Continued) Bid specifications—California Telecommunications Division.

RIDER H

SMALL BUSINESS PREFERENCE

I. NOTICE TO ALL BIDDERS

Section 14835 et seq. of the California Government Code requires that a 5% preference be given to bidders who qualify as a small business. The rules and regulations of this law, including the definition of a small business for procurements made by the Office of Procurement, are contained in Title 2, California Administrative Code, Section 1896, et seq. The Small Business Preference is for California based small businesses.

TO CLAIM SMALL BUSINESS PREFERENCE your firm must be approved by the Small and Minority Procurement Business Assistance Division. Questions regarding Small Business Preference approval and requests for a copy of the regulations should be directed to that office at (916) 322-5060.

II.

Bidders desiring to claim preference as a small business must complete the following and return with the Final Bid to be eligible for the Small Business Preference.

1. Are you claiming preference as a small business?

    YES _____ NO _____

    If yes, complete the following:

2. Non-Manufacturer _____ Manufacturer _____

3. Aggregate receipts for last three (3) years:_____

4. DECLARATION:
    By signing below with inclusion of the date and signature, the undersigned proposal DECLARES UNDER PENALTY OF PERJURY under the laws of the State of California that the information set forth above is true and correct.

Name and Title (Type or Print)	Firm Name
Signature	Date
Street Address	Telephone Number
City State Zip	

-33-

Figure 9.2 (Continued) Bid specifications—California Telecommunications Division.

CAPITOL LOOP FIBER MAINTENANCE

BID PROPOSAL FORM
FOR
MAINTENANCE OF CAPITOL FIBER OPTIC LOOP

All bidder responses shall be on these sheets. If additional sheets are needed for a complete response, they may be attached immediately after the sheet provided for the response. Failure to complete and return this form shall be cause for bid rejection.

Reference Paragraph 3.a - Bid Technical Requirements

a.   *Bidder must propose the necessary hardware to provide cable and jumper management within the fiber optic patch panels listed in Rider B. The Main objective of this requirement is the development of an orderly and systematic management scheme that will assist in developing system documentation and in readily identifying specific circuits. Vendor shall also periodically update current documentation to accurately reflect system configuration.*

Bidder must provide a description of the proposed methods and equipment that can be utilized to meet this objective, both initially, and throughout the duration of the contract.

BID PROPOSAL FORM - PAGE 1

Figure 9.2 (Continued)   Bid specifications—California Telecommunications Division.

CAPITOL LOOP FIBER MAINTENANCE

BID PROPOSAL FORM
FOR
MAINTENANCE OF CAPITOL FIBER OPTIC LOOP

All bidder responses shall be on these sheets. If additional sheets are needed for a complete response, they may be attached immediately after the sheet provided for the response. Failure to complete and return this form shall be cause for bid rejection.

Reference Paragraph 3.b - Bid Technical Requirements

b. *Bidder must maintain the lightwave terminal equipment and the fiber optic cable plant listed in Rider B. The main objective of system maintenance is to strive for system performance equal to the original system specifications. System maintenance will be considered acceptable if the following objectives are met:*

(1) *The transmission performance objective is error-free transmission in at least 99.999 percent of all one second intervals. If there are more than 30 error seconds per hour, the system service will be considered impaired. System performance is measured in "error seconds" , where error second is any 1 second interval during which one or more bit errors occurs.*

(2) *The reliability objective is that circuit integrity between stations remain intact and error free 99.9 percent of the time, based on a long term average. In terms of cumulative outage time, this objective equates to an average downtime of less than 8.75 hours per circuit per year.*

(3) *The availability objective is that no single outage exceed four hours. The objective recognizes the impact of long outages on the States's data networks and the perishable nature of some forms of data.*

Bidder must provide in the bid a statement confirming understanding of the State's objectives. Bidder must describe the preventative maintenance methods that shall be utilized to ensure the objectives continue to be met.

BID PROPOSAL FORM - PAGE 2

Figure 9.2 (Continued)   Bid specifications—California Telecommunications Division.

CAPITOL LOOP FIBER MAINTENANCE

*BID PROPOSAL FORM
FOR
MAINTENANCE OF CAPITOL FIBER OPTIC LOOP*

All bidder responses shall be on these sheets. If additional sheets are needed for a complete response, they may be attached immediately after the sheet provided for the response. Failure to complete and return this form shall be cause for bid rejection.

Reference Paragraph 3.c - Bid Technical Requirements

c.   *Bidder must have proven experience in work of this type. Bidder shall provide three (3) references in their response representing paying customers that ypify their experience with this type of work. Each reference must include the name of the contact person and phone number. Attach additional pages if necessary.*

BID PROPOSAL FORM - PAGE 3

Figure 9.2 (*Continued*) Bid specifications—California Telecommunications Division.

CAPITOL LOOP FIBER MAINTENANCE

BID PROPOSAL FORM
FOR
MAINTENANCE OF CAPITOL FIBER OPTIC LOOP

All bidder responses shall be on these sheets. If additional sheets are needed for a complete response, they may be attached immediately after the sheet provided for the response. Failure to complete and return this form shall be cause for bid rejection.

Reference Paragraph 3.c - Bid Technical Requirements

c.   Bidder must also list in the response the experience of person(s) assigned to perform the maintenance. The Bidders maintenance personnel must be certified by the manufacture to maintain the equipment, and have attended applicable schools and training. Copies of the certification or letter of credentials from the manufacture shall be included in the bid. Attach additional pages if necessary.

BID PROPOSAL FORM - PAGE 4

Figure 9.2 (Continued) Bid specifications—California Telecommunications Division.

CAPITOL LOOP FIBER MAINTENANCE

BID PROPOSAL FORM
FOR
MAINTENANCE OF CAPITOL FIBER OPTIC LOOP

All bidder responses shall be on these sheets. If additional sheets are needed for a complete response, they may be attached immediately after the sheet provided for the response. Failure to complete and return this form shall be cause for bid rejection.

Reference Paragraph 3.d - Bid Technical Requirements

d.   *Bidder must take responsibility for system maintenance by being able to respond to emergency requests for maintenance 24 hours a day, seven days a week. Bidder must have a 24 hour answering service to receive calls. Emergency maintenance will normally begin within two hours after notification. Upon arrival at the service site, the service personnel and the State personnel shall have the option to reschedule the service activity at a mutually agreeable time without additional cost. Bidder must explain how this requirement will be met. Attach additional pages if necessary.*

BID PROPOSAL FORM - PAGE 5

Figure 9.2 (*Continued*)   Bid specifications—California Telecommunications Division.

CAPITOL LOOP FIBER MAINTENANCE

BID PROPOSAL FORM
FOR
MAINTENANCE OF CAPITOL FIBER OPTIC LOOP

All bidder responses shall be on these sheets. If additional sheets are needed for a complete response, they may be attached immediately after the sheet provided for the response. Failure to complete and return this form shall be cause for bid rejection.

Reference Paragraph 3.f - Bid Technical Requirements

f. *Bidder shall list all test equipment and tools to be utilized for maintaining the system. Attach additional pages if necessary.*

BID PROPOSAL FORM - PAGE 6

Figure 9.2 (Continued) Bid specifications—California Telecommunications Division.

CAPITOL LOOP FIBER MAINTENANCE

BID PROPOSAL FORM
FOR
MAINTENANCE OF CAPITOL FIBER OPTIC LOOP

All bidder responses shall be on these sheets. If additional sheets are needed for a complete response, they may be attached immediately after the sheet provided for the response. Failure to complete and return this form shall be cause for bid rejection.

Reference Paragraph 4.b –

b.   *Conduct end-to-end proof-of-performance tests once a year, and on demand, to ensure the system meets the performance requirements of paragraph [3.b]. Bidder must provide a statement confirming his understanding of the State's requirement. Attach additional pages if necessary.*

BID PROPOSAL FORM – PAGE 7

Figure 9.2 (Continued)   Bid specifications—California Telecommunications Division.

CAPITOL LOOP FIBER MAINTENANCE

BID PROPOSAL FORM
FOR
MAINTENANCE OF CAPITOL FIBER OPTIC LOOP

All bidder responses shall be on these sheets. If additional sheets are needed for a complete response, they may be attached immediately after the sheet provided for the response. Failure to complete and return this form shall be cause for bid rejection.

Reference Paragraph 5.a - Spare Parts

a.   *Bidders must keep within 90 miles of Sacramento a stock of spare parts and modules necessary for maintaining the system. Bidder shall list these spare parts in the bid and provide an address for the local stocking point. The address may be a State facility if agreed to by the State.*

Facility Address: _____

_____

_____

_____

BID PROPOSAL FORM - PAGE 8

Figure 9.2 (*Continued*) Bid specifications—California Telecommunications Division.

CAPITOL LOOP FIBER MAINTENANCE

BID PROPOSAL FORM
FOR
MAINTENANCE OF CAPITOL FIBER OPTIC LOOP

All bidder responses shall be on these sheets. If additional sheets are needed for a complete response, they may be attached immediately after the sheet provided for the response. Failure to complete and return this form shall be cause for bid rejection.

Reference Paragraph 5.c - Spare Parts

c. *For the purposes of evaluation, the initial cost of the spare parts shall be considered a common cost among all bidders, and shall therefore not be considered part of the responses or quotations.*

Bidder must provide a statement confirming his understanding of the States requirements.

BID PROPOSAL FORM - PAGE 9

Figure 9.2 (Continued)   Bid specifications—California Telecommunications Division.

CAPITOL LOOP FIBER MAINTENANCE

BID PROPOSAL FORM
FOR
MAINTENANCE OF CAPITOL FIBER OPTIC LOOP

All bidder responses shall be on these sheets. If additional sheets are needed for a complete response, they may be attached immediately after the sheet provided for the response. Failure to complete and return this form shall be cause for bid rejection.

Reference Paragraph 6.a - Technical Standards

Bidder must provide in the response a statement confirming his understanding and compliance to standards referenced in this paragraph. Attach additional pages if necessary.

BID PROPOSAL FORM - PAGE 10

Figure 9.2 (Continued) Bid specifications—California Telecommunications Division.

CAPITOL LOOP FIBER MAINTENANCE

BID PROPOSAL FORM
FOR
MAINTENANCE OF CAPITOL FIBER OPTIC LOOP

All bidder responses shall be on these sheets. If additional sheets are needed for a complete response, they may be attached immediately after the sheet provided for the response. Failure to complete and return this form shall be cause for bid rejection.

Reference Paragraph 6.b –

The bidder shall provide a copy of his Standard Maintenance Procedures as part of this bid. Bids submitted without the Bidder's Maintenance Procedure Manual will not be considered for evaluation. Place Maintenance Procedure Manual after this sheet.

BID PROPOSAL FORM – PAGE 11

Figure 9.2 (*Continued*)   Bid specifications—California Telecommunications Division.

CAPITOL LOOP FIBER MAINTENANCE

BID PROPOSAL FORM
FOR
MAINTENANCE OF CAPITOL FIBER OPTIC LOOP

All bidder responses shall be on these sheets. If additional sheets are needed for a complete response, they may be attached immediately after the sheet provided for the response. Failure to complete and return this form shall be cause for bid rejection.

Reference Rider A, paragraph 2.k and 2.1 – Bidder shall list below the personnel to be assigned for the three levels of Technical Support.

FIRST LEVEL SUPPORT

NAME: _____

PHONE: _____

SECOND LEVEL SUPPORT

NAME: _____

PHONE: _____

THIRD LEVEL SUPPORT

NAME: _____

PHONE: _____

BID PROPOSAL FORM - PAGE 12

Figure 9.2 (Continued) Bid specifications—California Telecommunications Division.

CAPITOL LOOP FIBER MAINTENANCE

The preceding sheets constitute the complete bid response to IFB R-92-015

THIS FORM MUST BE COMPLETED, SIGNED, AND RETURNED. FAILURE TO SIGN/RETURN THIS
FORM SHALL BE CAUSE FOR REJECTION OF BID

BID PROPOSAL FORM
Capitol Fiber Loop Maintenance - Sacramento

BID NUMBER: R-92-015   BID DEADLINE: May 21, 1992 BY 2:00 p.m. PST
NOTE: YOUR SEALED BID MUST BE RECEIVED AT THE TELECOMMUNICATIONS DIVISION BY
         THIS DATE/TIME.

Address your BID PROPOSAL to: Department of General Services
                               Telecommunications Division
                               601 Sequoia Pacific Boulevard
                               Sacramento, CA 95814-0282
                               Attention Mark MacRae
and mark plainly
on the envelope: BID R-92-015 ENCLOSED -- DO NOT OPEN

Be sure to enclose ALL REQUIRED FORMS/MATERIALS WITH THIS COMPLETED/SIGNED BID
PROPOSAL FORM.

[ ]   I have read this BID PACKAGE carefully, and DO NOT wish to bid.

[ ]   I have read this BID PACKAGE carefully, and wish to bid in accordance
       with the BID SPECIFICATIONS:

Name of Company:_____

        Address:_____

City/State/Zip:_____

Telephone No.:_____

          Date:_____

  -   Signed:_____

         Title:_____

FEDERAL EMPLOYER ID NUMBER:_____

STATE EMPLOYER ID NUMBER:_____

If you do not have a Federal ID Number, please enter your Social Security Number.
Providing the Social Security Number is voluntary in accordance with the Privacy
Act of 1974.

PUBLIC BID OPENING WILL BE HELD ON May 21, 1992 at 2:00 p.m. PST

BID PROPOSAL FORM - PAGE 13

Figure 9.2 (Continued)   Bid specifications—California Telecommunications Division.

CAPITOL LOOP FIBER MAINTENANCE

*BID PROPOSAL FORM
FOR
MAINTENANCE OF CAPITOL FIBER OPTIC LOOP*

All bidder responses shall be on these sheets. If additional sheets are needed for a complete response, they may be attached immediately after the sheet provided for the response. Failure to complete and return this form shall be cause for bid rejection.

BID PROPOSAL FORM - PAGE 14

Figure 9.2 (Continued)   Bid specifications—California Telecommunications Division.

CAPITOL LOOP FIBER MAINTENANCE

BID PROPOSAL COST SHEET

YEAR ONE (estimated to be June 1, 1992 through May 31, 1993):

> BASIC MONTHLY MAINTENANCE CHARGE FOR 24 HOUR PER DAY, 7 DAY A WEEK, 2 HOUR RESPONSE, MAINTENANCE SERVICES ON EQUIPMENT LISTED ON RIDER B IN ACCORDANCE WITH THE TERMS AND CONDITIONS CONTAINED HEREIN.

> TOTAL MAINTENANCE COST FOR YEAR ONE:

> $_____$/MONTH x 12 MONTHS = $_____$
> [a]

****

YEAR TWO (estimated to be June 1, 1993 through May 31, 1994):

> My bid for the second year includes a _____% increase (this percentage CANNOT exceed 10%):

> TOTAL MAINTENANCE COST FOR YEAR TWO:

> $_____$/MONTH x 12 MONTHS = $_____$
> [b]

****

YEAR THREE (estimated to be June 1, 1994 through May 31, 1995):

> My bid for the third year includes a _____% increase (this percentage CANNOT exceed 10%):

> TOTAL MAINTENANCE COST FOR YEAR THREE:

> $_____$/MONTH x 12 MONTHS = $_____$
> [c]

Enter the total of [a], [b], and [c] onto the BID PROPOSAL SUMMARY, item [d].

BID PROPOSAL FORM - PAGE 15

Figure 9.2 (Continued) Bid specifications—California Telecommunications Division.

CAPITOL LOOP FIBER MAINTENANCE

BID PROPOSAL SUMMARY

YEAR ONE, TWO, AND THREE

GRAND TOTAL

[d] $_____
          [a] + [b] + [c]

BID PROPOSAL FORM - PAGE 16

Figure 9.2 (*Continued*)   Bid specifications—California Telecommunications Division.

(Date)

To: (Name)
    (Title)
    (Department) of Alcohol and Drug Programs
    (Address)

Re: Wire Specifications, (Location)

Dear (Name);

Enclosed please find the Final Wiring Specifications for your planned location at (Location), for an initial capacity of 320 work stations and 400 ultimate. Specifications are for wiring all Stations at initial installation due to (Source) average of $700.00 per station, plus lost time, for wiring individual locations after occupancy.

Additional time is anticipated for the following activities:

Utility Company/Building Owner/Vendor Coordination
On-Site Work activity Supervision
Cut-over Activity
Final Inspection and Report

The estimated Installed Cost for the Premise Wire specified in this document is $145,760.00. This estimate includes all material, labor and profit. The Estimated cost per additional Station is $387.00   .

Prior to Installation activities, the (Name of Your Company) would like to train 2 to 3 departmental employees on the Wire Specifications, Vendor Control and On-going maintenance. This training will take approximately 4 hours and will be team taught by Ed and myself. Should you have any questions, don't hesitate to call.

                                        Sincerely;

                                        (Name)
                                        (Title)

Figure 9.3 Wire specifications.

Premise Wiring Specifications - (Client name & Address)

WIRING SPECIFICATIONS

For

PREMISE DISTRIBUTION WIRING

at

(Address)

Prepared By:

(Company Name)
(Address)
(Phone Number)
(Date)

Figure 9.3 (*Continued*)   Wire specifications.

Premise Distribution Specifications - (Location)

GENERAL CONDITIONS

VENDOR RESPONSIBILITY SUMMARY

The successful vendor shall be responsible for the following:

a. Provide complete Premise Distribution System compatible
    wiring and cabling needs as provided in this document.   All
    cable and wire terminations will conform to AT&T wire
    termination configuration for both Voice and Data
    Terminations.

    b. Provide all labor and materials to pull, fasten, secure,
    install and designate all copper cable.   Terminate all
    conductors of all cable and wires onto terminal blocks,
    station jacks, and approved hardware.   Designate all circuit
    blocks with appropriate circuit numbers and station jacks
    with location numbers as designated on approved floor
    plans.   Mount all jacks as indicated on the approved floor
    plans.

c. Furnish and install all colored backboards, terminal
    connection blocks, distribution cables, station wire,
    connecting jacks, patch cords and all installation hardware.

d. Perform tests on all station and distribution wiring between
    the 1.1 terminal location and the end user station jack as
    described in the detailed specifications following and in
    accordance with IBM 3270 specifications.

e. Identify and label all circuits in the 1.1 Telephone Closet,
    Computer Room and Station locations in accordance with the
    labeling plan outlined in this document.

f. Remove all scrap to customer provided disposal containers.

g. Vendor shall provide an on-site supervisor to coordinate
    with other contractors on site.   On-site supervisor shall be
    a regular employee of the vendor with experience on similar
    size projects.

h. The vendor must comply with all federal, State and Local
    laws and ordinances, and must secure all permits and
    certificates required for any work to be performed under the
    requirements of this proposal.

i. Vendor to provide wire management plan and devices as
    required to facilitate minimum training and customer wire
    management certification.   Wire management plan to provide
    training module including system diagnostic processes.

i

Figure 9.3 (*Continued*)   Wire specifications.

Premise Distribution Specifications - (Location)

(GENERAL CONDITIONS - Continued)

j. Vendor shall warrantee materials and workmanship for a
    period of 1 (one) year from acceptance of work performed
    under this specification.

k. Title to all materials and/or equipment must pass to the
    (Client) upon acceptance of the System. Vendor to supply
    unconditional release of interest for all material and/or
    equipment installed or furnished from supplier, self or
    other interested parties prior to final payment.

Figure 9.3 (*Continued*) Wire specifications.

Premise Distribution Specifications - (Location)

1. VENDOR QUALIFICATIONS:

   In determining the capabilities of a vendor to perform the
work specified herein, the following are minimum qualifications
to be considered.   In submitting a bid, adequate information
regarding these capabilities must be provided by the Vendor:

   a. Valid C-61 Contractors License (C-7 if attained after
       11/89)

   b. Prior experience in the communications field.

   c. Prior experience in the installation of Voice and Data
       telecommunications wiring systems.

   All Vendors shall furnish a list of at least two (2) recent
distribution systems of similar size and complexity, that have
been operating for six months. For each system, state the user
capacity, level of involvement, year performed, location, owner,
telephone number and person who can be contacted.

2. APPLICABLE STANDARDS:

   It is required that the vendor be familiar with and perform
to the following standards and specifications:

   a. 1990 NEC
   b. 1989 State Wiring Management Plan
   c. National Electrical Safety Code
   d. National Fire Protection Standards
   e. State and Local Building Codes
   f. Occupational Safety and Health Code
   g. National Electrical Manufacturers Association

   Further, if an equivalent part is used, it must fully meet
or exceed the specifications and capabilities of the stated
parts listed.

3. CABLE AND WIRE SPECIFICATIONS:

3.1   HOUSE CABLE (From 2nd Floor Computer Room to Telephone
Closets)

CABLE CATEGORY   Multipair Cable, Unshielded
CABLE TYPE   24 Gauge AWG Solid Conductors,
                                  twisted pairs.

Figure 9.3 (*Continued*) Wire specifications.

Premise Distribution Specifications - (Location)

Individual Conductors

Gauge	24
Conductors	Bare Copper
Insulation	Semirigid Plastic
Jacket	Plastic (UL910 for Plenum)
Color Code	Standard Telco

Nominal Parameters: (24 Gauge)

Mutual Capacitance	<21 pf/ft
DC Resistance	<26 Ohms/1000 ft
Characteristic Impedance	600 Ohms @ 1 KHz
	92 Ohms @ 1 MHz
Attenuation	0.75 dB/1000 ft @ 1KHz
	8.00 dB/1000 ft @ 1MHz

Cable Sheath: AR Series Riser cable:

Composed of dual expanded
plastic insulated conductors
with Plastic over expanded
polyethylene, and sheathed with
a bonded aluminum plastic
(Alvyn)

Pairs   600

Sizes Available: 12 to 1800 Pairs

Approvals: UL Listed - Type CMR (UL 1666)
                                          - Type CMP (UL 910)
                               UL Classification 1666
                               NEC 800-50
                               IEEE 383 Vertical Flame Test

3.2   VOICE STATION WIRING:

WIRE CATEGORY	Hookup Wire (Station Wire)
WIRE TYPE	24 AWG Solid Conductors,
	Twisted Pairs

Individual Conductors:

Gauge	24
Conductors	Bare Copper
Pairs	8
Insulation	Semirigid Plastic Insulation
Jacket	N/A or UL 910 Equiv. for Plenum
Color Code	Standard Telco

Page 2

Figure 9.3 (*Continued*) Wire specifications.

Premise Distribution Specifications - (location)

Nominal Parameters:

Mutual Capacitance	< 20.0 Pf/ft
DC Resistance	25.7 Ohms/1000 ft
Characteristic Impedance	700 Ohms @ 1 KHz
	110 Ohms @ 1 MHz
Attenuation	0.50 dB/1000 ft @ 1 KHz
	6.6 dB/1000 ft @ 1 MHz
Pairs Available	2, 4, 6, 8, 12
Approvals	Utility Recognized - 1935
	90 Degrees C - 300 Volts
	Plenum - UL 910
	NEC 800-50

3.3   DATA GRADE STATION WIRE:

WIRE CATEGORY	Hookup Wire (Station Wire)
WIRE TYPE	22 AWG Solid Conductors,
	Twisted Pairs (IBM Type II)

Individual Conductors:

Gauge	22
Conductors	Bare Copper
Insulation	a. Foamed crosslinked polethylene with flame retardent skin.
	b. Foamed fluropolymer for Plenum.
Shield	a. Two pairs individually shielded with aluminum polyethylene teraphalate plus 4 pairs of 22 AWG solid conductors included inside the cable jacket.
	b. Overall Tinned Copper braided shield (90% Coverage over two pairs individually shielded)
Jacket	PVC (UL 910 for Plenum)
Color Code	Standard Telco

Nominal Parameters:

Mutual Capacitance	< 16 Pf/ft
Capacitance Unbalance	1500 Pf/KM @ 1 KHz

Page 3

Figure 9.3 (*Continued*) Wire specifications.

Premise Distribution Specifications - (Location)

DC Resistance   16.2 Ohms/1000 ft
Characteristic Impedance   150 Ohms 3 to 20 MHz Sweep
Attenuation   45 dB/KM Max. @ 16 MHz
                                           22 dB/KM Max. @ 4 MHz

Crosstalk (Near End)   40 dB Min. 12 to 20 MHz Sweep
                                           58 dB Min. 3 to 5 MHz Sweep

Pairs   6

Approvals   Utility Recognized - 1935
                                           90 Degrees C - 300 Volts

                                           (Plenum - UL Classified NEC
                                           Article 800-3(b)

4.   OPTICAL FIBER SPECIFICATIONS (2nd Floor Computer Room and
Telephone Closets)

4.1 **Minimum Specifications**

    Cladding Diameter   125.0 Microns
    Core Diameter   62.5 Microns
    Core Eccentricity   7.5% maximum (Typical - 1.5%)
    Core Ovality   20% Maximum (Typical 4%)
    Reflective Index delta   2.0%
    Numerical Aperture   0.29 (Theoretical)
    Attenuation Range   0.85 to 2.7 dB/km @ 1300 nm
                                4.0 dB/km @ 850 nm
    Bandwidth Range   300-700 MHz/km @ 1300 nm
                                160-300 MHz/km @ 850 nm
    Typical Field Splice Loss   0.10 dB (array)
                                0.20 dB (Mechanical)
    Coating Diameter   245 microns
    Type Sheath   Nonconductive, Optical Fiber
                                Plenum Sheath

    Approvals   NEC 800-3(B) (CMP
                                UL 910 STEINER TUNNEL TEST

4.2 OPTICAL CABLE DUCT MINIMUM SPECIFICATIONS
     (From Computer Room to Telephone Closets)

    o The Optical Fibers shall be enclosed in an extruded,
       flexible, annealed, polyethylene duct.

    o The duct shall conform to the applicable portions of ASTM
       Designations: D3485, D3035, D 2239 and D 2447, and

Page 4

Figure 9.3 (Continued) Wire specifications.

Premise Distribution Specifications - (Location)

applicable portions of NEMA TC7 and TC2.

o The duct shall be 1.90 inches outside diameter minimum, with a wall thickness of 0.20 inches minimum. The duct shall be designated for a pulling force of 1200 pounds, maximum. Pulling forces shall be taken by the duct, not the fiber optic cable.

o The Duct shall be imprinted at one meter intervals with

sequential numbers.

o The duct shall be Tamaqua Cable Products corporation; Integral Corporation or equal.

5.0 TERMINATION HARDWARE:

5.2 Computer Room:

a. Place   BLUE Backboards and terminate 4-600 Pair House Cables on PDS type 110-C-5, or equivalent, High Density (HD) connector blocks. Overhead access. Split 'D' Rings, 6" spacing, between columns.

b. The 62.5/125 Micron Multimode fibers will terminate on ST connectors and be mounted in interconnection units at the Computer Room.   Each   interconnection cabinet will be equipped with hinged doors to permit access to wiring aisles, fanout compartments,   mounting   connectors/couplers   and   connector enclosures.

5.3 Telephone Closets

a. Place BLUE Backboards and terminate 600 Pair House Cables on PDS type 110-C-5, or equivalent, High Density (HD) connector blocks.

b. The 62.5/125 Micron Multimode fibers will originate at this Computer Room and terminate on ST Connectors and be mounted in   interconnection units in the Telephone Closets.   Each interconnection cabinet will be equipped with hinged doors to permit access to wiring aisles, fanout compartments, mounting connectors/couplers and connector enclosures.

c. Place BLUE Backboards and terminate Voice Station Wiring on PDS type 110-C-4, or equivalent, High Density (HD) connector blocks.   Overhead   Access.   Allow 1 1/2" between rows and at least 4" between columns. Split 'D' rings, 6" spacing, between columns.

Page 5

Figure 9.3 (*Continued*) Wire specifications.

Premise Distribution Specifications - (Location)

d.   Allow 24" space between Voice Patch Panel Column and
Data Termination Block columns for the future mounting of
Telephone KTS equipment.

e.   Place BLUE backboards and Terminate Data Station Wiring
on Hinged IBM (or equivalent) Patch Panel. Mount 108 Position
IBM Type I terminations on IBM Type I connectors over 108
Position the RJ45 type connector termination for the 4 Pr. 22
Gauge, pairs in the IBM Type II station wire sheath. Allow 4'
slack in

Bonding of shields will be accomplished at the connector for the
IBM Type I connector to the Patch Panel Frame. The Frame is to
be grounded to the Building Ground. Split 'D' Rings at 6"
spacing to contain the station wiring bundle(s).

f.   Patching between the Voice CENTREX 66 Blocks and Voice
Station Wire 110 Type Patch panel will be accomplished by using
1 pair modular jumper wires at the patch panel and RJ 14 (2
Pair) Modular Patch Cords at the Station Location.

g.   Provide   200 Each Balun Connectors, Twisted Pair, 6
position/ 4 contact modular jack to male BNC. Active Pins 3 &
4.   Insertion Loss-Less than .5dB over the range of 100 KHz to
5MHz.   Maximum capacitance - .018 UF/1000 Feet of cable. Link
Isolation - 600 Volts.   Operation Humidity .0% to 95%
Temperature, 0 Degrees C to 55 Degrees C. Distance - 1000 Feet,
22, 24 AWG, Twisted Pair. Interface Compatible with IBM 3270
and 3174, Category A devices and plug compatibles. (i.e.
Controllers - 3174, 3271, 3272, 3274, 3275 & 3275. Muxes -
3299.   Terminals - 3178, 3277, 3278, 3279 & 3292. Printers -
3284, 3286, 3287, 3288 & 3289.) AMP RJ11 Jack (6 Pos.) to BNC
Plug Part Number 555329-1 or Equivalent. Patch between DCE and
Station Data Patch using Modular 4 Pair Patch Cords.

h   Provide 100 Each, subminiature RS232 Adapter interface
kits (RS232 to RJ45), and terminate on 4 Pair cable used for
data station terminations.

5.4 Station Location:

The work station location will employ the use of Modular
Furniture outlets and wall outlets.   In the Case of Wall
outlets, 1 - 3/4" conduit is provided to each location. Office
furniture will employ the use of Power poles. (Floor plans are
available through the agency or at the job site).

The Station location Jack will consist of a duplex, flush
mount faceplate equipped with 2 RJ14 (2-2 Pair Voice) modular
jacks and a duplex faceplate equipped with 1 IBM Type I
connector over an RJ45 Connector. Station wire connections will
be made using the "Direct Wire Connection (Insulation
Displacement) method.   Leave approximately 4" slack, each
bundle, at each station location.

Page 6

Figure 9.3 (*Continued*) Wire specifications.

Premise Distribution Specifications - (Location)

6.0 Hardware Performance Requirements:

All termination and modular equipment specified in this specification, unless otherwise stated, will be warranted for a minimum of 100 insertions and extractions. Direct Wire Connection components will provide solder equivalent electrical connectivity.

7.0 LABELING:

7.1   1.1 Telephone Closet: Permanently mark House cable designation on backboard above connector blocks. Permanently label connector blocks in 5 pair increments.

7.2 Computer Room.

o   Permanently label House cable designation on backboard above connector blocks. Permanently label connector blocks in 5 pair increments.

o   Permanently label Fiber Optic Cable indicating fiber Number and Closet location.

7.3 Telephone Closets

o   Permanently label House cable designation on backboard above connector blocks. Pewrmanently label connector blocks in 5 pair increments.

o   Permanently label Fiber Optic Cable, indicating fiber number and closet reference.

o   Permanently label each station connector block as to its unique number according to the station designation shown on the office floor plan drawing for the site, for Voice and Data Terminations.

o   Voice Patch Panel.   Permanently label each connector block and corresponding modular patch jack according to its unique number as described above. Label the equipment side to the equipment assignment specifications.

7.3   Station Location.   Permanently label station location as shown on the office floor plan drawings for the site.

Page 7

Figure 9.3 (*Continued*) Wire specifications.

Room to all Station Jacks. The vendor shall perform Time Domain Reflectometer (TDR), or equivalent tests, to assure conformance with the Data Grade Wire specifications indicated in Section 3 above. All applicable tests will be made through the connector

Page 8

Figure 9.3 (*Continued*) Wire specifications.

Premise Distribution Specifications - (Location)

8.0 <u>CONFORMANCE TEST REQUIREMENTS</u>:

8.1   The vendor shall provide a DC (Direct Current) Conformance test report on each twisted Cable Pair (Voice and Data) in every cable sheath and station wire application to indicate and clear the following defective pair conditions:

  o <u>Opens</u> - condition indicating that one or both wires in a
     pair of twisted pair have a break in continuity.

  o <u>Shorts</u> - condition that occurs when tip and ring wire of
     a cable or twisted pair come in contact with each other.

  o <u>Grounds</u> - condition which occurs when either wire of a
     twisted pair come in contact with a ground potential.

  o <u>Crosses</u> - condition that occurs when one side of a cable
     pair comes in contact with another cable pair.

  o <u>Splits</u> - condition which occurs when the tip and ring of
     a cable pair are not terminated on adjacent and proper
     pins on a connector block.

8.2   All IBM Type I circuits will be tested in accordance with IEEE 802.5 specifications.

8.3   The Vendor shall perform the following Multimode Optical Fiber Cable Tests:

  o Optic continuity tests at 850 and 1300 nm.

  o Optic Attenuation or loss at 850 and 1300 nm.

The following optional tests may be performed at the discretion of the cable installation vendor.   The performance of these tests will serve to indemnify the placement vendor against any claims   made   by   subsequent   equipment vendors as to the performance characteristics of the individual fibers:

  o Optical Time Domain Reflectometry - Return loss/errors
     per km.

  o Cable dispersion or bandwidth - measured in
     nsec/km.

All fibers are to guarantee good in accordance with the technical   specifications   listed   under   Fiber   Optics Specifications.

8.4 <u>Terminal and Computer Room Building Ground</u>:

  o <1.0 volt AC ground potential difference between serving
rooms.

     The vendor shall also perform resistance and dB loss tests on each pair (Voice and Data). Tests will be performed from the 3.2 Telephone Closet to the Computer Room and from the Computer

Figure 9.3 (*Continued*) Wire specifications.

Premise Distribution Specifications - (Location)

blocks, patch cords and station jacks for each circuit.

Each pair will be classified according to its condition based upon the testing phase, as follows:

o Good - No defects.

o Defective (Unrecoverable): - Indicate fault (i.e. factory created shorts and grounds would generally be the only valid reason for this classification).

All information will be thoroughly documented and presented to the Client Agency prior to system acceptance.

9. GENERAL CONSTRUCTION NOTES:

a. The Pacific Bell Entrance facilities are located in each floors Telephone Closet. The cable count is available at the site.

b. The premise will consist of three types of offices;

    o Private
    o Semi Private
    o Large rooms with modular partitions

c. The prints showing power poles serving modular and outside wall locations are available upon request from the bid issuing agent.

d. While all indications point to a Plenum environment for the placement of communications facilities, the vendor must assure themselves of the placement environment prior to placement of non-Plenum rated cable. Secure Station Wires in such a way as to avoid pinching, crimping and over stress conditions. Wire Wraps are not allowed.

e. CONTACT LIST

    Telecommunications Design: (Name)
                                  (Name of Company)
                                  (Address)
                                  (Phone)

    Real Estate Design   (Name)
                                  (Address)
                                  (Phone)

Page 9

Figure 9.3 (Continued) Wire specifications.

Premise Distribution Specifications - (Location)

> Telephone Utility   (Name of Person)
>   (Title)
>   (Phone)
>
> Communications Consultant - (Name of Company)
>   (Phone)

10.0 BUILDING OWNER SUPPLIED MATERIAL AND FACILITIES:

The building owner or Electrical Contractor is to provide the following facilities prior to the vendor installing the low-voltage wiring system described above:

1.   Place 3-4" metallic conduits from the 2.1 Telephone Closet to the Computer Room, with a pull wire enclosed from point to point in each conduit. No more than 180 Degrees cumulative radius from point to point in any conduit run. Minimum 8' Radius or curve.   Total degrees to be calculated horizontally and vertically.

2.   Place 4'x8'x3/4" construction grade plywood at the following locations:

> a. Computer Room - 5 Each, South Wall
> b. 1.1 Closet   - 2 Each, West Wall
> c. 2.1 Closet   - 2 Each, West Wall
> d. 3.1 Closet   - 2 Each, West Wall
> e. 4.1 Closet   - 2 Each, West Wall
>
> NOTE: Minimum 42' in front of and 30" lateral board clearance from power equipment/facilities required for low-voltage installation.

3. Owner to provide suitable building Ground Bus in accordance with NEC article 250 at the 1.1 Telephone closet and the Computer Room .

4. Owner to provide Fire Wall penetrations for Station Wire routing. Fire Stop Provision applies.

11.0 WORK ITEM LIST:0

Location	Item	$ Estimate
Telco Closets		
Place BLUE Backboards		
Place Termination Blocks		
Place 4-600 Pair Cables		
Terminate 2400 Pair		
Place 40' #6 Ground Wire		
Ground Cable Sheath		
Ground Metallic Conduit		
Conformance Test 2400 Pair		
Permanently Label Facilities		
Fire Stop 4 Conduit		
Place 16-FO Cables		
Place FO Panel		

Page 10

Figure 9.3 (Continued) Wire specifications.

Premise Distribution System - (Location)

    Closets (Continued)
                        Place Data Patch Panel
                        Terminate 320 Voice Stations
                        (Average Length 155') (4Pr)
                        Terminate 320 Data Stations
                        (Average Length 155'((6Pr)

    Computer Room   Place BLUE Backboards (All)
                        Place 110 Type Connector Blks
                        Place FO Panel
                        Terminate 2400 Pairs (Cable)
                        Terminate 4 FO Cables
                        Ground Metallic Conduit
                        Place 40' #6 Ground Wire
                        Ground Cable Sheaths
                        Permanently Label Facilities
                        Conformance Test FO Cables
                        Provide Voice Patch Cords
                        Provide 200 Baluns
                        Provide 100 Sub-mini RS232/RJ45

    Station Location   Place 320 Duplex Voice
                        Modular Faceplates & Jacks
                        Place 320 Voice Station Cords
                        Place 320 Duplex DATA
                        Place 320 Data Station Cords
                        Modular Faceplates & Jacks
                        Terminate Station Wire
                        Permanently Label Facilities
                        Conformance Test Station Wire
                        Fire Stop Wall Penetrations

    TOTAL ESTIMATED COST AS SPECIFIED

    COST PER ADDITIONAL STATION (Wire, Jacks, Labor)_____

12.0 SCHEDULING

        IN SERVICE DATE
        Vendor Site Walk-thru
        CENTREX Service Ready Date
        Material On-Site (This Specification)
        Installation Start Date
        Modular Furniture Install Complete
        Furniture Move Complete
        Conformance Test Complete
        Voice & Data Equipment Move Date
        System Acceptance

Page 11

Figure 9.3 (*Continued*) Wire specifications.

Premise Distribution Specifications - (Location)

13.0   <u>LOCATION DIAGRAMS</u>:

<u>Computer Room South Wall Termination Detail</u>

(Not to Scale - Does not represent Quantities)

FUT
<u>KTS EQU.</u>   <u>2400 PAIR HOUSE</u>   <u>DATA Patch</u>

<u>STATION TERMINATION</u>

RJ14   IBM TYPE I
RJ14   RJ45

<u>VOICE CIRCUIT DIAGRAM</u>

<u>Computer Rm.</u>   <u>Telco Closet</u>   <u>Station Loc.</u>

VOICE Circuit (Twisted Pairs)

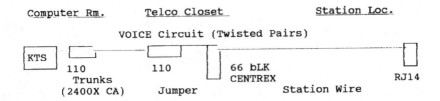

| KTS |
110   110   66 bLK
Trunks   CENTREX   RJ14
(2400X CA)   Jumper   Station Wire

Page 12

Figure 9   (1) Computer room south wall termination detail; (2) station termination; (3) voice circuit diagram.

Premise Distribution Specifications - (Location)

DATA Circuit

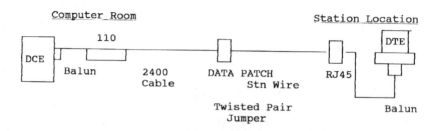

CABLE PLACEMENT SKETCH

(Address)
(Face West)

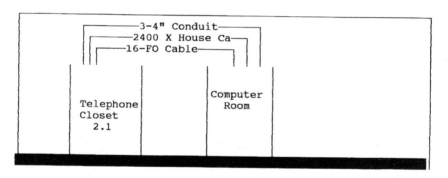

NOTE: Take 1 600 Pair House Cable to Each Closet on Each Floor. Take 4 Fibers to Each Closet on Each Floor.

Page 13

Figure 9 **(Continued)** (4) Data circuit; (5) cable placement sketch.

10

Job Scheduling

10.1 Purpose

Job scheduling requires close coordination of all related functions. Each function must meet its deadline or the work of the various parties will not be successful, and the client will suffer a penalty. Job scheduling must be planned so that each job component has adequate time for completion, and once these schedules have been established, they should be met with a minimum of exceptions. All management personnel having a part in the scheduling process have a combined responsibility for the overall administration of job scheduling. They must provide the discipline to see that the elements of the procedure are adhered to.

This chapter is provided to acquaint the user with information adequate to coordinate multiagency and vendor schedules and assure timely provision of service.

10.2 Coordination Responsibility

The scheduler should be knowledgeable about the functions involved in project implementation in order to identify possible road blocks to meeting the schedule. The scheduler should be the agency or company telecommunication manager for the work-requesting party. This individual must be able to interact with the various public and private sector representatives in order to effect timely installation of telecommunication facilities and services.

Experience has taught that vendors seem to be absorbed primarily in completing their task in a least-time, least-expense manner. The scheduler must be able to confront these individuals on objective levels without causing delay. It is therefore recommended that the individual assigned this responsibility be completely familiar with the component requirements of each job and have superior communication and diplomatic skills. There is a popular saying among the younger members of the business community today to the effect that it is not important that people be familiar with the tasks they supervise, only that they be skilled in people management techniques. This adage does not hold true for the scheduler position.

10.3   Scheduler Authority and Responsibility

The scheduler must be given complete overall responsibility for project completion. This means that the individual must be in charge of project development and implementation through all phases of construction and system acceptance. To attempt to fragment the process will lead to time loss, confusion in issue resolution, and, inevitably, cost overruns. The scheduler must be aware of the magnitude of the responsibility and the consequences of missed or delayed due dates.

Once the responsibility is assigned, the scheduler must also be given the authority to make decisions, effect changes, and halt work should critical standards appear to be in jeopardy. This authority can have limits only if appropriate escalatory limits are imposed and known to affected parties. Higher management must express confidence and support for the scheduler and expect timely and periodical formal feedback on total or phased project work, depending on the complexity of the task, from this individual.

10.4   Scheduling Techniques

The job scheduler will be required to assure that responsibility is assigned for the functions involved in any particular project. This is not a job to be assigned to junior members of the staff or one to be given as a rotational assignment, since this individual will require ultimate authority in project decisions, as discussed above. The job scheduler will be responsible for identifying and maintaining a working relationship with key personnel in several internal and external operations, such as:

- Office/building planning counterparts
- Equipment and service order origination organization
- Facilities design/engineering department or vendor

- Procurement department or services
- Public sector utility companies
- Vendor and contractor schedules

It is important to note that the wiring installation project is usually a subfunction of a larger implementation project. The job scheduler, as referred to in this work, is the individual responsible for making sure that the communication systems are in place when requested. The overall project will involve interaction with many organizations to accurately establish when the scheduling or communications and wiring can realistically begin and end. It is important to distinguish between a project superintendent and scheduler in this manner. To that end, the following responsibility descriptions are presented. The reader may not encounter these distinct divisions of responsibility, but will find these functions involved in any undertaking.

10.4.1 Office/building planning counterparts

The planning organization negotiates the space requirements for the various organizational clients. A representative of the planning organization normally chairs the joint meeting between the prime, or user, organization and external and private sector entities involved with moves and additional space acquisitions. In the best of circumstances, the planning organization also accumulates the voice and data requirements for agency and departmental needs and updates in service dates.

In the case of new or remodeled office space, the scheduler must know when the space to be occupied will be ready for system installations. Normally, the support structure is installed first, wiring second, and equipment last. In the case of new or remodeled space, the scheduler must obtain a firm commitment from the space planning authority for when the support structure will be ready. Once that commitment is obtained, the scheduler must assure that the facilities are installed as planned and ready for the next phase. Should there be inadequate or improper design, the group responsible must be notified and correction commitments obtained before work groups under the scheduler's control are committed. Since the design of facilities is not usually the responsibility of the scheduler, assistance from the designer in determining proper facility compliance and acceptance should be required. If this proves impractical, the scheduler may use the bid specifications as a guide in conjunction with a vendor inspection.

10.4.2 Equipment and service
order origination organization

It is recommended that each group affected by project work prepare its own need document, demand forecast, and detailed engineering plan. Each group should also prepare a budget and receive approval prior to issuing orders for service. Once this is accomplished, the scheduler should be in the loop before firm in-service commitment dates are issued and service is requested from external sources (e.g., telephone companies and vendors). Once the service orders are received, the scheduler will match the request against the job schedule to assure that a realistic in-service date is supplied.

10.4.3 Facilities design/engineering
department or vendor

The engineering/design organization or vendor is normally available to assist the scheduler in arranging for the proper design change of the new building space. This is the source of the bid specifications for entrance facilities, closet design, and site specifications. As such, this organization must be involved with the scheduler up to structure turnover and acceptance.

10.4.4 Procurement department
or services

The procurement department will normally arrange for the ordering of material or process the contractor bid for the installation of cable and wire. There are usually internal procedures for the processing of bids and material orders. The scheduler must assure that the engineering design for cable and wire is provided to the procurement department and that all updates and changes are processed in a timely manner to assure complete material availability and schedule adherence. It may be necessary for the scheduler to perform this function in the absence of such a processing unit.

10.4.5 Public sector utility companies

The utility company will have to be apprised of service needs at the particular location at the earliest possible date. If new facilities will have to be engineered by them, time frames of 6 months to a year may be required. The scheduler should be sure that utility companies have been alerted to the client's needs and that facilities will be available on the in-service date.

10.4.6   Vendor and contractor schedules

Equipment and cable vendor delivery and installation schedules are susceptible to equipment availability and manpower constraints. Therefore, system specifications and vendor selection functions must be front-end–loaded into the scheduling process to assure timely installation and turn-up of equipment.

10.5   Performance Control System

As referred to here, performance control pertains to those functions directly under the control of the scheduler. Project tasks are divided into these functions:

- Cable placement
- Termination activity
- Testing
- System acceptance

10.5.1   Cable placement activity

AT&T developed a very detailed and precise unit productivity model that, when applied across a broad universe of work, became a fairly accurate method of forecasting placement time. However, in the micro world of an individual job, the model deteriorated as the impact of individual work effort imposed substantial variances to the average. To overcome this problem for the individual scheduler, it has been most effective to develop placement time frames by using the contractor's or vendor's estimate of labor. If the contract is on a cost-plus basis (time and material plus extras), the time should be multiplied by 1.2. If the installation agreement is lump sum (not to exceed price, no labor given), the scheduler must either develop a material price, deduct it from the total price, and divide by the prevailing labor rate, or request an installation schedule from the placement supervisor. Multiply this amount by 1.35.

The multiplication factor recommended is based on the tendency of cost-plus contracts to be more reliable, in terms of estimated time, than lump sum. The lump sum contract is normally selected on least-cost totals, which most contractors hedge on. After they receive the contract, it is the rule to then add in the components that were left out of the original bid, thus requiring more time and expense. These are basic rules to go by in an unknown vendor situation. If the vendor is known through previous work association, the scheduler will develop multiplication factors appropriate to that particular vendor's performance. It is always best to overestimate time frames than to miss a due date!

10.5.2 Cable termination activity

This activity involves the termination of individual wires, coax cables, or optical fibers on access blocks. This work is normally performed in apparatus closets and major access points. Again, the general rules which apply for placement apply here, with a few exceptions.

Placement activity normally occurs prior to the installation of carpeting, paneling, office partitions, and furniture. Station terminations, however, cannot be completed prior to the placement of office cubicles and walls. The scheduler must assure that all conduits and support structures to the end-user location are in place and ready for service before termination work begins. Occasionally, closet and access room terminations can be completed before the office space is complete, but vendors usually like to work start to finish without interruption for efficiency considerations.

Once the scheduler is satisfied that the support structure for all block placements is ready, the termination labor may be scheduled. Before committing forces to this task, it is critical that the scheduler ensure that all trades that may conflict with the placement and termination activity are not scheduled to be in the same place at the same time. This is a very difficult task and requires a good relationship with the on-site construction supervisors. There will always be minor conflicts, but common sense and patience will normally resolve all conflicts.

If, by some circumstance, irreconcilable conflicts develop, make sure that a swift resolution mechanism is in place. Know the names of every top manager and have their phone numbers readily available. Do not be afraid to be a name dropper.

10.6 Inspection Requirements

Inspection of support structure, electrical wiring, and building and office partitions will be accomplished by local building code enforcement authority. The scheduler should insist that such inspections are complete prior to commitment of work forces under the scheduler's control to effectively eliminate responsibility for infractions.

Should work under the control of the scheduler require permit application, the scheduler will ensure that the vendor or contractor performing the work has obtained the necessary permits and that appropriate inspection documentation is completed prior to acceptance of work.

When wire/cable work is complete, the scheduler will arrange for the system design engineer to inspect the work for adherence to the bid specification. Discrepancies will be formally recorded and presented to the responsible vendor/contractor for correction. This work will be scheduled and time limits for its completion established and mutually agreed to. It will also be necessary to carefully determine if the corrective action is necessitated by a modification of the original plan or by an oversight by the placing vendor.

If it is determined that additional work operations from the original plan have been authorized by the design engineering forces, the scheduler must assure that the appropriate budgeting organization is aware of the situation and has appropriated additional funds to cover the cost of the additional work operations.

Should the corrective action be a result of oversight by the responsible vendor, additional funding should not be authorized. Many times an attempt will be made by the vendor to capture additional funds in this manner, and disputes may ensue. It is important to adhere to the bid specification in these instances and ensure that time delay responsibility and potential penalties are made clear to the vendor.

10.7   Conformance Testing

This section describes the conformance tests required on all new cables after all installation work has been completed, but before the facilities are turned over to the client organization. These tests are necessary in order to identify any defective pairs or facilities that may exist within the new placement.

Economic studies indicate that the dollar savings is high when conformance testing is done. The major portion of the savings comes from an increase in usable facilities.

After all conformance tests have been completed, the vendor should provide a complete set of cable and/or facility records which reflect the condition of each pair or circuit on every facility placed. End-to-end tests are the most reliable standard to follow and should be called for in the bid specifications.

10.7.1   Types of defects

There are certain types of defects which are commonly associated with the installation of a new cable. These defects are listed below with a brief description of the most common causes:

- *Open*—when one or both wires in a cable pair have a break in continuity

- *Short*—when the tip and ring wire of a cable pair come in contact with each other

- *Ground*—when either wire of a cable pair comes in contact with a ground potential

- *Crosses*—when one side of a cable pair comes in contact with another cable pair

- *Splits*—when the tip and ring of a cable pair are not terminated on adjacent termination pins

Conformance tests can be classified as either AC or DC. Since the client will normally be installing nonloaded cables, it will be necessary to address only the dc type of defects for voice circuits and the ac tests for data circuits. The DC designation comes from the use of direct-current–type test instruments to identify and locate these defects.

Once a defective pair is identified, it should be further classified into one of the following categories:

- Type A—defective pair that is not suitable for any grade of service, such as opens and shorts.

- Type B—pair which meets minimum voice-grade requirements. These pairs should not be used for data or other special service requirements.

10.7.2   Data cable tests

Since data cables are shielded and the shielding is grounded in the serving closet, it becomes vital to test each pair or circuit for ground potential. Should the shield not be grounded at one end, the shield will become an AM/FM antenna and interfere with most data transmission signals.

The vendor should also test for ground potential difference between serving closets. There should not be a reading of more than 1.0 Vac. Another test to be performed is to measure the grounding terminal of the AC power receptacle. This should not measure more than 3.5 Ω.

Decibel and resistance loss measurements, along with time-domain reflectometer tests, may be desired when specific vendor equipment is planned for use. These tests are more exacting and will at times require the assistance of the design engineering and equipment vendor forces. The scheduler should assure that the proper forces and equipment are available to assure system acceptability.

10.7.3 Conformance tests on interbuilding or campus distribution systems

Basically, the tests for these cables are the same. Since these cables require special engineering and may include loaded or aerial cables with splices, special consideration will have to be given to the conformance tests made on these facilities.

When dealing with these types of loops, the scheduler must make sure that appropriate directions are given to the placing force, that the workers understand the testing methodology, and that proper documentation is prepared. The scheduler must then arrange for the responsible design organization to review and accept the testing results.

10.8 System Acceptance

This is the point at which the rubber meets the road. The client organization will normally be anxious that the facilities meet expectations (which are normally far in excess of physical reality), the engineering group is ready to forget the whole project, and the vendor wants compensation.

To satisfy all the interests involved and assure that the scheduler is not put in the middle of disputes, the following documentation must be complete:

- Project schedule with start and complete dates
- Correspondence concerning all work changes and delays with responsible party identification
- Complete system end-to-end conformance testing results, signed off by appropriate engineering representatives
- Location records updated to reflect new facilities
- Logs of interim payments to the vendors/contractors
- Signed warranties and maintenance agreements

Once this information is compiled, the scheduler may want to schedule a final walk-through with the client agency representative and the appropriate vendor and engineering representatives. At this time, the system can be explained to the client agency representative, minor additions or changes arranged for, and acceptance communicated, face to face. This meeting should be documented and formally filed with the project papers.

Once this process is complete, the scheduler will authorize final payment to the vendors and contractors.

Glossary of Telecommunications Technologies

A

A and B signaling—a procedure used in most T1 transmission facilities where one bit, robbed from each of the 24 subchannels in every sixth frame, is used for carrying dial and control information. A type of in-band signaling used in T1 transmission.

A-law encoding—encoding according to CCITT Recommendation G711, used with European 30-channel pulse code modulation (PCM) systems that comply with CCITT Recommendation G732. Employs non-uniform quantizing to obtain the desired compression characteristic.

abandoned call—call in which a caller cancels the call after a connection has been made but before conversation takes place, for example, after hearing a recorded announcement.

abbreviated dialing—enables a caller to dial a frequently used number by means of a few digits instead of the entire telephone number. Also called speed dialing and short-code dialing.

access charge—cost assessed to communications users for access to the interexchange, interstate message toll telephone network to originate and receive interstate toll calls, as well as access to the customer's local access and transport area (LATA).

access code—the digit, or digits, that a user must dial to be connected to an outgoing trunk facility.

access control—hardware, software, and administrative tasks that monitor system operation, perform user identification, and record system accesses and changes.

access line—Connection from the customer to the local telephone company for access to the telephone network; also represents the connection between the serving toll center and the serving office of the interexchange carrier used for access to public switched network services. Also known as local loop.

accounting codes (voluntary or forced)—Entered by a caller by dialing the appropriate digits for the call being placed followed by the digits composing the accounting codes. With forced accounting codes, if the caller fails to dial a valid accounting code within a specific timeframe, an intercept tone (or as a customer option, a recorded message) can be played to the caller and the call will be terminated.

accounting rate—charge per traffic unit, which can be a unit of time or information content, covering communications between zones controlled by different telecommunications authorities; used to establish international tariffs.

ACK—"acknowledge" character. A transmission control character transmitted by a station as an affirmative response to the station with which a connection has been set up. An acknowledge character may also be used as an accuracy control character. See also *NAK (negative acknowledgement character)*.

ACK/NAK/END—acknowledged/not acknowledged/inquiry.

acoustic coupler—device that allows a telephone handset to be used for access to the switched telephone network for data transmission. Digital signals are modulated as sound waves, and data rates are typically limited to about 300 bps, some up to 1.2K bps.

ACTGA—see *attendant control of trunk group access.*

ACU—see *automatic calling unit.*

A/D—Analog/Digital.

adapter—a connective device designed to link different parts of one or more systems and/or subsystems

adaptive differential pulse code modulation (ADPCM)—encoding technique (CCITT) that allows analog voice signals to be carried on a 32K bps digital channel. Sampling is done at 8kHz with 3 or 4 bits used to describe the difference between adjacent samples.

Glossary of Telecommunications Technologies

adaptive equalization—line equalization, used for optimizing signal transmission, that is adjusted while signals are being transmitted in order to adapt to changing line characteristics.

adaptive routing—routing that automatically adjusts to network changes such as changes of traffic pattern or failures.

ADC—analog-to-digital conversion. A method of sampling and costing an analog signal to create a digital signal.

ADCCP—see *advanced data communications control procedure.*

add-on conference—sometimes referred to as "three-way calling." Used in association with consultation hold features, it is a conference facility that allows the station user to add another internal party to the existing conversation.

add-on data modules—plug-in circuit boards that allow a PBX to receive and transmit both digital and analog signals.

address—1) coded representation of the destination of data, or of its originating terminal. Multiple terminals on one communications line, for example, each must have a unique address. 2) group of digits that makes up a telephone number. Also known as the called number. 3) in software, a location that can be specifically referred to in a program.

ADPCM—see *adaptive differential pulse code modulation.*

advanced data communications control procedure (ADCCP)—a bit-oriented protocol developed by ANSI.

advanced mobile phone service—AT&T-developed analog cellular radio standard, adopted in the USA and in many other parts of the world.

advanced private line termination (APLT)—provides the PBX user with access to all the services of an associated Enhanced Private Switched Communications Services (EPSCS) network. It also functions when associated with a Common Control Switching Arrangement (CCSA) network.

AFIPS—see *American Federation of Information Processing Societies.*

agent sign-on/sign-off—enables any ACD agent to take ACD calls from any ACD-assigned set.

AIOD—see *automatic identification of outward dialing.*

alarm display—indicators on attendant console that show the status of the system. Usually two alarms are included: a minor alarm to signal a malfunction in a part of the system and a major alarm to indicate that system will not work at all.

alarm signals to stations—ability of a system to interface with various types of signaling systems and to transmit such alarms in the form of coded tones to all, or preselected, stations.

all number calling (ANC)—calling by means of seven digits instead of previously used two letters plus five digits.

all trunks busy (ATB)—condition in which all trunks in a group are engaged.

alpha test—the stage during new product research and development when a prototype is operated to ascertain that the system concept and design are functional and to identify areas that need further development and/or enhancement.

alternate mark inversion—digital signaling method in which the signal carrying the binary value alternates between positive and negative polarities; zero and one values are represented by the signal amplitude at either polarity, while no-value "spaces" are at zero amplitude. Also called *bipolar.*

alternate routing—the routing of a call or message over a substitute route when an established route is busy or otherwise unavailable for immediate use.

alternate voice/data (AVD)—method of deriving telegraph-grade lines from unused portions of a voice circuit.

AM—see *amplitude modulation.*

AMA—automatic message accounting. See also *station message detail recording.*

ambient noise—communications interference present in a signal path at all times.

American Bell—name of ATTIS before AT&T was forced to drop the Bell name by the divestiture decree; see also *ATTIS.*

American Federation of Information Processing Societies (AFIPS)—an organization of computer-related societies; its members include The Association for Computer Machinery, The Institute of Electrical and

SEPTEMBER 1989

Glossary of Telecommunications Technologies

Electronic Engineers Computer Group, Simulation Councils, Inc., and American Society for Information Science.

American National Standards Institute (ANSI)—a standards-forming body affiliated with the International Organization for Standardization (ISO) that develops standards for transmission codes protocols, media, and high level languages, among other things.

American Standard Code for Information Interchange (ASCII)—1) eight-bit code yielding 128 characters, both displayed and nondisplayed (for device control); used for text transmission. 2) a widely used asynchronous protocol based on ASCII code.

amplitude modulation (AM)—in communications technology, a transmission method where the height of the carrier wave is modified to satisfy the height of the signal wave; contrast with *frequency modulation.*

AMPS—see *advanced mobile phone service.*

analog—referring or pertaining to a signaling technique in which a transmission is conveyed by modulating (varying) the frequency, amplitude, or phase of a carrier. Contrast with *digital.*

analog loopback—technique for testing transmission equipment and devices that isolate faults to the analog signal receiving or transmitting circuitry, where a device, such as a modem, echoes back a received (test) signal that is then compared with the original signal. Compare with *digital loopback.*

analog signal—signal in the form of a continuous wavelike pattern representing a physical quantity, such as voltage, which reflects variations in some quantity, such as loudness of the human voice. Analog signaling is generic to the public switched telephone network (PSTN), as well as to certain other audio-frequency and radio-frequency facilities. A digital baseband signal generated by a business machine must be converted to analog form in order to transmit that signal over an analog facility, e.g., a voice-grade telephone line.

analog transmission—transmission of a continuously variable signal as opposed to a discretely variable signal. Also called analog signaling.

ANC—see *all-number calling.*

ancillary equipment—terminal or communications devices not required for the provision of basic telephone service. Answering machines, conferencing devices, and automatic dialers are types of ancillary equipment.

angle modulation—a form of modulation in which the angle of a sine wave carrier is varied.

ANI—see *automatic number identification.*

anisochronous data channel—communications channel capable of transmitting data but not timing information. More commonly known as *asynchronous data channel.*

ANS—answer.

ANSI—see *American National Standards Institute.*

answer signal—supervisory signal (usually in the form of a closed loop) from the called telephone to the exchange and back to the calling telephone (usually in the form of a reverse battery) when the called number answers. Signal can also initiate call charging.

answerback—a manually or automatically initiated response from a terminal that usually includes the terminal address to verify that the correct terminal has been reached and that it is operational.

APLT—see *advanced private line termination.*

append—to change or alter a file or program by adding to the end of the file or program.

application layer—the top of the seven-layer OSI model, generally regarded as offering an interface to, and largely defined by, the network user; in IBM's SNA, the end-user layer.

application program—software programs in a system are usually known as "application programs" and "supervisory programs." Application programs contain no input/output coding (except in the form of macroinstructions that transfer control to the supervisory programs) and are usually unique to one type of application.

architecture—the physical inter-relationship between the components of a computer.

archive—a procedure for transferring information from an online storage diskette or memory area to an off-line storage medium.

area code—synonymous with numbering plan area (NPA); A three-digit code designating a "toll" center not in the NPA of the calling party. The first digit is any number from 2 through 9. The second digit is always a "1" or "0."

Glossary of Telecommunications Technologies

area code restriction—Ability of the switching equipment to selectively identify three-digit area codes, and either allow or deny passage of long distance calls to those specific area codes.

ARL—see *attendant release loop.*

ARO—after receipt of order.

ARPA—the generic name for services designed by the Department of Defense in the 1970s; it is now a de facto standard for networking multivendor computers running on differing operating systems. Includes FTP, SMTP, and TELNET.

ARQ—see *automatic repeat request.*

ARS—see *automatic route selection.*

artificial intelligence—the capability of a computer to perform such functions that are associated with human logic as reasoning, learning, and self-improvement.

artificial language—a convention based on a set of rules established prior to its usage and without a precise relationship to the user applications it will be used for. Contrast with *natural language.*

ARU—see *audio response unit.*

ASR—see *automatic send/receive.*

asynchronous—occurring without a regular or predictable time relationship to a specified event, e.g., the transmission of characters one at a time as they are keyed. Contrast with *synchronous.*

asynchronous transmission—transmission in which each information character, or sometimes each word or small block, is individually synchronized, usually by the use of start and stop elements. Also called start-stop transmission.

ATB—see *all trunks busy.*

ATND—see *attendant.*

ATS—see *applications technology satellite.*

attached resource computer (ARC)—the local networking products and philosophy developed by Datapoint Corporation.

attendant (ATND)—usually refers to a local switchboard operator, for example, on a PBX. See also *Operator.*

attendant console—centralized operator position, either desktop or floor-mounted, that uses pushbutton keys for all control and call connecting functions.

attendant monitor—special attendant circuit that allows listening in on all circuits with the console handset/headset transmitter deactivated.

attendant override—allows an attendant to enter a busy trunk connection and key the trunk number within the PBX. A warning tone will be heard by the connected parties, after which the connected parties and the attendant will be in a three-way connection.

attendant supervisory console—special attendant console used by the chief operator of large PBX systems. It provides, in addition to standard console operation, certain monitoring and control functions over other consoles.

attended operation—situation in which both data stations require attendants to establish the connection and transfer the data sets from talk (voice) mode to data mode.

attenuation—a decrease in the power of a received signal due to loss through lines, equipment, or other transmission devices. Usually measured in decibels.

ATTIS—AT&T Information Systems; division of AT&T Technologies that supplied and manufactured customer premises equipment ca. 1984-86.

audible ringing tone—tone received by the calling telephone indicating that the called telephone is being rung (formerly called ringback tone).

audio frequencies—frequencies that correspond to those that can be heard by the human ear (usually 30 Hz to 20,000 Hz).

audio response unit (ARU)—device that provides prerecorded spoken responses to digital inquiries from a telephone caller once the connection is established. Requires use of a pushbutton telephone; callers are instructed to press a certain number to obtain a certain type of information.

audiotex—communications system that allows a host computer to pass data to a voice mail computer, a store-and-forward mechanism for digitized voice, where it is interpreted and delivered over the telephone.

authorization code—code allowing a station user to override the restriction level associated with that user's station line or incoming trunk.

SEPTEMBER 1989

Glossary of Telecommunications Technologies

auto answer—automatic answering; capability of a terminal, modem, computer, or similar device to respond to an incoming call on a dial-up telephone line and to establish a data connection with a remote device without operator intervention. Allows unattended operation for incoming dial-up calls.

auto call—automatic calling; a machine feature that allows a transmission control unit or a station to automatically initiate access to a remote system over a switched line.

auto dial—automatic dialing; the capability of a terminal, modem, computer, or a similar device to place a call over the switched telephone network and establish a connection without operator intervention.

automated attendant system—processor controlled system that performs most attendant functions, such as answering calls, extending them to station users, taking messages, and providing assistance. The system provides voice prompts, to which users reply via any standard 12-button dialpad.

automatic call distributor (ACD)—system designed to distribute a large volume of incoming calls uniformly to a number of operators or agents, e.g., for airline reservations.

automatic calling unit (ACU)—a unit, which may or may not be integrated within a modem, that automatically dials calls based on digits supplied by the attached business machine.

automatic circuit assurance (ACA)—assists the customer in identifying possible trunk malfunctions. The PBX maintains a record of the performance of individual trunks relative to short hold time (SHT) and long hold time (LHT) calls. A significant increase in the number of short calls or a single long call can indicate a trunk failure. When a possible failure is detected, a referral call is initiated to the attendant.

automatic exclusion—first station accessing a line automatically prevents any other station from gaining access to that line.

automatic identification of outward dialing (AIOD)—hardware system or PBX feature that automatically obtains the identity of a calling station over a separate data link for the purpose of automatic message accounting.

automatic message accounting (AMA)—automatic recording system that documents all the necessary billing data of subscriber-dialed long distance calls. See also *station message detail recording.*

automatic message-switching center—in a communications network, location at which data is automatically routed according to its destination.

automatic number identification (ANI)—feature that automatically identifies a calling station; for use in message accounting.

automatic repeat request (ARQ)—an error control technique that requires retransmission of a data block that contains detected errors. A special form, called "go-back-n," allows multiple blocks to be acknowledged with a single response. "Stop and wait" requires an acknowledgment after each block.

automatic route selection (ARS)—provides automatic routing of outgoing calls over alternative customer facilities based on the dialed long-distance number. The station user dials either a network access code or a special ARS access code followed by the number. The PBX routes the call over the first available special trunk facility, checking in a customer-specified order. Alternative routes can also include tie trunks to a distant PBX. When such routing is used, the restriction level associated with the call can be transmitted to the distant PBX as a traveling class mark. Incoming tie trunks from other locations (e.g., main or satellite) can be arranged to have automatic access to ARS. This allows station users at these systems to dial a single access code to use the ARS feature at a distant PBX.

AUTOVON—automatic voice network (U.S. military).

Aviane—a heavy launch vehicle produced by the European Space Agency (ESA).

AVD—see *alternate voice/data.*

AVD circuits—alternate voice/data circuits that have been conditioned to handle both voice and data traffic.

B

backbone network—the portion of a communications facility that connects primary nodes; a primary shared communications path that serves multiple users via multiplexing at designated jumping-off points.

backhaul—the terrestrial link between an earth station and a switching or data center.

backup—the provision of facilities, logical or physical, to speed the process of restart and recovery following failure.

Glossary of Telecommunications Technologies

balanced-to-ground—with a two-wire circuit, the impedance-to-ground on one wire equals the impedance-to-ground on the other wire. Compare with *unbalanced-to-ground*, a preferable condition for data transmission.

balun—balanced/unbalanced. An impedance-matching device used to connect balanced twisted-pair cabling with unbalanced coaxial cable.

band—1) the range of frequencies between two defined limits. 2) in relation to WATS service, the specific geographical area which the customer is entitled to call.

bandpass filter—circuit designed to allow a single band of frequencies to pass, neither of the cut-off frequencies being zero or infinity.

bandwidth—the range of frequencies, expressed in hertz (Hz), that can pass over a given transmission channel. The bandwidth determines the rate at which information can be transmitted through the circuit. The greater the bandwidth, the more information that can be sent through the circuit in a given amount of time.

baseband signaling—transmission of a digital or analog signal at its original frequencies, i.e., a signal in its original form, not changed by modulation. It can be an analog or digital signal.

base station—within a mobile radio system, a fixed radio station providing communication with mobile stations and, where applicable, with other base stations and the public telephone network.

basic rate interface—in ISDN, the interface to the basic rate CCITT 2B + D, 2 channels + 1 signaling channel. See *integrated services digital network*.

BASR—buffered automatic send/receive; see *automatic send/receive*.

battery reserve power—provides an alternative, independent source of power to maintain PBX service during a power failure or "brownout" at the customer location. The power supply consists of storage batteries and permanently connected battery chargers operating from a commercial AC power source.

baud—a measure of data rate, often used to denote bits per second (bps). A baud is equal to the number of discrete conditions or signal events per second. There is disagreement over the appropriate use of this word since, at speeds above 2400 bps, the baud rate does not equal the data rate in bits per second.

baudot code—a data code using a five-bit structure used on vintage teleprinter (e.g., Telex) terminals. Two of the possible 32-bit patterns are used to indicate case shift, which doubles the interpretations of the remaining bit patterns. An additional four characters are reserved for control functions, limiting the number of data characters to 52.

bay—see *rack*.

BCD—see *binary coded decimal*.

BDC—see *block down conversion*.

beam width—angular width of an antenna radiation pattern, or beam, within which the radiation exceeds some proportion of the maximum.

Bell Operating Company (BOC)—any of 22 local telephone companies spun off from AT&T as a result of divestiture, such as Bell of Pennsylvania, New Jersey Bell, and Southern Bell. Does not include Southern New England Telephone or Cincinnati Bell. The 22 operating companies are divided into seven regions and are held by seven RBHCs (Regional Bell Holding Company).

Bellcore—Bell Communications Research; organization established by the AT&T divestiture, representing and funded by the RBHCs, for the purpose of establishing telephone-network standards and interfaces; includes much of former Bell Labs.

benchmark—a point of reference from which measurements can be made; involves the use of typical problems for comparing performance and is often used in determining which computer can best serve a particular application.

BER—see *bit error rate*.

BERT—see *bit error rate test/tester*.

beta test—the stage at which a new product is tested under actual usage conditions.

binary phase-shift keying—see *phase shift keying*.

binary synchronous communications (BSC)—bisync; a half-duplex, character-oriented data communications protocol originated by IBM in 1964. It includes control characters and procedures for controlling the establishment of a valid connection and the transfer of data. Although still enjoying widespread usage, it is being replaced by IBM's more efficient protocol, Synchronous Data Link Control (SDLC).

SEPTEMBER 1989

bipolar—1) the predominant signaling method used for digital transmission services, such as DDS and T1, in which the signal carrying the binary value successfully alternates between positive and negative polarities. Zero and one values are represented by the signal amplitude at either polarity, while no-value "spaces" are at zero amplitude. 2) a type of integrated circuit that uses both positively and negatively charged currents, characterized by high operational speed and cost. Also called *alternate mark inversion*.

bipolar violation—modified bipolar signaling in which a control code is inserted into the original digital format.

BISDN—broadband ISDN. See *integrated services digital network*.

bisync—see *binary synchronous communications*.

bit duration—equivalent to the time that it takes one encoded bit to pass a point on the transmission medium. In serial communications, a relative unit of time measurement used for comparison of delay times (e.g., propagation delay, access latency), where the data rate of a transmission channel can vary.

bit error rate (BER)—in data communications testing, the ratio between the total number of bits transmitted in a given message and the number of bits in that message received in error. A measure of the quality of a data transmission, usually expressed as a number referred to a power of 10; e.g., 1 in 10^5.

bit error rate test/tester (BERT)—a test conducted by transmitting a known pattern of bits (commonly 63, 511, or 2047 bits in length), comparing the pattern received with the pattern transmitted, and counting the number of bits received in error. Also see *bit error rate*. Contrast with *block error rate test/tester (BLERT)*.

bit-oriented—describes a communications protocol or transmission procedure where control information is encoded in fields of one or more bits. Oriented toward full-duplex link operation.

bit rate—the speed at which bits are transmitted, usually expressed in bits per second (bps).

bits per second (bps)—basic unit of measurement for serial data transmission capacity; abbreviated as K bps, or kilobit/s, for thousands of bits per second; M bps, or megabit/s, for millions of bits per second; G bps, or gigabit/s for billions of bits per second; T bps, or terabit/s, for trillions of bits per second.

blanking interval—the area in a video signal that falls between frames. It is often used to accommodate data including synchronizing information.

BLERT—see *block error rate test/tester*.

BLF—see *busy lamp field*.

block—a string of records, words, or characters treated as a logical entity. Blocks are separated by interblock gaps, and each block may contain one or more records.

block down conversion—the conversion of a full satellite band to a lower frequency, (e.g., from SHF to UHF or VHF).

block error rate—in data communications testing, the ratio between the total number of blocks transmitted in a given message and the number of blocks in that message received in error; a measure of the quality of a data transmission.

block error rate test/tester (BLERT)—a test conducted by transmitting a known blocked bit pattern, comparing the pattern received with the pattern transmitted, and counting the number of blocks containing errored bits. Also see *block error rate*. Contrast with *bit error rate test/tester (BERT)*.

block length—a measure of the size of a block, usually specified in units such as records, words, computer words, or characters.

board—the circuit card on which integrated circuits are mounted.

BOC—Bell Operating Company.

bottleneck—the operation with the least capacity in a total system with no alternative routings; the total system can be effectively scheduled by simply scheduling the limiting operation.

bps—bits per second.

break—an interruption to a transmission; frequently a provision permitting a controlled terminal to interrupt the controlling computer.

breakout box—a device that allows access to individual points on a physical interface connector (e.g., EIA RS-232-C) for testing and monitoring.

breakout panel—a breakout box mounted as a component in some larger device.

Glossary of Telecommunications Technologies

bridge—1) to connect a load across a circuit. 2) the connection itself that allows the interconnection of LANs, permitting communication between devices on separate networks.

bridged ringing—system in which ringers on a line are connected across that line.

bridge lifter—device that removes, either electrically or physically, bridged telephone pairs. Relays, saturable inductors, and semiconductors are used as bridge lifters.

broadband—1) transmission equipment and media that can support a wide range of electromagnetic frequencies. 2) any voice communications channel having a bandwidth greater than a voice grade telecommunications channel; sometimes used synonymously with wideband. 3) typically the technology of CATV (q.v.) transmission, as applied to data communications; employs coaxial cable as the transmission medium and radio frequency carrier signals in the 50M Hz to 500M Hz range.

broadcast—a transmission to multiple receiving locations simultaneously. A broadcast can be made, for example, over a multipoint line to all terminals that share the line, or over a radio or television channel to all receivers tuned to that channel.

broadcasting service—radio communications service of transmission to be received directly by the general public. It can consist of sound, video, facsimile, or other one-way transmission.

brownout operation—in response to heavy demand, main system voltages are sometimes lowered leading to brownouts, where power is not lost but reduced. Although conventional PBX equipment is relatively immune to brownouts, the computer controlling the system is very sensitive to voltage variations and could fail under these conditions. Most PBXs today have the capability to cope with these reductions, or a heavy-duty power supply can be furnished as an option. An uninterruptible power supply (UPS) can be installed to ensure continued service during prolonged outages.

BSC—see *binary synchronous communications.*

BT—see *busy signal.*

buffer—storage device used to compensate for a difference in rate of data flow, or time of occurrence of events, when transmitting data from one device to another.

bundled—a pricing strategy in which a computer manufacturer includes all products—hardware, software, services, training, etc.—in a single price.

burst—in data communications, a sequence of signals counted as one unit in accordance with some specific criterion or measure.

bus—1) physical transmission path or channel. Typically an electrical connection, with one or more conductors, wherein all attached devices receive all transmissions at the same time. 2) local network topology, such as used in Ethernet and the token bus, where all network nodes listen to all transmissions, selecting certain ones based on address identification. Involves some type of contention-control mechanism for accessing the bus transmission medium.

business television—a form of business communications that enables organizations to transmit corporate television programs to reach audiences in many dispersed locations.

busy (BY)—describes a line or trunk that is in use.

busy hour (BH)—continuous one-hour period that has the maximum average traffic intensity.

busy lamp—visual indicator on a piece of telephone equipment that indicates the associated line is in use.

busy lamp field (BLF)—panel providing the attendant with visual indications of either busy or idle conditions for a particular group of one hundred station lines selected by the attendant.

busy override or intrude—see *busy verification of station lines; executive busy override.*

busy signal—1) audible and/or flashing signal that indicates that the called number is unavailable. Also called engaged tone. 2) signal that indicates all voice paths are temporarily unavailable.

busy tone (BT)—see *busy signal.*

busy verification of station lines—attendant confirmation that a line is actually in use by establishing a connection to an apparently busy line. Prior to the attendant's connection the PBX sends a burst of tone to the line to alert the talking parties.

bypass—the name given to the method of establishing a communication link without using the facilities of the local exchange carrier (telephone company).

byte—small group of bits of data that is handled as a unit.

SEPTEMBER 1989

Glossary of Telecommunications Technologies

C

C band—portion of the electromagnetic spectrum, approximately 4G Hz to 6G Hz, used primarily for satellite and microwave transmission.

C conditioning—type of line conditioning that controls attenuation, distortion, and delay distortion so they lie within specific limits.

cable—assembly of one or more conductors within a protective sheath; constructed to allow the use of conductors separately or in groups.

CACS—see *customer administration center system.*

CALC—customer access line charges (Centrex).

call—any demand to set up a connection. Also used as a unit of telephone traffic.

call accounting system—device that tracks outgoing calls and records data for reporting. See *station message detail recording.*

call detail recording (CDR)—see *station message detail recording.*

call duration—interval of time between the establishment of the connection between the calling and called stations and the termination of the call.

call holding time (CHT)—total length of time that a circuit is held busy.

call information logging—automatic recording of chargeable calls made on a PBX system, with details of extension number, exchange line number, time, call duration, and digits dialed. This can be used for call accounting or billing.

call processing—sequence of operations performed by a switching system from the acceptance of an incoming call through the final disposition of the call.

call progress tones—audible signals returned to the station user by the switching equipment to indicate the status of a call; dial tones and busy signals are common examples.

call record—all recorded data pertaining to a single call.

called party—the subscriber requested by the calling subscriber. Also known as *called subscriber.*

called subscriber—see *called party.*

calling relay—relay that is controlled via a subscriber's line or trunk line. Also called a *line relay.*

CAMA—see *centralized automatic message accounting.*

camp-on—feature whereby a subscriber calling a busy number is placed in a waiting condition; both phones ring automatically when the called party hangs up.

CAP—see *customer administration panel.*

card (circuit)—the individual boards that carry the necessary circuits for particular functions; these cards (or boards) are designed to fit expansion slots provided by many computer and communications manufacturers.

carrier (CXR)—a signal of known characteristics (for example, frequency) that is altered (modulated) to transmit information. Knowing the expected signal, the receiving terminal interprets any change in signal as information. Changes to the signal made by outside influences (noise) can cause the receiving terminal to misinterpret the information transmitted. See also *common carrier* and *specialized carrier.*

carrier frequency—frequency of the carrier wave that is modulated to transmit signals.

carrier sense multiple access (CSMA)—a local area network access technique in which multiple stations connected to the same channel are able to sense transmission activity on that channel and to defer the initiation of transmission while the channel is active. Similar to contention access.

carrier sense multiple access with collision detection (CSMA/CD)—a refinement of CSMA in which stations are able to detect the interference caused by simultaneous transmissions by two or more stations (collisions) and to retransmit colliding messages in an orderly manner.

carrier signaling—any of the signaling techniques used in multichannel carrier transmission. The most commonly used techniques are in-band signaling, out-of-band signaling, and separate channel signaling.

carrier system—means of obtaining a number of channels over a single path by modulating each channel on a different carrier frequency and demodulating at the receiving point to restore the signals to their original frequency.

Glossary of Telecommunications Technologies

Carterfone decision—landmark 1968 FCC decision that first permitted the electrical connection of customer-owned terminal equipment to the telephone network.

CAS—see *centralized attendant service.*

CATV—see *community antenna television.*

CBX—1) computerized branch exchange. 2) centralized branch exchange. See *private automatic branch exchange.*

CCIR—Comité Consultatif International de Radiocommunication. Technical committee set up under the international telecommunications union (ITU) with responsibility for radio communications.

CCIS—see *common channel interoffice signaling.*

CCITT—Comité Consultatif International de Téléphonie et de Télégraphie. An advisory committee to the international telecommunications union (ITU) whose recommendations covering telephony and telegraphy have international influence among telecommunications engineers, manufacturers, and administrators.

CCS (hundred call seconds)—unit of telephone traffic load calculated by multiplying the number of calls per hour by the average call duration in seconds and dividing the result by one hundred (e.g., 10 CCS = 1,000 seconds).

CCSA—see *common-control switching arrangement.*

CCSA access—provision of inward and outward service between the PBX and the CCSA (common-control switching arrangement) network.

CCSR—see *centrex customer station rearrangement.*

CCTV—see *closed circuit television.*

cellular radio—technology employing low-power radio transmission as an alternative to local loops for accessing the switched telephone network. Differs from standard mobile telephony in that service is provided through a large number of areas or cells which are served by a low-power transmitter in each cell, rather than through a single high-power transmitter for the entire region.

CEN—European Standards Institute (Comite Europeen de Normalisation).

CENLEC—European Electrical Standards Institute (Comite Europeen de Normalisation Electrique).

central office (CO)—the physical location where communications common carriers terminate customer lines and locate the switching equipment which interconnects those lines. See also *exchange.*

centralized automatic message accounting (CAMA)—automatic message accounting system that is located at an exchange but that serves various adjacent exchanges. Calls not processed by ANI (Automatic Number Identification) must be routed through an operator who dials the calling number into the equipment.

Centrex—central exchange; the telephone company's termination point for multiple lines from customers. Switching facilities at the central exchange allow interconnection of the various lines or circuits.

Centrex customer station rearrangement (CCSR)—feature that allows Centrex users to make their own moves and changes. This feature requires the use of customer premises equipment (CPE).

channel—1) in data communications, a one-way path along which signals can be sent between two or more points. Contrast with *circuit.* 2) in telecommunications, a transmission path (may be one-way or two-way, depending on the channel) between two or more points provided by a common carrier. See also *link, line, circuit,* or *facility.*

channel bank—equipment typically used in a telephone central office that performs multiplexing of lower speed, digital channels into a higher speed composite channel. The channel bank also detects and transmits signaling information for each channel and transmits framing information so that time slots allocated to each channel can be identified by the receiver.

channel capacity—an expression of the maximum data traffic that can be handled by the channel.

channel service unit (CSU)—a component of customer premises equipment (CPE) used to terminate a digital circuit, such as a DDS or T1 facility, at the customer site. Performs certain line-conditioning functions, ensures network compliance per FCC rules, and responds to loopback commands from the central office. Also, ensures proper 1's density in a transmitted bitstream and performs bipolar violation correction.

channel, voice grade—channel suitable for transmission of speech, digital or analog data, or facsimile, generally with a frequency range of about 300 Hz to 3000 Hz.

character-oriented—describes a communications protocol or transmission procedure that carries control information encoded in fields of one or more bytes.

SEPTEMBER 1989

Glossary of Telecommunications Technologies

circuit—1) means of two-way communication between two or more points. 2) a group of electrical/ electronic components connected to perform a specific function. See also *channel*.

circuit board—a flat card with connections for integrated circuits.

circuit, four-wire—communications path in which four wires (two for each direction of transmission) are connected to the station equipment.

circuit grade—the data-carrying capability of a circuit; the grades of circuit are broadband, voice, subvoice, and telegraph.

circuit switching—temporary direct connection of two or more channels between two or more points in order to provide the user with exclusive use of an open channel with which to exchange information. A discrete circuit path is set up between the incoming and outgoing lines, in contrast to message switching and packet switching, in which no such physical path is established. Also called *line switching*.

circuit terminating arrangements (CTA)— arrangement for terminating activated circuits at customers' premises. May involve changes to presentation, e.g., 4-wire to 2-wire on wideband circuits.

circuit, two-wire—circuit formed by two conductors insulated from each other that can be used as a oneway transmission path, a half-duplex path, or a duplex path.

cladding—in fiber optic cable, a colored, low refractive index material that surrounds the core and provides optical insulation and protection to the core.

class of exchange—ranking assigned to switching point in the telephone network determined by its switching functions, interrelationships with other exchanges, and transmission requirements. Also called *class of office*.

class of service—1) in X.25 networks, refers to the speed at which the data circuit terminating equipment (DCE) and the data terminal equipment (DTE) communicate. 2) in telephony, it refers to the categorization of telephone subscribers according to specific type of telephone usage. Telephone service distinctions that include, for example, rate differences between individual and party lines, flat rate and message rate, and restricted and extended area service.

class of service display to attendant—shows the class of service for calls to the attendant console from inside

extensions. The class of service shows which trunks or lines the extension is restricted from accessing.

class of service intercept—station is automatically routed to the attendant if it attempts to place a call that is not authorized by its class of service.

clear-forward/clear-back signal—signal transmitted from one end of a subscriber line or trunk, in the forward/backward direction, to indicate at the other end that the established connection should be disconnected. Also called *disconnect signal*.

clear to send—a control circuit that indicates to the data terminal equipment that data can or cannot be transmitted.

clipping—loss of initial or final parts of words or syllables due to operation of voice-actuated devices.

closed circuit television (CCTV)—television transmission via direct link between two points, as opposed to broadcast transmission to many receiving locations.

closed user groups (CUG)—restricts access to and from one or more terminals to other members of the CUG (found on packet switched systems, E-mail, etc.).

cluster—two or more terminals connected to a single point or node.

cluster controller—a device that handles the remote communications processing for multiple (usually dumb) terminals or workstations.

CMOS—complementary metal-oxide semiconductor (logic circuit).

CO—see *central office*.

coaxial cable—a popular transmission medium consisting of one or more central wire conductors, surrounded by a dielectric insulator, and encased in either a wire mesh or extruded metal sheathing. There are many varieties, depending on the degree of EMI shielding afforded, voltages, and frequencies accommodated. Common CATV transmission cable typically supports RF frequencies from 50 to about 500M Hz.

COCOT—customer-owned, coin-operated telephone.

code—the conventions specifying how data may be represented in a particular system.

code call access—feature that allows attendants and station users to dial an access code and a two- or threedigit called party code to activate signaling devices

Glossary of Telecommunications Technologies

throughout a customer's premises with a coded signal corresponding to the called code (either audible or visible). The called or "paged" party can then be connected to the calling party by dialing a "meet-me" answering code from any station within the PBX system.

code character—set of conventional elements established by the code to enable the transmission of a written character (letter, figure, punctuation sign, arithmetical sign, etc.) or the control of a particular function (spacing, shift, line-feed, carriage return, phase corrections, etc.).

code restriction—denies selected station lines completion of dialed outgoing exchange network calls to selected exchange and area codes, both local and distant.

codec—coder-decoder device used to convert analog signals, such as speech, music, or television, to digital form for transmission over a digital medium, and back again to the original analog form. One is required at each end of the channel.

coin box—telephone, usually public, requiring insertion of coins before it can be used.

coin-value tones—tones produced in multislot coin telephones when different coins are deposited. The tones are detected and transmitted to the operator so that the correct amount can be checked. Also called *coin-denomination tones.*

collect call—see *reverse charge call.*

collision—overlapping transmissions that interfere with one another. Occurs when two or more devices attempt to transmit at or about the same instant.

Comité Consultatif International de Téléphonie et de Télégraphie (CCITT)—International Consultative Committee for Telephone and Telegraph; an advisory committee within the International Telecommunications Union (ITU) that recommends data communications standards.

command—a signal or group of signals which causes a computer to execute an operation or series of operations.

command-driven—programs requiring that the task to be performed be described in a special language with strict adherence to syntax. Compare to menu-driven.

common battery—DC power source in the exchange that supplies power to all subscriber stations and exchange office switching equipment.

common-battery signaling—method by which supervisory and telephone address information is sent to an exchange by depressing and releasing the switch on the cradle of the handset.

common bell—capability of an individual station ringer to respond to all incoming calls on all lines terminated on that instrument.

common carrier—an organization in the business of providing regulated telephone, telegraph, telex, and data communications services.

common channel interoffice signaling (CCIS)—an electronic means of signaling between any two switching systems independent of the voice path. The use of CCIS makes possible new customer services, versatile network features, more flexible call routing, and faster call connections.

common channel signaling system number 7—a CCITT-specified signaling protocol used in high-speed digital networks to provide communication between intelligent network nodes.

common control—automatic switching arrangement in which the control equipment necessary for the establishment of connections is shared, being associated with a given call only during the period required to accomplish the control function.

common control switching arrangement (CCSA)—switching facilities connected by the telephone company to incorporate tie line networks. Switching of the leased lines in the organization's network is accomplished by common control exchange switching equipment. All stations in the network can then dial one another regardless of distance and without using exchange facilities. They can also dial outside the network via local and/or foreign exchange lines.

communications channel—see *channel.*

communications controller—dedicated computer with special processing capabilities for organizing and checking data and handling information traffic to and from many remote terminals or computers, including functions such as message switching.

communications processor—see *communications controller.*

communications satellite—earth satellite designed to act as a telecommunications radio relay and usually positioned in geosynchronous orbit 35,800 kilometers (23,000 miles) above the equator so that it appears from earth to be stationary in space.

SEPTEMBER 1989

Glossary of Telecommunications Technologies

community antenna television (CATV)—1) where television reception is poor, signals can be received at a selected site by sensitive, directional antennas and then transmitted to subscribers via a cable network. Additional channels, not normally available in that area, can also be transmitted. 2) data communications based on radio frequency (RF) transmission, generally using 75-ohm coaxial cable as the transmission medium. CATV offers multiple frequency-divided channels, allowing mixed transmissions to be carried simultaneously.

compandor—combination of a compressor at one point in a communications path for reducing the volume range of signals, followed by an expandor at another point for restoring the original volume range. Designed to improve the ratio of the signal to the interference entering in the path between the compressor and expandor.

compression—the application of any of several techniques that reduce the number of bits required to represent information in data transmission or storage, therefore conserving bandwidth and/or memory.

compressor—electronic device that compresses the volume range of a signal.

computer database—a collection of facts in machine-readable form. Computer databases vary according to type (e.g., numeric, bibliographic, full-text) and to availability (proprietary or public).

computerized branch exchange (CBX)—see *private automatic branch exchange.*

concatenation—the linking of transmission channels (phone lines, coaxial cable, optical fiber) end-to-end.

concentrator—device that connects a number of circuits that are not all used at once to a smaller group of circuits for economical transmission. A telephone concentrator achieves the reduction with a circuit-switching mechanism.

conditioning—procedure to make transmission impairments of a circuit lie within certain specified limits and typically used on telephone lines leased for data transmission to improve the possible transmission speed. Two types are used: C conditioning and D conditioning. Also called *line conditioning.*

Conference of European Posts and Telecommunications (CEPT)—an organization formed by the European PTTs for the discussion of operational and tariff matters.

connect time—the amount of time that a circuit, typically in a circuit-switched environment, is in use; see also, *holding time.*

console—the part of a computer used for communicating between the operator or maintenance engineer and the computer. A CRT terminal or a typewriter console are the most common.

console-less operation—allows internal stations to answer all incoming calls and process them to the proper stations.

consultation hold—incoming call is automatically placed on hold and the station user is given a PBX system dial tone. The station user proceeds to establish connection with another station. The original station can return to the incoming call.

consultation hold, all calls—similar to the consultation hold facility, but not restricted to incoming calls.

contact—part of a switch designed to touch a similar contact to allow current to flow and to break this union to cause a current to cease.

contention—method of line control in which the terminals request to transmit. If the channel in question is free, transmission proceeds; if it is not free, the terminal will have to wait until it becomes free. A queue of contention requests can be built up by a computer, and this queue can either be organized in a prearranged sequence or in the sequence in which requests are made.

continuity check—in common-channel signaling, a test performed to check that a path exists for speech or data transmission.

control character—a character inserted into a datastream for signaling the receiving station to perform a function or to identify the structure of the message. Newer protocols are getting away from character-oriented control procedures into bit-oriented control procedures.

control station—the network station that supervises procedures such as polling, selecting, and recovery. It also is responsible for maintaining order in the event of any abnormal situation.

controlled station-to-station restriction—allows the attendant to inhibit station-to-station calling. When activated, attempted station calls are rerouted to attendant intercept.

conversion—the process of changing from one method to another; may refer to changing processing methods, data, or systems.

Glossary of Telecommunications Technologies

COS—see *class of service*. Also, Corporation for Open Systems.

CPE—see *customer premises equipment*.

CPI—computer to PBX interface. Gateway providing 24 digital PCM channels at an aggregate rate of 1.544M bps. Developed by Northern Telecom. See also *DMI*.

cps—characters per second.

CPT—Conference of European Posts and Telecommunications. An organization formed by the European PTTs for the discussion of operational and tariff matters.

CPU—central processing unit.

CRC—see *cyclic redundancy check*.

critical path scheduling—a project planning and monitoring system used to check progress toward completion of the project by scheduling events, activities, milestones, etc.

cross-modulation—interference caused by two or more carriers in a transmission system interacting through nonlinearities in the system.

cross section—signal transmission capacity of a transmission system, usually measured in terms of the number of two-way voice channels.

crossbar switch—switch having multiple vertical and horizontal paths and an electromagnetically operated mechanical means for interconnecting any one of the vertical paths within any one of the horizontal paths.

crosspoint—1) switching element in an exchange that can be mechanical or electronic. 2) two-state semiconductor switching device having a low transmission system impedance in one state and a very high one in the other.

crosstalk—interference or an unwanted signal from one transmission circuit detected on another, usually parallel, circuit.

crosstalk attenuation—extent to which a communications system resists crosstalk.

CRR—customer service record.

CRT—cathode-ray tube.

CSMA—see *carrier sense multiple access*.

CSMA/CD—see *carrier sense multiple access with collision detection*.

CSR—Customer Service Record.

CSU—see *channel service unit*.

CTA—see *circuit terminating arrangements*.

CT2—second generation (digital) cordless telephone specification.

C/T—carrier-to-noise-temperature. Expressed in degrees Kelvin.

CUG—see *closed user groups*.

current loop—a transmission technique that recognizes current flows, rather than voltage levels. It has traditionally been used in teletypewriter networks incorporating batteries as the transmission power source.

cursor keys—keys that control the movement of the position indicator in the CRT; usually designated with arrows; see also *cursor*.

custom key set—specialized multibutton telephones designed expressly for a particular PBX. Unlike the locking buttons on normal key telephones, the buttons on a custom key set are used to communicate with the system and are typically nonlocking buttons. Custom key set buttons can be arranged to activate specific features such as speed dialing and executive override as well as to select lines.

customer access line charges (CALC)—basic rates paid for lines from customer premises to central office.

customer administration center system (CACS)—allows the customer to administer station and electronic tandem switching features as well as obtain traffic measurements and recent circuit assurance data from one or more PBX switching locations. The user operates the CACS system via an interactive terminal. The following features can optionally be provided: station rearrangement and change, facilities administration and control, traffic data to customer, and facilities assurance reports.

customer administration panel (CAP)—provides a panel that is a simplified alternative to the customer administration center system. With this panel, the customer has plug-in access to a PBX for the purpose of performing administration of station features and/or electronic tandem switching capabilities.

SEPTEMBER 1989

Glossary of Telecommunications Technologies

customer premises equipment (CPE)—terminal equipment, supplied by either the telephone common carrier or by a competitive supplier, that is connected to the telephone network.

customer service record (CSR)—detailed description of a subscriber's service and equipment charges.

cutover—physical changing of lines from one system to another, usually at the time of a new system installation.

CVSD—Continuous Variable Slope Delta Modulation. Speech encoding and digitizing technique that uses 1 bit to describe the change in the slope of the curve between 2 samples, rather than the absolute change between the samples. Sampling is usually done at 32K bps.

CXR—see *carrier*.

cyclic redundancy check (CRC)—a powerful error detection technique. Using a polynomial, a series of two 8-bit block check characters are generated that represent the entire block of data. The block check characters are incorporated into the transmission frame, then checked at the receiving end.

D

d conditioning—type of conditioning that controls harmonic distortion and signal-to-noise ratio so that they lie within specified limits.

DA—don't answer.

DAA—see *data access arrangement*.

DACS—see *digital access and cross-connect system*.

daisychaining—the connection of multiple devices in a serial fashion. An advantage of daisychaining is a savings in transmission facilities. A disadvantage is that if a device malfunctions, all of the devices daisychained behind it are disabled.

DAMA—see *demand assigned multiple access*.

DASS—a public network standard for Digital Access Signaling System, developed by the United Kingdom telecommunications industry. The present version is DASS2, a message-based signaling system based on the ISO model, developed by British Telecom and its suppliers for multiline integrated digital access to the public network.

data access arrangement (DAA)—an interface device used to allow interconnect customer-owned data transmission equipment (DTE) to the telephone network; now generally integrated into such directly attached devices. DAAs were required in the 1970s; they are no longer mandatory.

data center—common designation for the computer-equipped, central location within an organization. The center processes and converts information to a desired form such as reports or other types of management information records.

data circuit—a communications facility that allows transmission of information in digital form.

data circuit terminating equipment (DCE)—in a communications link, equipment that is either part of the network, an access point to the network, a network node, or equipment at which a network circuit terminates. In the case of an RS-232-C connection, the modem is usually regarded as DCE while the user device is DTE, or data terminal equipment; in a CCITT X.25 connection, the network access and packet-switching node is viewed as the DCE.

data compression—a technique that saves storage space by eliminating gaps, empty fields, redundancies, or unnecessary information to shorten the length of records or blocks.

data encryption standard (DES)—a cryptographic algorithm designed by the National Bureau of Standards (now the National Institute of Standards and Technology) to encipher and decipher data using a 64-bit key.

data line interface (DLI)—point at which a data line is connected to a telephone system.

data line privacy—critical system extension lines, used with such devices as facsimile machines and computer terminals, are very sensitive to extraneous noise. Data privacy prohibits activities that would insert tones on the station line while it is in use. Data lines can then be connected through the PBX without danger of losing data through interference.

data link—any serial data communications transmission path, generally between two adjacent nodes or devices and without any intermediate switching nodes.

data link control—1) procedures to ensure that both the sending and receiving devices agree on synchronization, error detection and recovery methods, and initialization and operation methods for point-to-point or multipoint configurations. 2) the second layer in the ISO Reference Model for Open Systems Interconnection.

Glossary of Telecommunications Technologies

data PBX—a switch that allows a user on an attached circuit to select from other circuits, usually one at a time and on a contention basis, for the purpose of establishing a through connection. A data PBX is distinguished from a PBX in that data transmission, and not voice, is supported.

data rate—the speed at which a channel carries data, measured in bits per second (bps).

data service unit (DSU)—a device that connects a digital terminal to a digital communications facility, converting between incompatible digital formats. DSUs can be considered as modem replacements in digital networks.

data set—infrequently used term for a modem.

data terminating equipment (DTE)—generally end-user devices, such as terminals and computers, that connect to DCE, which either generate or receive the data carried by the network. In RS-232-C connections, designation as either DTE or DCE determines signaling role in handshaking; in a CCITT X.25 interface, the device or equipment that manages the interface at the user premises. Compare with *DCE*.

data transmission—the process of transmitting information from one location to another by means of some form of communications media.

Dataphone—an AT&T trademark. 1) an AT&T modem family. 2) an AT&T digital data service offering.

datel—data transmission services offered by European PTTs, using public switched telephone networks (PSTNs).

dB—see *decibel*.

DB/DC systems—database/data communications systems.

dBm—decibels relative to one milliwatt.

dBr—relative level in decibels.

DCE—see *data circuit terminating equipment*.

DCS—see *digital cross connect system*.

DDC—see *direct department calling*.

DDCMP—see *digital data communications message protocol*.

DDD—see *direct distance dialing*.

DDI—direct dialing in. See *direct inward dialing*.

DDS—see *digital data service*.

decibel (dB)—unit for measuring relative strength of a signal parameter such as power or voltage. The number of decibels is 10 times the logarithm (base 10) of the ratio of the power of two signals, or ratio of the power of one signal to a reference level.

dedicated attendant link—assures the availability of an intercom link for the attendant announcing incoming calls to the station within the system.

dedicated circuit—end-to-end communications line used exclusively by one organization.

delay announcements—associated with Automatic Call Distribution capability, provides a recorded announcement to incoming calls that are delayed and placed in queue. The user can vary the time interval between the first and second delay announcements.

delay distortion—the change in a signal from the transmitting end to the receiving end resulting from the tendency of some frequency components within a channel to take longer to be propagated than others.

delimiter—in data communications, a character that separates and organizes elements of data.

delta modulation—method of representing a speech waveform (or other analog signal) in which successive bits represent increments of the waveform. The increment size is not necessarily constant.

deluxe queuing—allows station users, tie trunks and attendants, or attendant-assisted calls to be placed in a queue whenever all routes for completing a particular call are busy. The queue can be a Ringback Queue (RBQ), in which case a user goes on-hook and is called back when a trunk becomes available, or an Off-Hook Queue (OHQ), in which case the user remains off-hook and is connected to a trunk when it becomes available. Four types of specialized queuing arrangements are available; these are combinations of RBQ and OHQ with one or two queues.

demand assigned multiple access (DAMA)—allocation of communication satellite time to earth stations as the need arises.

demarc—demarcation point between carrier equipment and customer-premises equipment (CPE); usually a terminal block.

SEPTEMBER 1989

Glossary of Telecommunications Technologies

demodulation—the opposite of modulation; the conversion of a signal from analog to its original (e.g., digital) form.

derivation equipment—equipment used to produce narrow band facilities from a wider band facility. Commonly used on alternate voice/data (AVD) circuits to derive telegraph-grade lines from unused portions of a voice circuit.

DES—see *data encryption standard.*

destination field—a field in a message header that contains the address of the station to which a message is being directed.

diagnostics—software routines used to check equipment malfunctions and to pinpoint faulty components.

dial—device that transmits a coded signal to actuate the exchange switching equipment according to the digit dialed.

dial pulse (DP)—current interruption in the DC loop of a calling telephone produced by the breaking and making of the dial pulse contacts of a calling telephone when a digit is dialed.

dial speed—the number of pulses that a rotary dial can transfer in a given amount of time, typically 10 pulses per second.

dial tone (DT)—signal sent to an operator or subscriber indicating that the switch is ready to receive dial pulses.

dial-up—the process of, or the equipment or facilities involved in, establishing a temporary connection via the switched telephone network.

dial-up line—a circuit that is established by a switched circuit connection; generally refers to the AT&T telephone network.

DID—see *direct inward dialing.*

digital—referring to communications procedures, techniques, and equipment whereby information is encoded as either binary "1" or "0"; the representation of information in discrete binary form, discontinuous in time, as opposed to the analog representation of information in variable, but continuous, waveforms.

digital access and cross-connect system (DACS)—an AT&T term for a digital cross-connect system (DCS). See *digital cross-connect system.*

digital cross-connect system (DCS)—a computerized facility allowing DS1 lines (1.544M bps) to be remapped electronically at the DS0 level (64K bps) meaning that DS0 channels can be individually rerouted and reconfigured into different DS1 lines.

digital data service (DDS)—a digital transmission service supporting speeds up to 1.544M bps or higher.

digital loopback—technique for testing the digital processing circuitry of a communications device. It can be initiated locally, or remotely via a telecommunications circuit; the device being tested will echo back a received test message, after first decoding and then reencoding it, the results of which are compared with the original message—compare with *analog loopback.*

digital network—network incorporating both digital switching and digital transmission.

digital signal—discrete or discontinuous signal; one whose various states are discrete intervals apart.

digital speech interpolation (DSI)—when speech is digitized, it can be cut into slices such that no bits are transmitted when a person is silent. As soon as speech begins, bits flow again.

digital switching—the process of establishing and maintaining a connection, under stored program control, where binary-encoded information is routed between an input and an output port. Generally, a virtual through circuit is derived from a series of time slots (time-division multiplexing), which is more efficient than requiring dedicated circuits for the period of time that connections are set up.

DIP switch—a dual in-line switch; allows the user to set current paths on or off; they are frequently used to reconfigure microcomputer components and peripherals.

direct attendant signaling lines—station user (momentarily) activates the pushbutton to signal the attendant, and when time permits, the attendant rings the station and verbally seeks instructions.

direct department calling (DDC)—incoming calls on a specific trunk or group of trunks are routed to specific stations or groups of stations.

direct distance dialing (DDD)—telephone exchange service that enables the telephone user to call long distances without operator assistance.

direct-in lines—allow direct termination of separate exchange lines to station instruments, bypassing the attendant console.

Glossary of Telecommunications Technologies

direct inward dialing (DID)—incoming calls from the exchange network can be completed to specific station lines without attendant assistance. Also called *direct dialing in (DDI)*.

direct inward system access (DISA)—feature that allows an outside caller the ability to dial directly into a PBX system, without attendant intervention, and gain complete access to PBX system facilities and outgoing trunk circuits and tie line circuits.

direct outward dialing (DOD)—allows a PBX, Centrex, or hybrid system user to gain access to the exchange network without the assistance of the attendant.

direct station selection (DSS)—ability to call any station within a PBX by means of a single pushbutton associated with that station number. A DSS panel is usually part of an attendant console.

direct trunk group selection—allows the attendant to access an outgoing trunk from a particular group by pushing a single button associated with the group rather than dialing an access code.

directory number—the full complement of digits (or letters and figures) associated with the name of a subscriber in the directory.

DISA—see *direct inward system access.*

disconnect signal—see *clear-forward clear-back signal.*

discriminating ringing—different types of station ringing are provided to give audible indication of internal or external incoming calls and other special calls.

distortion—the modification of the waveform or shape of a signal caused by outside interference or by imperfections of the transmission system. Most forms of distortion are the result of the varying responses of the transmission system to the different frequency components of the transmission signal.

distribution frame—structure (typically wall-mounted) for terminating telephone wiring, usually the permanent wires from, or at, the telephone exchange, where cross-connections are readily made to extensions. Also called *distribution block.*

diversity—provision of more than one communications channel to improve the reliability of a service.

divestiture—the breakup of AT&T in the U.S., mandated in 1984 by the federal courts based on an anti-trust accord reached between AT&T and the U.S. Department of Justice. The decision included the sep-aration of 22 AT&T-owned local Bell Operating Companies (BOCs) into seven independent Regional Bell Holding Companies (RBHCs).

DMI—Digital Multiplexed Interface. Gateway providing 23 digital PCM channels + 1 signaling channel at an aggregate rate of 1.544M bps. Developed by AT&T. See also *CPI.*

DOD—see *direct outward dialing.*

downlink—the portion of a satellite circuit extending from the satellite to the earth. Compare to *uplink.*

download—the process of loading software into the nodes of a network from one node over the network media.

downtime—the total time a system is out of service due to equipment failure.

DP—1) dial pulse (signaling). 2) distribution point.

DPC—data processing center.

DPNSS—a digital private network signaling system standard developed by British Telecom and other United Kingdom PBX suppliers.

drop—a connection point between a communicating device and a communications network.

drop, subscriber's—line from a telephone cable to a subscriber's building.

DS-C1—digital signal level 1C; telephony term describing a 3.152M bps digital signal carried on a T1 C facility.

DS0—digital signal level 0; telephony term for a 64K bps standard digital telecommunications signal or channel.

DS1—digital signal level 1, a digital transmission format in which 24 voice channels are multiplexed into one 1.544M bps (U.S.) or 2.108M bps (Europe) T1 digital channel.

DS2—digital signal level 2; telephony term describing the 6.312M bps digital signal carried on a T2 facility.

DS3—digital signal level 3; telephony term describing the 44M bps digital signal carried on a T3 facility.

DS4—digital signal level 4; telephony term describing the 273M bps digital signal carried on a T4 facility.

DSI—see *digital speech interpolation.*

Glossary of Telecommunications Technologies

DSS—see *direct station selection.*

DSU—see *data service unit.*

DT—see *dial tone.*

DTE—see *data terminal equipment.*

DTMF—see *dual tone multifrequency signaling.*

dual tone multifrequency signaling (DTMF)—basis for operation of pushbutton telephone sets. A method of signaling in which a matrix combination of two frequencies, each from a group of four, is used to transmit numerical address information. The two groups of four frequencies are 697 Hz, 770 Hz, 852 Hz, and 941 Hz, and 1209 Hz, 1336 Hz, 1477 Hz, and 1633 Hz.

duplex circuit—circuit used for transmission in both directions at the same time. It can be called "full-duplex" to distinguish it from "half-duplex." See *full-duplex and half-duplex.*

duplex signaling (DX)—signaling system that occupies the same cable pair as the voice path yet does not require filters.

duplex transmission—simultaneous, two-way, independent transmission. Also called *full-duplex transmission.*

E

E&M signaling—signaling arrangement that uses separate paths for signaling and voice signals. The M lead (derived from "mouth") transmits ground or battery to the distant end of the circuit, while incoming signals are received as either a grounded or open condition on the E (derived from "ear") lead. E&M is the traditional, inband, send and receive signaling method used by the North American Public Switched Telephone network. Some PBXs use E&M signaling, also. Most telcos are replacing E&M with computer-controlled CCIS (out-of-band) signaling, which is required for ISDN implementation.

EAROM—electrically alterable read-only memory.

earth station—ground-based equipment used to communicate via satellites. See also *ground station.*

EAS—see *extended area service.*

EAX—electronic automatic exchange.

EBCDIC (extended binary coded decimal interchange code)—a coded character set comprising 8-bit coded characters, developed by IBM and a de facto industry standard.

echo—wave that has been reflected or otherwise returned with sufficient magnitude and delay for it to be perceived as a wave distinct from that directly transmitted.

echo canceller—a device used to reduce or eliminate echo. It operates by placing a signal that is equal and opposite to the echo signal on the return transmission path. This feature is necessary for voice transmission but often interferes with data transmission.

echo return loss (ERL)—attenuation of echo currents in one direction caused by telephone circuits operating in the other direction.

ECL—emitter-coupled logic. Type of ALU technology.

ECMA—European Computer Manufacturers Association.

ECS—European fixed-service satellite system.

EDC—see *error detecting code* and *error correcting code.*

EDI—see *electronic data interchange.*

EEPROM—electrically erasable programmable read-only memory. A memory that can be electronically programmed and erased, but which does not require a power source to retain data.

EIA—Electronic Industries Association.

EIA interface—a standardized set of signal characteristics (time duration, voltage, and current) specified by the Electronic Industries Association.

EIRP—effective isotropic radiated power. The combination of transmitted power and antenna gain.

EKTS—electronic key telephone system.

electromagnetic interference (EMI)—the energy given off by electronic circuits and picked up by other circuits; based on type of device and operating frequency, EMI can be reduced by shielding; minimum acceptable levels are detailed by the FCC.

electromagnetic spectrum—entire range of wavelengths or frequencies of electromagnetic radiation extending

Glossary of Telecommunications Technologies

from gamma rays to the longest radio wave and including visible light. See also *infrared* and *radio frequency*.

electromechanical ringing—bell or buzzer provided in the station instrument to give audible incoming call indication.

electronic automatic exchange (EAX)—term used by the General Telephone Company for electronic exchange equipment.

electronic data interchange (EDI)—paperless electronic exchange of trading documents, such as invoices and orders.

electronic mail (E-mail)—the transmission of letters, messages, and memos from one computer to another without using printed copies.

electronic switching system (ESS)—system using computer-like operations to switch telephone calls.

electronic tandem networking—operation of two or more switching systems in parallel.

elevation—the angle between the horizon and an antenna's beam.

emergency access—emergency alarm system integrated into the PBX switching system that can be implemented by uniquely ringing all station instruments to indicate an emergency condition.

EMI—see *electromagnetic interference*.

emulate—to imitate one system with another, so that the imitating system accepts the same data, executes the same computer programs, and achieves the same results as the imitated system.

EN—equipment number.

encryption—the process of systematically encoding a bit stream before transmission so that an unauthorized party cannot decipher it.

end office—class 5 central office, at which subscriber loops terminate.

end-to-end test—as used by the Bell System, a method of exercising a communications link; it requires Bell maintenance personnel at each end of the circuit.

energy communications—provides the ability for the PBX to communicate with energy-consuming devices throughout a hotel or business complex. Audio signals are sent over the telephone wiring to power unit consumption devices in the building. This feature is designed to use existing telephone wiring wherever possible. A separate interface unit is required for each power unit. Specific operating modes include:

* Energy consumption and demand monitoring,
* Guestroom cycling function,
* Individual load cycling function,
* Peak demand shedding function, and
* Vacant room energy function.

engaged tone—see *busy signal*.

enhanced private switched communications service (EPSCS)—AT&T Communications service providing a private interLATA electronic tandem tie-line network linking several of a customer's locations together.

ENQ (Enquiry)—in data communications, a request for a response from another terminal; it is used to obtain identification and/or an indication of the other station's status.

envelope delay—an analog line impairment that conditioning is meant to overcome. A variation of signal delay with frequency across the data channel bandwidth.

EPABX—electronic private automatic branch exchange. See *exchange, private automatic branch*.

EPROM—erasable and programmable read-only memory.

EPSCS—see *enhanced private switched communication service*.

equal access—Department of Justice ruling (effective 9/84) that requires all RBHCs with ESS systems using SPC technology, and serving a market of at least 10,000 access lines, to offer the same quality of connection at the same rates to all common carriers. Under this arrangement, subscribers choose a primary long-distance carrier that they access by dialing "1" + area code + telephone number. Other long-distance carriers are reached by dialing "10" plus a three-digit access code unique to each carrier, then dialing "1" + area code + telephone number.

equalization—the introduction of components to an analog circuit by a modem to compensate for signal attenuation and delay distortion. Generally, the higher the transmission rate, the greater the need for equalization.

ergonomics—a discipline that promotes the consideration of human factors in the design of the working environment and its components (heat, light, sound, equipment, etc.).

ERL—see *echo return loss*.

SEPTEMBER 1989

Glossary of Telecommunications Technologies

Erlang (E)—unit of traffic intensity. One erlang is the intensity at which one traffic path would be continuously occupied, e.g., 1 call-hour per hour, 1 call-minute per minute, etc., generally referred to as 36 hundred call seconds (ccs).

Erlang B—traffic engineering formula used when traffic is random and there is no queuing. Erlang B assumes that blocked callers either automatically use another route, or blocked calls disappear entirely.

Erlang C—traffic engineering formula used when traffic is random and queuing is provided. Erlang C assumes that all callers will wait indefinitely until a line becomes available.

ERMES—program sponsored by the Commission of the European Communities for a pan-European radio(-paging) message system.

error—in data communications, any unwanted change in the original contents of a transmission.

error burst—a concentration of errors within a short period of time as compared with the average incidence of errors. Retransmission is the normal correction procedure in the event of an error burst.

error control—a process of handling errors, which includes the detection and correction of errors.

error correcting code—code incorporating sufficient additional signaling elements to enable the nature of some or all of the errors to be indicated and corrected entirely at the receiving end.

error detecting code—code in which each telegraph or data signal conforms to specific rules of construction so that departures from this construction in the received signals can be automatically detected.

error rate—the ratio of the amount of data incorrectly received to the total amount of data transmitted.

ESPRIT—European Strategic Program for Research and Development in Information Technologies.

ESS—see *electronic switching system*.

ETACS— extended TACS. See *TACS*.

Ethernet—a local area data network, developed by Xerox Corporation and supported by Intel Corp., Digital Equipment Corp., and Hewlett-Packard.

ETN—see *electronic tandem networking*.

ETSI—European Telecommunications Standards Institute.

ETX—end-of-text (indicator).

even parity check (odd parity check)—tests whether the number of digits in a group of binary digits is even (even parity check) or odd (odd parity check).

EUTELSAT—European Telecommunications Satellite Organization. Operates the Eutelsat services of satellites.

exchange—assembly of equipment in a communications system that controls the connection of incoming and outgoing lines and includes the necessary signaling and supervisory functions. Different exchanges, or switches, can be co-sited to perform different functions, e.g., local exchange, trunk exchange, etc. See *class of exchange*. Also known as *central office*.

exchange area—a calling area served by a local telco central office switch, may comprise one or more NXX codes.

exchange line socket—2- or 4-wire socket providing physical access to the PSTN.

exchange, private (PX)—exchange serving a particular organization and having no means of connection with a public exchange.

exchange, private automatic (PAX)—dial telephone exchange that provides private telephone service to an organization and that does not allow calls to be transmitted to or from the public telephone network.

exchange, private automatic branch (PABX)—private automatic telephone exchange that provides for the transmission of calls internally and to and from the public telephone network.

exchange, private branch (PBX)—see *private automatic branch exchange (PABX)*.

exchange, private digital—see *PDX*.

exciter—the transmitter components that modulate and drive a signal. Used in uplinks.

exclusive hold—only the station that has placed a line circuit on hold is capable of breaking the hold condition and reestablishing conversation.

executive busy override—allows preselected stations to dial a single digit and gain access to the conversation taking place upon encountering a busy signal.

Glossary of Telecommunications Technologies

expandor—transducer that, for a given amplitude range of input voltages, produces a larger range of output voltages.

extended area service (EAS)—option whereby the telephone subscriber can pay a higher flat rate in order to obtain wider geographical coverage without additional per-call charges.

extended binary-coded decimal interchange code (EBCDIC)—a coded character set consisting of 8-bit coded characters. EBCDIC is the usual code generated by synchronous IBM devices.

F

facilities administration and control—allows customer to administer the assignment of parameters that determine user calling privileges, such as restriction levels and authorization codes. Manual control (override) of time of day routing is provided. Activation and deactivation of trunk group queues are also provided.

facilities assurance reports—allows customer to obtain an audit trail of the data generated by the automatic circuit assurance feature. The audit trail indicates the identity of the trunk circuit, time of referral, date, nature of the referral (e.g., short-holding-time failure or long-holding-time failure), and whether a test was performed in response to the referral.

facility—1) any or all of the physical elements of a plan used to provide communications services. 2) a component of an operating system. 3) transmission path between two or more points, provided by a common carrier.

facsimile—system for the transmission of images, usually over the public telephone network. The image is scanned at the transmitter, reconstructed at the receiving station, and duplicated on some form of paper.

fading—a phenomenon, generally of microwave or radio transmission, where atmospheric, electromagnetic, or gravitational influences cause a signal to be deflected or diverted away from the target receiver. The reduction in intensity of the power of a received signal.

fail softly—when a piece of equipment fails, the programs let the system fall back to a degraded mode of operation rather than let it fail completely.

far-end crosstalk—crosstalk that travels along a circuit in the same direction as the signals in that circuit. Compare with *near-end crosstalk*.

fault—a condition that causes any physical component of a system to fail to perform in acceptable fashion.

fault tolerance—the ability of a program or system to operate properly even if a failure occurs.

fax—see *facsimile*.

FCC—see *Federal Communication Commission*.

FDDI—see *fiber distributed data interface*.

FDM—see *frequency-division multiplexing*.

FDMA—see *frequency-division multiple access*.

FD or FDX—see *full-duplex*.

feature group—under the provisions of Equal Access, three types of local-line access must presently be provided by the BOCs to long-distance carriers:

—Feature Group A access, which amounts to dialing the carrier's 7-digit access number, as well as an account code and a 10-digit long-distance number (you could use an automatic dialer for this purpose).

—Feature Group B access, which is a trunk-side connection accessed by dialing 950-XXXX plus a 10-digit long-distance number. Feature Group B provides better quality but is more expensive.

—Feature Group D access, which is similar to FGB, permits 1 + 10-digit dialing of the primary long-distance carrier and can be considered the most complete form of access.

Federal Communications Commission (FCC)—Washington, DC regulatory agency established by the Communications Act of 1934, charged with regulating all electrical and radio communications in the U.S.

federal telecommunications system (FTS)—a government communications system administered by GSA; it covers 50 states, plus Puerto Rico and the Virgin Islands, and provides services for voice, teletypewriter, facsimile, and data transmission.

feedback—the return of part of the output of a machine, process, or system, to the computer as input from another phase; also refers to system messages that keep a user informed of system activities during a process.

feed horn—a small antenna at the focal point of a reflector antenna that gathers the signal.

SEPTEMBER 1989

Glossary of Telecommunications Technologies

FEP—front-end processor.

fiber distributed data interface (FDDI)—an ANSI standard specifying a packet switched LAN-to-LAN backbone for transporting data at high throughput rates over a variety of multimode fibers.

fiber optic waveguides—thin filaments of glass through which a light beam can be transmitted for long distances by means of multiple internal reflections. Occasionally other transparent materials, such as plastic, are used.

fiber optics—a technology that uses light as a digital information carrier. Fiber optic cables (light guides) are a direct replacement for conventional coaxial cables and wire pairs. The glass-based transmission facilities occupy far less physical volume for an equivalent transmission capacity, which is a major advantage in crowded underground ducts. The fibers are immune to electrical interference. Also called *lightwave communications*.

file server—in local networks, a station dedicated to providing file and mass data storage services to the other stations on the network.

filter—circuit designed to transmit signals of frequencies within one or more frequency bands and to attenuate signals of other frequencies.

firmware—permanent or semipermanent control coding implemented at a micro-instruction level for an application program, instruction set, operating routine, or similar user-oriented function.

fixed night service—routes incoming exchange calls to preselected stations within a PBX system when the attendant is not on duty.

flat rate—fixed payment for service within a defined area, independent of use, with an additional charge for each call outside the area.

flexible intercept—allows completely flexible and instantly changeable intercept conditions to the attendant, such as unassigned number, temporary disconnect, etc.

flexible line ringing—indicates the ability to arrange each station within the system with complete flexibility in regard to the ringing on incoming outside calls.

flexible night selection—allows the attendant to "set up" night connections in accordance with day-to-day requirements.

flexible numbering of stations—station numbers can be assigned to lines at installation based on customer specifications and can easily be changed thereafter. Some PBXs allow the user to perform the modifications while others require the attention of a service person. This feature is frequently used in hotels where the extension number is the same as the room number.

flexible release—ability of the switching system to effect connection release either by the calling party or by the first party to hang up.

flexible route selection (FRS)—used with Centrex service to achieve least-cost routing.

flow control—the use of buffering and other mechanisms, such as controls that turn a device on and off, to prevent data loss during transmission.

FM—see *frequency modulation.*

foreign exchange (FX)—connects a customer's location to a remote exchange. This service provides the equivalent of local service from the distant exchange.

forward channel—the communications path carrying voice or data from the call initiator to the called party.

FOTS—fiber optic transmission systems. See *fiber optic waveguides.*

four-wire channel—a circuit containing two pairs of wire (or their logical equivalent) for simultaneous (i.e., full-duplex) two-way transmission. Contrast with *two-wire channel.*

four-wire circuit—a circuit in which two pairs of conductors, one for the inbound channel and one for the outbound channel, are connected to the station equipment. Contrast with *two-wire circuit.*

frame—1) in high-level data-link control (HDLC), the sequence of contiguous bits bracketed by and including opening and closing flag (01111110) sequences. 2) in data transmission, the sequence of contiguous bits bracketed by and including beginning and ending flag sequences. 3) a set of consecutive digit time slots in which the position of each time slot can be identified by reference to a frame alignment signal. 4) in a time-division multiplexing (TDM) system, a repetitive group of signals resulting from a signal sampling of all channels, including any additional signals for synchronizing and other required system information.

frame-grabber—device that can seize and record a single frame of video information out of a sequence of many frames.

Glossary of Telecommunications Technologies

framing—a control procedure used with multiplexed digital channels, such as T1 carriers, where bits are inserted so that the receiver can identify the time slots that are allocated to each subchannel; framing bits may also carry alarm signals indicating specific alarm conditions.

frequency—an expression of how frequently a periodic (repetitious) wave form or signal regenerates itself at a given amplitude. It can be expressed in hertz (Hz), kilohertz (KHz), megahertz (MHz), etc.

frequency coordination—international procedure to prevent interference between new and existing radio communications services. Under International Telecommunication Union (ITU) Radio Regulations, potential operators must consult countries and administrations whose services might be affected.

frequency-division multiple access (FDMA)—communicating devices at different locations sharing a multipoint or broadcast channel by means of a technique that allocates different frequencies to different users.

frequency-division multiplexing (FDM)—a technique of dividing a single communications line into several data paths of different frequencies, each supporting an independent information stream. Contrast with *time-division multiplexing.*

frequency modulation (FM)—one of three ways of modifying a sine wave signal to make it carry information. The sine wave or "carrier" has its frequency modified in accordance with the information to be transmitted. The frequency function of the modulated wave may be continuous or discontinuous.

frequency shift keying (FSK)—a method of modulation that uses two different frequencies to distinguish between a mark (digital 1) and a space (digital 0) when transmitting on an analog line. Used in modems operating at 1200 bps or slower.

front-end processor (FEP)—a dedicated communications system that intercepts and handles activity for the host. It may perform line control, message handling, code conversion, error control, and such applications functions as control and operation of special-purpose terminals; see also *communications controllers.*

FRS—see *flexible route selection.*

FSK—see *frequency shift keying.*

full-duplex (FD, FDX)—refers to a communications system or equipment capable of transmission simultaneously in both directions. Contrast with *half-duplex.*

fully restricted stations—feature that denies selected station lines the ability to place or receive any but internal station-to-station calls.

functional split—division within an automatic call distributor (ACD) that allows incoming calls to be directed from a specific group of trunks to a specific group of agents.

functional test—test carried out under normal working conditions to verify that a circuit or a particular part of the equipment functions correctly.

FX—foreign exchange.

G

gain—denotes an increase in signal power in transmission from one point to another, usually expressed in dB.

GAP—an ad hoc working group, Groupe d'Analysis et Prevision.

GaAs FETs—Gallium Arsenide Field Effect Transistors—a form of generating low-power but reliable microwave energy up to 50 Hz.

gate assignments—used in context of ACD equipment. Gates are made up of trunks that require similar agent processing. Individual agents can be reassigned from one gate to another gate by the customer via the supervisory control and display station. Also called *splits.*

gateway—a conceptual or logical network station that serves to interconnect two otherwise incompatible networks, network nodes, subnetworks, or devices. Gateways perform a protocol-conversion operation across a wide spectrum of communications functions or layers.

GHz—gigahertz (one billion cycles per second).

gigabyte—one billion bytes.

gigahertz (GHz)—a frequency unit equal to one billion cycles per second.

GOSIP—U.S. and U.K. government OSI procurement specification.

SEPTEMBER 1989

Glossary of Telecommunications Technologies

grade of service—quality of telephone service provided by a system described in terms of the probability that a call will encounter a busy signal during the busiest hour of the day.

grd—ground.

ground circuit—1) circuit in which energy is carried one way over a metallic path and returned through the earth. 2) circuit connected to earth at one or more points.

ground start—telephony term describing a signaling method whereby one station detects that a circuit is grounded at the other end.

ground station—assemblage of communications equipment, including signal generator, transmitter, receiver, and antenna, that receives (and usually transmits) signals to/ from a communications satellite. Also called *earth station*.

group band modem—modem designed for CCITT V.35, V.36, or V.37 standard for transmission over group band (60-108 KHz) circuits.

group call—special type of station hunting requiring a special access number that will allow a call to the special access number and ring the first available number in that group.

GAPS—made up of industrial, Member State, and Commission representatives in the European Community for the study and agreement of common approaches to various communication issues. Reports by the group have covered ISDN, broadband technologies and networks, mobile communications, and Open Network Architecture (ONA).

grps—groups.

Groupe Speciale Mobile (GSM)—formed by the CEPT to define a pan-European digital cellular mobile standard.

H

half-duplex (HD or HDX)—a circuit designed for transmission alternately in either direction but not both directions simultaneously. Contrast with *full-duplex*.

handoff—transfer of duplex signaling as a mobile terminal passes to an adjacent cell in a cellular radio network.

hands-free station—capability of the station user to have speakerphone operation on all calls.

handshaking—exchange of predetermined signals for control when a connection is established between two modems or other devices.

hard wired—1) referring to a communications link, whether remote phone line or local cable, that permanently connects two nodes, stations, or devices. 2) descriptive of electronic circuitry that performs fixed logical operations by virtue of unalterable circuit layout, rather than under computer or stored-program control.

harmonic distortion—a waveform distortion, usually caused by the nonlinear frequency response of a transmission.

HD or HDX—see *half-duplex*.

HDLC—see *high-level data-link control*. Bit-oriented communication protocol developed by the ISO.

header—the initial portion of a message, which contains any information and control codes that are not part of the text (e.g., routine, priority, message type, destination addressee, and time of origination).

header record—data containing common, constant, or identifying information for a group of records that follow.

headset—operator or attendant telephone set that consists of a telephone transmitter, a receiver, and cord and plug, designed to leave the operator's hands free.

headset/recorder jack—allows a headset or input plug from a recorder to be connected to the talk circuit on the station instrument.

hertz (Hz)—unit of electromagnetic frequency equal to one cycle per second.

heterogeneous (computer) network—a system of different host computers, such as those of different manufacturers. Compare with *homogeneous network*.

heuristic—exploratory methods of problem solving in which solutions are arrived at by an interactive, self-learning method.

HF—high frequency.

hierarchical (computer) network—a system in which processing and control functions are performed at several levels by computers specially suited in capability for the functions performed.

Glossary of Telecommunications Technologies

high frequency (HF)—portion of the electromagnetic spectrum, typically used in short-wave radio applications; frequencies approximately in the 3 MHz to 30 MHz range.

high-level data-link control (HDLC)—a bit-oriented protocol developed by the ISO.

high-usage (HU) route—direct trunks provided, where traffic volume warrants, to bypass a part of the DDD switching network. Also called *high-usage trunk.*

histogram—a graph of contiguous vertical bars representing a frequency distribution in which the groups or classes of items are marked at equal intervals in ascending order on the x axis, and the number of items in each class is indicated by a horizontal line segment drawn above the x axis at a height equal to the number of items in the class.

hold—feature whereby a station user can remain connected to a line while not off-hook to the line. PBXs have provided a new range of hold features that are more easily implemented by means of their programmed intelligence than through mechanical arrangements.

holding time—length of time a communications channel is in use for each transmission. Includes both message time and operating time. Also called *connect time.*

homogeneous (computer) network—a system of similar computers such as those of one model by the same manufacturer; contrast with Heterogeneous Network.

hookswitch—see *switchook.*

hoot and holler circuit—four-wire, private line circuit using manual or automatic voice signaling or ringdown to transmit voice information among dispersed sites, such as branch offices. Used primarily in financial institutions to provide timely communication of investment information among traders. Derived from the lively conversational mode of the investment trading environment.

host computer—the primary or controlling system in a multiple computer network operation.

host interface—the link between a communications processor or network and a host computer.

hot line—line serving two telephone sets exclusively, on which one set will ring immediately when the receiver of the other set is lifted.

hot standby—alternate equipment ready to take over an operation quickly if the equipment on which it is being performed fails.

house phone—allows certain stations to reach the attendant or another station by merely going off-hook.

howler tone—tone used to alert a subscriber when the handset is off-hook.

hum—spurious electrical interference, usually picked up from a conventional alternating current power supply.

hunting—movement of a call as it progresses through a group of lines. Typically, the call will try to be connected on the first line of the group; if that line is busy it will try the second line, and then the third, etc.

hybrid circuit—circuit having four sets of terminals arranged in two pairs designed so that there is high loss between the two sets of terminals of a pair but low loss between the terminals of a pair and the terminals of the other pair are suitably terminated. Commonly used to couple four-wire circuits to two-wire circuits.

hybrid network—a network composed of both public and private facilities.

hyper-density—1-inch magnetic tape with a density of 3,200 characters per inch.

Hz—see *hertz.*

I

I/O—see *input/output.*

I/O bound—refers to programs with a large number of I/O (input/output) operations, which slow the CPU.

IBS—see *international business service.*

ICA—International Communications Association.

ICC—International Control Center.

ICI—see *incoming call identification.*

ICPT—intercept tone. See *vacant number intercept.*

IDD—see *international direct dialing.*

identified outward dialing (IOD)—see *automatic identification of outward dialing (AIOD).*

IDF—see *intermediate distributing frame.*

SEPTEMBER 1989

Glossary of Telecommunications Technologies

IDN—see *integrated digital network.*

IDR—see *international digital route.*

IEC—1) interexchange carrier. 2) International Electrotechnical Commission. Affiliated to the ISO and responsible for international standards in the electrotechnical field. Some work relates to telecommunications, particularly in the area of wires, cables, waveguides, and CATV systems, but concentrating on stadards for materials, components, and methods of measurement.

IFRB—see *International Frequency Registration Board.*

IMTS—see *international message telephone service.*

IMV/VS—see *information management system/ virtual storage.*

INC—incoming.

induction coil—apparatus for obtaining intermittent high voltage consisting of a primary coil through which the direct current flows, an interrupter, and a secondary coil of a larger number of turns in which the high voltage is induced.

information feedback—see *message feedback.*

information management system/virtual storage (IMS/ VS)—a common IBM host operating environment, usually under the MVS operating system, oriented toward batch processing and telecommunications-based transaction processing.

information path—the functional route by which information is transferred.

information systems network—a network of multiple operational-level systems and one management-oriented system (centered around planning, control, and measurement processes). The network retrieves data from databases and synthesizes that data into meaningful information to support the organization.

infrared—pertaining to the frequency range in the electromagnetic spectrum that is higher than radio frequencies but below the range of visible light. See also *electromagnetic spectrum.*

input/output (I/O) channel—a component in a computer system, controlled by the central processing unit, that handles the transfer of data between main storage and peripheral equipment.

insertion loss—difference in the amount of power received before and after a device is inserted into a circuit or a call is connected.

Institute of Electrical and Electronics Engineers (IEEE)—a group involved in recommending standards for the computer and communications field.

integrated adapter—provides for the direct connection of an external device and uses neither a control unit nor the standard I/O interface.

integrated circuit—a combination of the interconnected circuit elements inseparably associated on or within a continuous substrate; see also *substrate.*

integrated digital network (IDN)—network employing both digital switches and digital transmission.

integrated services digital network (ISDN)—project underway within the CCITT for the standardization of operating parameters and interfaces for a network that will allow a variety of mixed digital transmission services to be accommodated. Access channels under definition include a basic rate defined by CCITT 2B + D (64K + 64K + 16K bps, or 144K bps) and a primary rate that is DS1 (1.544M bps in the U.S., Japan, and Canada, and 2.048M bps in Europe).

integrated voice/data terminal (IVDT)—devices that feature a keyboard/display and a voice telephone instrument; many contain varying degrees of local processing power, ranging from full PC capacity to directory storage for automatic telephone dialing. They may be designed to work with a specific customer-premises PBX or be PBX-independent.

Intelsat—International Telecommunications Satellite Consortium, formed in 1964 with the purpose of creating a worldwide communications satellite system.

inter-PBX call transfer—part of a main/satellite configuration. An incoming exchange call to a main PBX or a satellite PBX can be put in a three-way conference mode. In addition, an incoming exchange call to a main PBX can be transferred over a tie trunk to a satellite PBX station and vice versa.

inter-PBX coordinated station numbering—component of main/satellite configurations. Stations at the main and satellite can dial each other without intervening dial tone. The dialing plan for an inter-PBX call is the same as for an intra-PBX (station-to-station) call.

interchange point—a location where interface signals are transmitted.

Glossary of Telecommunications Technologies

intercom—internal communications system that allows calling generally within the same building, but not outside the system. Key systems are frequently provided with intercom lines that allow quick communication between stations on the key system. The PBX has features that can replace or enhance the familiar key system intercom functions on custom key sets.

interconnect company—organization, other than the serving telephone company, that supplies telephone equipment by sale, rental, or leasing.

interexchange channel (IXT)—direct circuit between exchanges.

interface—boundary between two pieces of equipment across which all the signals that pass are carefully defined. The definition includes the connector signal levels, impedance, timing, sequence of operation, and the meaning of signals.

interface EIA standard RS-232C—a standardized method adopted by the Electronic Industries Association to insure physical and signal uniformity of interface between data communication equipment and data processing terminal equipment.

interLATA—between different Local Access and Transport Areas.

intermediate distributing frame (IDF)—frame having distributing blocks on both sides, permitting connection of any telephone number with any line circuit.

international business service—a satellite-based service at up to 8M bps. Services include data, facsimile, digital voice, and video- and audioconferencing.

international digital route—a proposed digital service intended to replace the current analog system. It is designed to allow access to the public switched network at rates up to 45M bps.

international direct dialing (IDD)—cooperative service enabling subscribers to place international calls without operator assistance.

International Frequency Registration Board (IFRB)—within the International Telecommunication Union (ITU), the IFRB is responsible for the maintenance of an international list of radio frequency usage and the allocation of new frequencies.

international message telephone service—a public switched voice service.

international number—digits that have to be dialed after the international prefix to call a subscriber in another country; i.e., the country code followed by the subscriber's national number.

International Organization for Standardization (ISO)—an organization established to promote the development of standards to facilitate the international exchange of goods and services, and to develop mutual cooperation in areas of intellectual, scientific, technological, and economic activity.

international record carrier (IRC)—term for group of common carriers that until recently provided data and text service between certain U.S. gateway cities and locations abroad.

international switching center (ISC)—exchange used to switch traffic between different countries over international circuits.

international subscriber dialing (ISD)—see *international direct dialing.*

International Telecommunication Union (ITU)—telecommunications agency of the United Nations, established to provide standardized communications procedures and practices including frequency allocation and radio regulations on a worldwide basis. Parent group of the CCITT.

interoffice trunk—direct trunk between local central offices (Class 5 offices), or between Class 2, 3, or 4 offices; also called *intertoll trunk.*

interposition calling—one attendant in a multiposition system can call an attendant at another position for consultation.

interposition transfer—an operator at one console can transfer a call to an operator at another position. Used where certain positions are assigned to handle specific types of calls.

intraLATA—within a single Local Access and Transport Area.

intraoffice trunk—trunk connection within the same central office.

INTUG—see *International Telecommunications Users Group.*

inverter—used to convert a direct current into a higher voltage direct current or an alternating current.

inward restriction—blocks selected extension lines from receiving incoming exchange network calls and CCSA calls, completed via either DID or the attendant. Calls can be given any intercept treatment.

SEPTEMBER 1989

Glossary of Telecommunications Technologies

IOD—identified outward dialing (may use operator).

IPM—impulses per minute (interruption rate for call progress tones).

IPSS—international extension of British Telecom's PSS.

IRC—see *international record carrier*.

ISC—see *international switching center*.

ISD—international subscriber dialing. See *international direct dialing*.

ISDN—see *integrated services digital network*.

ISO—see *International Organization for Standardization*.

ISPABX—integrated services PABX, compatible with ISDN standards.

ITU—see *International Telecommunication Union*.

IVDT—see *integrated voice/data terminal*.

IXC—interexchange carrier.

IXSD—international telex subscriber dialing.

IXT—see *interexchange channel*.

J

jack—a device used generally for terminating the permanent wiring of a circuit, access to which is obtained by the insertion of a plug.

jitter—slight movement of a transmission signal in time or phase that can introduce errors and loss of synchronization for high-speed synchronous communications. See also *phase jitter*.

Josephson junction—a type of circuit capable of switching at very high speeds when operated at a temperature approaching absolute zero.

jumper—patch cable or wire used to establish a circuit, often temporarily, for testing or diagnostics.

junctor—in crossbar systems, a circuit extending between frames of a switching unit and terminating in a switching device on each frame.

K

K bps—kilobits per second.

Ka-Band—portion of the electromagnetic spectrum allotted for satellite transmission; frequencies are approximately in the 20 to 30GHz range.

key illumination: incandescent lamp behind key—circuit status lamp indication behind circuit pushbutton with flashing incoming, steady busy, and "wink" hold visual indications.

key pulsing (KP)—manual method of sending numerical and other signals by the operation of nonlocking pushkeys. Also called *key sending*.

key service unit (KSU)—central operating unit of a key telephone system.

keyboard perforator—perforator provided with a bank of keys, the manual depression of any one of which will cause the code of the corresponding character or function to be punched in a tape.

KHz—see *kilohertz*.

kilobyte—1,024 bytes, or 2^{10} bytes.

kilohertz (KHz)—one thousand cycles per second.

KSU—key service unit.

KTU—key telephone unit.

ku band—portion of the electromagnetic spectrum being used increasingly for satellite communications. Frequencies approximately in the 12GHz to 14GHz range.

KW—kilowatt.

KWH—kilowatt hour.

L

L-Band—portion of the electromagnetic spectrum commonly used in satellite and microwave applications, with frequencies approximately in the 1GHz region.

LAMA—see *local automatic message accounting*.

LAN—see *local area network*.

laser—acronym for "light amplification by stimulated emission of radiation." Lasers convert electrical en-

Glossary of Telecommunications Technologies

ergy into radiant energy in the visible or infrared parts of the spectrum, emitting light with a small spectral bandwidth. For this reason, they are widely used in fiber optic communications, particularly as sources for long-haul links.

last trunk busy (LTB)—condition in which the last-choice trunk of a given group is busy.

LATA—see *local access and transport area.*

layer—in the OSI reference model, referring to a collection of related network-processing functions that comprise one level of a hierarchy of functions.

LCD—see *liquid crystal display.*

LCR—see *least cost routing, automatic route selection.*

LCU—see *line control unit.*

LD—see *long-distance.*

LDN—see *listed directory number.*

leaky PBX—a PBX that allows incoming private-line (tie line) traffic to access the local exchange.

leased line—a communication channel contracted for exclusive use from a common carrier, frequently referred to as a private line.

least cost routing (LCR)—see *automatic route selection.*

least squares method—a method of smoothing or curve fitting which selects the fitted curve so as to minimize the sum of squares of deviations from the given points.

LEC—local exchange carrier; see *local exchange.*

LED—see *light-emitting diode.*

level—1) in data management structures or communication protocols, the degree of subordination in a hierarchy. 2) measurement of signal power at specific point in a circuit.

LF—see *low frequency.*

light-emitting diode (LED)—semiconductor junction diode that emits radiant energy and is used as a light source for fiber optic communications, particularly for short-haul links. Also used in alphanumeric displays in electronic telephones and calculators.

lightwave communications—term sometimes used in place of "optical" communications to avoid confusion

with visual information and image transmission such as facsimile or television; see also *fiber optics.*

limited-distance modem—a device that translates digital signals into analog signals (and vice versa) for transfers over limited distances; some operate at higher speeds than modems that are designed for use over analog telephone facilities.

limiting operation—see *bottleneck.*

line—1) a communications path between two or more points, including a satellite or microwave channel. See *channel.* 2) in data communications, a circuit connecting 2 or more devices. 3) transmission path from a nonswitching subscriber terminal to a switching system. 4) in management structures, an authority relationship in an ·organization where one person (manager) has responsibility for the activities of another person (subordinate).

line balancing—the assignment of tasks to multiple workstations so as to minimize the number of workstations and the total amount of unassigned time at all stations.

line conditioning—see *conditioning.*

line control unit (LCU)—in data communications, a special-purpose processor that controls input/output of communication lines not directly accessed by the computer.

line discipline—see *control procedure.*

line driver—communications transmitter/receiver which is used to extend the transmission distance between terminals and computers that are directly connected; acts as an interface between logic-circuits and a 2-wire transmission line.

line finder—switch that finds a calling line among a group of lines and connects it to another device. Used typically in step-by-step (S × S) switching.

line hit—electrical interference causing the introduction of undesirable signals on a circuit.

line loading—the process of installing loading coils in series with each conductor on a transmission line. Usually 88 milliHenry coils installed at 6,000 foot intervals.

line load control—equipment in a telephone system that provides a means by which essential paths may be assured continuity of service under over-loaded condi-

SEPTEMBER 1989

Glossary of Telecommunications Technologies

tions, generally accomplished by temporarily denying originating service to some or all of the nonessential lines.

line lockout with warning—telecommunications system feature which provides 10 seconds of intercept tone and then holds the line out of service when the station line remains off-hook for longer than a predetermined time.

line number—the identification number sequence of an instruction or statement in a sequential program.

line of sight—characteristic of some open-air transmission technologies where the area between a transmitter and a receiver must be clear and unobstructed; said of microwave, infrared, and open-air laser-type transmission; a clear, open-air, direct transmission path free of obstructions such as buildings but in some cases impeded by adverse weather or environmental conditions.

line preference—telecom, system feature in which the user selects a line by pressing the button associated with that line.

line speed—maximum data rate that can be reliably transmitted over a line.

line switching—see *circuit switching.*

link—1) physical circuit between two points. 2) conceptual (or virtual) circuit between two users of a packet switched (or other) network that allows them to communicate, even when different physical paths are used.

link layer—the logical entity in the OSI model concerned with transmission of data between adjacent network nodes; it is the second layer processing in the OSI model, between the physical and the network layers; see also *OSI.*

link redundancy level—the ratio of actual number of paths to the minimum number of paths required to connect all nodes of a network.

liquid crystal display (LCD)—a graphic display on a terminal screen using an electroluminescent technology to form symbols or shapes.

LLC—see *logical link control.*

load—in computer operations, the amount of scheduled work, generally expressed in terms of hours of work. In programming, to feed data or programs into the computer.

load balancing for station/trunk lines—traffic balancing may be required when system traffic limits are approached. This feature provides the capability, during installation or on an in-service system to change specific station and trunk terminations on the PBX system switching network with minimum installer effort and without requiring number changes for the purpose of balancing the traffic load on the switch network.

load center—a group of workstations that can be considered together for purposes of scheduling.

loading—adding inductance (load coils) to a transmission line to minimize amplitude distortion.

loading coil—induction device employed in local loops, generally those exceeding 5,500 meters in length, that compensates for the wire capacitance and serves to boost voice grade frequencies. They are often removed for new generation, high-speed, local-loop data services as they can distort data signals at higher frequencies than those used for voice.

local access and transport area (LATA)—geographic regions within the U.S. that define areas within which the Bell Operating Companies (BOCs) can offer exchange and exchange access services (local calling, private lines, etc.).

local area network (LAN)—a system for linking terminals, programs, storage, and graphic devices at multiple workstations over relatively small geographic areas.

local automatic message accounting (LAMA)—combination of automatic message accounting equipment, automatic number accounting equipment, and automatic number identification equipment in the same office. In such a system, a subscriber-dialed call can be automatically processed without operator assistance.

local call—any call for a destination within the local service area of the calling station.

local call billing—telecom system feature used in hotel applications which computes the charges for local calls placed by guests based on total message units and, optionally, service charges stored for each guestroom telephone via the station message register service feature and the hotel local call billing rate parameter.

local exchange—switching center in which subscribers' lines terminate. The exchange has access to other exchanges and to national trunk networks. Also called *local central office, end office.*

Glossary of Telecommunications Technologies

local loop—a line connecting a customer's telephone equipment with the local telephone company exchange.

local service area—area within which the telephone operating company uses local rates for calling charge.

local systems environment—those resources which exist in a real, open system but which are outside the scope of the OSI environment.

local trunk—trunks between local exchanges.

logic circuit—an electronic path that performs information processing; generally encoded on a chip, it is composed of logic gates which are the boolean logic building blocks. See *logic gate.*

logic design—the specification of the working relations between the parts of a system in terms of symbolic logic.

logical data independence—the capacity to change the overall logical structure of the data without changing the application programs.

logical link control (LLC)—a protocol developed by the IEEE 802, common to all of its local network standards, for data link-level transmission control. The upper sublayer of the IEEE layer-2 (OSI) protocol that complements the MAC protocol (IEEE 802.2).

logical record—a collection of items independent of their physical environment; portions of the same logical record may be located in different physical records.

long-distance—any telephone call to a destination outside the local service area of the calling station. Also called *toll call.*

long-haul—long distance, describing (primarily) telephone circuits that cross out of the local exchange.

longitudinal balance—measure of the electrical balance between the two conductors (tip and ring) of a telephone circuit. Specifically, the difference between the tip-to-ground and ring-to-ground AC signal voltages, expressed in decibels.

longitudinal redundancy check—a data communications error trapping technique in which a character is accumulated at both the sending and receiving stations during the transmission and is compared for an equal condition, which indicates a good transmission of the previous block.

loop—1) local circuit between an exchange and a subscriber telephone station. Also called *subscriber loop*

and *local line.* 2) in programming, a sequence of computer instructions that repeats itself until a predetermined count or other test is satisfied.

loop checking—see *message feedback.*

loop circuit—generally refers to the circuit connecting the subscriber's equipment with local switching. Also called *metallic circuit* and *local loop.*

loop signaling systems—any of three methods of transmitting signaling information over the metallic loop formed by the trunk conductors and the terminating equipment bridges. Transmission of the loop signals can be accomplished by 1) opening and closing the DC path around the loop, 2) reversing the voltage polarity, or 3) varying the value of the equipment resistance.

loop start—most commonly used method of signaling an off-hook condition between an analog phone set and a switch, where picking up the receiver closes a wire loop, allowing DC current to flow, which is detected by a PBX or local exchange and interpreted as a request for service.

loopback—a diagnostic procedure used for transmission devices; a test message is sent to a device being tested, which then sends the message back to the originator for comparison with the original transmission. Loopback testing may be performed within a locally attached device or conducted remotely over a communications circuit. See *analog loopback, digital loopback.*

loss (transmission)—decrease in energy of signal power in transmission along a circuit due to the resistance or impedance of the circuit or equipment.

loudspeaker paging access—allows the PBX/key system attendant and station users to have access to paging equipment for the purpose of voice paging. All voice paging facilities make use of the telephone transmitter as the microphone. Available to station users on a class-of-service basis, this function allows direct connection to the paging system by dialing a unique number for each zone or for all zones simultaneously.

low frequency (LF)—generally indicates frequencies between 30 KHz and 300 KHz.

low-level language—a programming language in which instructions have a 1-to-1 relationship with machine code; see also *computer language.*

LRC—see *longitudinal redundancy check.*

LTB—see *last trunk busy.*

SEPTEMBER 1989

Glossary of Telecommunications Technologies

LTE—local telephone exchange; see *local exchange.*

M

M—mega; 1,048,576 bytes (1,024K squared) or roughly 1 million.

Mbps—megabits per second.

macro—a subroutine that retrieves frequently used phrases and formats with a few keystrokes.

main distribution frame (MDF)—a wiring arrangement that connects outside lines on one side, and internal lines from exchange equipment on the other.

main/satellite service—a PBX networking feature in which multilocation PBX customers can concentrate their attendant positions at one location, referred to as the main. Other unattended locations, equipped with dial switching equipment, are referred to as satellites. Only one listed directory number (LDN) is provided per complex, and all attendant-handled calls are switched through the main PBX over tie trunks to and from satellite PBX locations.

main station—subscriber's instrument (e.g., telephone or terminal) connected to a local loop, used for originating calls and answering incoming calls from the exchange.

maintenance time—the interval required for hardware correction and enhancement; contrast with *available time.*

managed object—a data processing or data communications resource that may be managed through the use of an OSI Management protocol. The resource itself need not be an OSI resource. A managed object may be a physical item of equipment, a software component, some abstract collection of information, or any combination of all three.

management information—information, associated with a managed object that is required to control and maintain that object.

management information base (MIB)—a conceptual composite of information about all managed objects in an open system.

manual exchange—exchange in which connections are made by an operator.

manual exclusion—method by which a PBX station user, by entering a special code, can block all other stations on that line from entering the call, assuring privacy on the line.

manual hold—method of placing a line circuit on "hold" by activating a nonlocking common hold button.

manual originating line service—provides station lines that require the attendant to complete all outgoing calls. All nonattendant-handled call attempts are intercepted. This arrangement can be used for lobby phones or emergency telephones to minimize system abuse.

manual signaling—depressing a specific button on a telephone to send an audible signal to a predetermined station.

manual terminating line service—provides extension lines that require all terminating calls to be completed by the attendant. Intercepts all call attempts not handled by the attendant. This feature can be activated, for example, on patient phones in a hospital to prevent disturbance.

manufacturing automation protocol (MAP)—application-specific protocol based on Open Systems Interconnection (OSI) standards. It is designed to allow communication among competing vendors' digital products in the factory.

MAP—see *manufacturing automation protocol.*

mapping—in network operations, the logical association of one set of values, such as addresses on one network, with quantities or values of another set, such as devices on a second network (e.g., name-address mapping, internetwork-route mapping).

mark—the signal (communications channel state) corresponding to a binary 1. The marking condition exists when current flows (current-loop channel) or when the voltage is more negative than -3 volts (EIA RS-232-C channel).

marker—wired-logic control circuit that, among other functions, tests, selects, and establishes paths through a switching state(s) in response to external signals.

master number hunting—an incoming call routing pattern in which the specific number is assigned as the first number in the hunt group, rather than the more traditional procedure of using the first station within the group.

Glossary of Telecommunications Technologies

master station—unit having control of all other terminals on a multipoint circuit for purposes of polling and/or selection.

master terminal—a reserved workstation used by the system manager that can initiate conversations for network management tasks.

matrix—1) an arrangement of elements (numbers, characters, dots, diodes, wires, etc.) in perpendicular rows. 2) in switch technology, that portion of the switch architecture where input leads and output leads meet, any pair of which may be connected to establish a through circuit. Also called a *switching matrix.*

MDF—see *main distribution frame.*

MDNS—managed data network services. A pan-European public VANs operator initiated by the Commerical Action Committee of CEPT.

mean—the arithmetical average value of a group of values; see also *median, mode.*

mean-time-between-failure (MTBF)—average length of time for which the system, or a component of the system, works without fault.

mean-time-to-repair (MTTR)—the average time required to perform corrective maintenance on a failed device.

measured rate—message rate structure which includes payment for a specified number of calls within a defined area, plus a charge for additional calls.

median—the middle value in a group.

medium—1) the material on which data is recorded; for example, magnetic tape, diskette. 2) any material substance that is, or can be, used for the propagation of signals, usually in the form of modulated radio, light, or acoustic waves, from one point to another, such as optical fiber, cable, wire, dielectric slab, water, air, or free space (ISO).

medium frequency (MF)—frequencies in the range between 300 KHz and 3 MHz.

meet-me conference—conference circuit on a PBX given a single-digit access code. All stations dialing that code at a predetermined time (or upon direction by the operator) are connected in conference.

megabyte—approximately 1,000,000 bytes.

megahertz (MHz)—a unit of frequency equal to 1 million cycles per second.

memory—area of a computer system that accepts, holds, and provides access to information.

menu-driven—set of instructions using a list of commands and available options; the user only has to select the desired option; compare to *command-driven.*

message—sequence of characters used to convey information or data. In data communications, messages are usually in agreed format with a heading, which establishes the destination of the message, and text, which consists of the data being sent.

message feedback—method of checking the accuracy of data transmission in which the received data are returned to the sending end for comparison with the original data, which are stored there for this purpose. Also called *information feedback* and *loop checking.*

message format—rules for the placement of such portions of a message as message heading, address, text, end-of-message indication, and error-detecting bits.

message numbering—identification of each message within a communications system by the assignment of a sequential number.

message registration—a hotel telecommunications system feature provides an electronic or mechanical readout of outgoing local and long-distance calls from each guest station. This information can be displayed at a hotel console or on mechanical counters.

message relay—a feature that allows a PBX extension user to record and store a message for transmission to a given extension by a given time, after which it is relayed to the attendant.

message switching—a technique that transfers messages between points not directly connected. The switching facility receives messages, stores them in queues for each destination point, and retransmits them when a facility becomes available. Synonymous with *store-and-forward.*

message unit (MU)—unit of measure for charging local calls that details the length of call, distance called, and time of day.

message waiting—feature that allows an attendant or voice messaging system to light an indicator lamp on a user's telephone to show that a message is waiting.

MF—1) medium frequency. 2) multifrequency. See *dual tone multifrequency signaling (DTMF).*

MHz—see *megahertz.*

SEPTEMBER 1989

Glossary of Telecommunications Technologies

MIB—see *management information base.*

microsecond—one-millionth of a second.

microwave—1) portion of the electromagnetic spectrum above about 760 MHz. 2) describing high-frequency transmission signals and equipment that employ microwave frequencies, including line-of-sight open-air microwave transmission and, increasingly, satellite communications.

millisecond—one-thousandth of a second.

mixed station dialing—indicates the ability of the switching system to accommodate both rotary dial and pushbutton dial stations.

mobile earth station—radio transmitter and/or receiver situated on a ship, vehicle, or aircraft and used for satellite communications.

mobile satellite communications—satellite-based services to serving ships, motor vehicles, and aircraft.

mobile telephone exchange (MTE)—exchange providing service to mobile telephone subscribers.

mobile telephone service (MTS)—telephone service provided between mobile stations and the public switched telephone network; radio transmission provides the equivalent of a local loop. See also *cellular radio.*

mode—1) the most common or frequent value in a group of values. 2) one of many energy propagation patterns through a waveguide.

model—an approximate mathematical representation that simulates the behavior of a process device or concept so that an increased understanding of the system is attained.

modem—a contraction of *mo*dulate and *dem*odulate; a conversion device installed in pairs at each end of an analog communications line. The modem at the transmitting end modulates digital signals received locally from a computer or terminal; the modem at the receiving end demodulates the incoming signal, converting it back to its original (i.e., digital) format, and passes it to the destination business machine.

modem sharing unit (MSU)—a device that permits two or more terminals to share a single modem.

modular—a design technique that permits a device or system to be assembled from interchangeable components; permits the system or device to be expanded or modified simply by adding another module.

modulation—the process of converting voice or data signal for transmission over a network.

modulation, amplitude (AM)—form of modulation in which the amplitude of the carrier is varied in accordance with the instantaneous value of the modulating signal.

modulation, frequency (FM)—form of modulation in which the instantaneous frequency of a sine wave carrier is caused to depart from the carrier frequency by an amount proportional to the instantaneous value of the modulating signal.

modulation, pulse amplitude (PAM)—form of modulation in which the amplitude of the pulse carrier is varied in accordance with successive samples of the modulating signal.

modulation, pulse code (PCM)—digital transmission technique that involves sampling of an analog information signal at regular time intervals and coding the measured amplitude value into a series of binary values, which are transmitted by modulation of a pulsed, or intermittent, carrier. A common method of speech digitizing using 8-bit code words, or samples, and a sampling rate of 8KHz.

modulation, pulse width (PWM)—process of encoding information based on variations of the duration of carrier pulses. Also called *pulse duration modulation* or *PDM.*

modulator—device that converts a signal (voice or other) into a form that can be transmitted.

module—a hardware or software component that is discrete and identifiable.

monitor—a software tool used to supervise, control, or verify the operations of a system.

monitor (display)—a device used to display computer-generated information.

monitoring key—feature button that allows an attendant to listen on a circuit without sensibly affecting the transmission quality of that circuit.

moving average—the mean of the most recent observations; as each new observation is added, the oldest one is dropped.

msec—millisecond.

MSU—see *modem sharing unit.*

MTBF—see *mean-time-between-failure.*

Glossary of Telecommunications Technologies

MTE—see *mobile telephone exchange.*

MTS—see *mobile telephone service.*

MTTR—see *mean-time-to-repair.*

MTX—see *mobile telephone exchange.*

MU—see *message unit.*

mu-law encoding—encoding according to CCITT recommendation G.711, used with 24-channel PCM systems in the U.S. It is similar to a-law encoding, but the two differ in the size of the quantizing intervals.

multiaccess—the ability for several users to communicate with a computer at the same time.

multidrop—a communications arrangement where multiple devices share a common transmission channel, though only one may transmit at a time. See also *multipoint.*

multidrop line—see *multipoint line.*

multileaving—the transmission of a variable number of datastreams between user devices and a computer, usually via bisync facilities and using bisync protocols.

multimode fiber—a fiber supporting propagation of multiple modes. See *modes.*

multiple—system of wiring so arranged that a circuit, a line, or a group of lines are accessible at a number of points. Also called *multipoint.*

multiple console operation—PBX supporting more than one attendant's position to handle heavy traffic. Call traffic is distributed evenly among consoles in use.

multiple customer group operation—PBX that can be shared by several different companies, each having separate consoles and trunks. Stations are assigned to one company or the other and are then able to reach only that company's trunks and attendants.

multiple listed directory number service—allows more than one listed directory number to be associated with a single PBX installation. Each listed number can be assigned a unique incoming call identification.

multiple trunk groups—indicates that the switching system is capable of being equipped for more than one group of trunk circuits.

multiplex—to interleave or simultaneously transmit two or more messages on a single channel.

multiplex hierarchy—12 channels = 1 group; 5 groups (60 channels) = 1 supergroup; 10 supergroups (600 channels) = 1 mastergroup (U.S. standard); 5 supergroups (300 channels) = 1 mastergroup (CCITT standard); 6 U.S. mastergroups = 1 jumbo group.

multiplexer (mux)—a device that enables more than one signal to be sent simultaneously over one physical channel. It combines inputs from two or more terminals, computer ports, or other multiplexers, and transmits the combined datastream over a single high-speed channel. At the receiving end, the high-speed channel is demultiplexed, either by another multiplexer or by software.

multipoint—pertaining or referring to a communications line to which three or more stations are connected. It implies that the line physically extends from one station to another until all are connected. Contrast with *point-to-point.*

multiprocessing system—a computer capable of performing more than one task at a time.

multiprogramming—a technique allowing several programs to run on one computer system at the same time.

multitasking—two or more program segments running in a computer at the same time.

multithreading—concurrent processing of more than one message (or similar service request) by an application program.

multiuser—a computer that can support several workstations operating simultaneously.

music on hold, trunk—availability of audio source input for all "held" conditions placed on trunk circuits in the system.

mux—see *multiplexer.*

N

nailed-up connection—slang term for permanent, dedicated path through a switch; often used for lengthy, regular data transmissions going through a PBX.

nak—see *negative acknowledgement.*

nanosecond—one-billionth of a second.

narrowband channels—subvoice grade pathways characterized by a speed range of 100 to 200 bps.

SEPTEMBER 1989

Glossary of Telecommunications Technologies

NARUC—National Association of Regulatory Utility Commissioners.

NATA—North American Telecommunications Association.

national number—digits that have to be dialed following the trunk prefix to call a subscriber in the same country but outside the local numbering area. These digits uniquely identify a station in an area identified by a country code.

NBS—see *NIST*.

near-end crosstalk (NEXT)—unwanted energy transferred from one circuit usually to an adjoining circuit. It occurs at the end of the transmission link where the signal source is located, with the absorbed energy usually propagated in the opposite direction of the absorbing channel's normal current flow. Usually caused by high-frequency or unbalanced signals and insufficient shielding.

negative acknowledgement (NAK)—an indication that a previous transmission was in error and the receiver is ready to accept retransmission.

neper—basic unit of a logarithmic scale used for the expression of ratios of voltages, currents, and similar quantities.

NETS—Normes Europeans de Telecommunications. Pan-European communications equipment type approval standards agreed to by the European Telecommunications Standards Institute (ETSI).

network—1) a series of points connected by communications channels. 2) switched telephone network is the network of telephone lines normally used for dialed telephone calls. 3) private network is a network of communications channels confined to the use of one customer.

network architecture—the philosophy and organizational concept for enabling communications between data processing equipment at multiple locations. The network architecture specifies the processors and terminals, and defines the protocols and software that must be used to accomplish accurate data communications.

network database—an organization method that allows information relationships to be expressed.

network inward/outward dialing (NIOD)—ability to provide dialing both ways between a toll network and a local network.

network layer—in the OSI model, the logical network entity that services the transport layer. It is responsible for ensuring that data passed to it from the transport layer is routed and delivered through the network.

network management center (NMC)—center used for control of a network. May provide traffic analysis, call detail recording, configuration control, fault detection and diagnosis, and maintenance.

network redundancy—a communications pathway that has additional links to connect all nodes.

network security—measures taken to protect a communications pathway from unauthorized access to and accidental or willful interference of regular operations.

network terminating unit (NTU)—the part of the network equipment that connects directly to the data terminal equipment.

network topology—describes the physical and logical relationship of nodes in a network; the schematic arrangement of the links and nodes of a network, typically either a star, ring, tree, or bus topology, or some hybrid combination thereof.

network virtual terminal—a communications concept wherein a variety of data terminal equipment (DTEs), with different data rates, protocols, codes, and formats, are accommodated in the same network. This is done as a result of network processing, where each device's data is converted into a network standard format, then converted into the format of the receiving device at the destination end.

NEXT—see *near-end crosstalk*.

night audit—provides automatic printout of message registration data for all guestrooms by key operation at front desk console.

night console position—provides an alternate attendant position that can be used at night in lieu of the regular console.

NIOD—see *network inward/outward dialing*.

NIST—National Institute of Standards and Technology. Prior to 1988, was called the National Bureau of Standards.

NMC—see *network management center*.

NMOS—n-channel metal oxide semiconductor.

NND—national number dialing; see *national number*.

Glossary of Telecommunications Technologies

NNX—older form of central office code where N is any digit Room 2 to 9 and X is any digit from 0 to 9.

node—a termination point for two or more communications links. The node can serve as the control location for forwarding data among the elements of a network or multiple networks, as well as performing other networking and, in some cases, local processing functions. A node is usually connected to the backbone network and serves end points and/or other nodes.

noise—unwanted electrical signals, introduced by circuit components or natural disturbances, that tend to degrade the performance of a communications channel. Contrast with *signal*.

nonblocking—describes a switch where a through traffic path always exists for each attached station. Generically, a switch or switching environment designed to never experience a busy condition due to traffic volume.

nonconsecutive hunting—nonconsecutive station numbers can be "searched" by the switching equipment upon dialing the initial number in the hunting group to find connection to the first nonbusy station.

nonvolatile storage—a storage medium whose contents are not lost when the power is removed.

NPA—see *numbering plan area*.

NRZ—non-return to zero (magnetic tape format).

ns—nanosecond (also nsec).

NTSC—National Television Systems Committee. Television broadcasting system using 525 picture lines and a 60Hz field frequency. Developed by the Committee, and used primarily in the U.S., Canada, Mexico, and Japan. See also *PAL* and *SECAN*.

NTU—see *network terminating unit*.

number-unobtainable tone— audible signal received by the caller indicating that the attempted call cannot be completed due to faulty equipment or lines, invalid number dialing, or because access to that number is barred.

numbering plan area (NPA)—geographic subdivision of the territory covered by a national or integrated numbering plan. An NPA is identified by a distinctive area code (NPA code).

O

OCB—see *outgoing calls barred*.

OCC—Other Charges or Credits (on phone bill), also Other Common Carrier (MCI, U.S. Sprint, etc.).

octet-8-bit byte. Term used in preference to byte when talking about packet services.

ODA—office document architecture. An OSI standard intended to provide parameters for electronically transmitted documents.

ODD—see *operator distance dialing*.

OEM—see *original equipment manufacturer*.

off-hook—telephone set in use; handset is removed from its cradle.

off-line—1) pertaining to equipment or devices not under direct control of the central processing unit. 2) used to describe terminal equipment that is not connected to a transmission line.

off-premises extension (OPX)—telephone extension located other than where the main switch is.

office classification—see *class of exchange*.

OGT—see *outgoing trunk*.

OHQ—off-hook queue. See *queue*.

on-hook—telephone set not in use; handset resting in cradle.

on-hook dialing—station user can dial a number and listen to the call's progress over the set's speaker, leaving the receiver on-hook until the call goes through and conversation begins.

on-line—1) connected to a computer so that data can pass to or from the computer without human intervention. 2) directly in the line loop. 3) in telegraph usage, transmitting directly onto the line rather than, for example, perforating a tape for later transmission.

online services—computer functions offered to end users not owning a host computer; includes timesharing, archival storage, and prepared software programs.

one-way splitting—when the attendant is in connection with an outside trunk and an internal station, activation of a key allows the attendant to speak privately with the internal station.

SEPTEMBER 1989

Glossary of Telecommunications Technologies

one-way trunk—trunk between a switch (PBX) and an exchange, or between exchanges, where traffic originates from only one end.

ONI—see *operator number identification*.

ONP—see *open network provision*.

open network provision (ONP)—a developing pan-European standard for ensuring the provision of the network infrastructure by European telecommunications administrations to users and competitive service providers on terms equal to those for the administrations themselves.

open systems interconnection (OSI)—referring to the reference model, OSI is a logical structure for network operations standardized within the OSI; a seven-layer network architecture being used for the definition of network protocol standards to enable any OSI-compatible computer or device to communicate with any other OSI-compliant computer or device for a meaningful exchange of information.

operating time—the time required for dialing the call, waiting for the connection to be established, and coordinating the transaction with the personnel or equipment at the receiving end. Also known as *call setup time*.

operator distance dialing (ODD)—establishment of long-distance calls requiring the intervention of an intermediate operator.

operator number identification (ONI)—at an exchange, a feature that allows the operator to come in long enough to acquire the calling number so that it can be keyed into CAMA equipment.

OPS—off-premises station; see *off-premises extension*.

optical fiber—any filament or fiber, made of dielectric materials, that is used to transmit laser- or LED-generated light signals, usually for digital communications. An optical fiber consists of a core, which carries the signal, and cladding, a substance with a slightly higher refractive index than the core, which surrounds the core and serves to reflect the light signal back into it. Also called *lightguide* or *fiber-optic waveguide*.

OPX—see *off-premises extension*.

OR—originating register (crossbar switching).

original equipment manufacturer (OEM)—maker of equipment that is marketed by another vendor, usually under the name of the reseller. The OEM may only manufacture certain components, or complete computers, which are then often configured with software, and/or other hardware, by the reseller.

originating restriction—station line with this restriction cannot place calls at any time. Calls directed to the station, however, will be completed normally.

OS—see *operating system*.

OSI—see *open systems interconnection*.

OSI environment—those resources that enable information processing systems to communicate openly, that is, to conform to the services and protocols of Open Systems Interconnection (OSI).

OSI management—the facilities to control, coordinate, and monitor the resources which enable communications in an OSI environment.

OSITOP—a promotions group for OSI standards.

other common carrier (OCC)—specialized common carriers (SCC), domestic and international record carriers, and domestic satellite carriers engaged in providing services authorized by the Federal Communications Commission.

OTQ—see *outgoing trunk queuing*.

outgoing calls barred (OCB)—prevention of calls to distant addresses; in particular, preventing PBX users from making calls outside the PBX system.

outgoing line restriction—ability of the system to selectively restrict any outside line to an "incoming only" line.

outgoing station restriction—ability of the system to restrict any given station from originating outgoing calls.

outgoing trunk (OGT)—one-way trunk that carries only outgoing traffic.

outgoing trunk queuing (OTQ)—extensions can dial a busy outgoing trunk group, be automatically placed in a queue, and then be called back when a trunk in the group is available. This feature allows more efficient use of expensive special lines as, instead of having to redial the trunk access code until a line is free, the caller can activate OTQ, followed by the trunk access code, and take care of other affairs until a line is free.

outward restriction—station lines within the PBX can be denied the ability to access the exchange network without the assistance of the attendant. Restricted calls are routed to intercept tone.

Glossary of Telecommunications Technologies

overflow—excess traffic, on a particular route, that is offered to another (alternate) route.

override—seizure of a circuit even though the circuit is already occupied.

P

PABX—private automatic branch exchange; see *private branch exchange (PBX)*.

packet—group of binary digits, including data and call control signals, that is switched as a composite whole. The data, call control signals, and error control information are arranged in a specified format.

packet assembler/disassembler (PAD)—a protocol conversion device or program that permits end-user devices (e.g., terminals) to access a packet switched network. See also *packet switching*.

packet overhead—a measure of the ratio of the total packet bits occupied by control information to the number of bits of data, usually expressed as a percent.

packet switched network—a network designed to carry data in the form of packets. The packet and its format is internal to that network. The external interfaces may handle data in different formats, and conversion is done by an interface computer.

packet switching—a data communications technique where a message is broken down into fixed-length units which are then transmitted to their destination through the fastest route; although all units in a message may not travel the same pathway, the receiving station ascertains that all units are received and in proper sequence before forwarding the complete message to an addressee.

PAD—see *packet assembler/disassembler*.

paging by zone—by dialing the appropriate access code, any station is able to selectively page groups of predesignated stations or speakers.

paging speakers—inclusion of such speakers within the station instrument. Also includes external units located in larger areas.

paging, total system—upon dialing the appropriate special code, any station can originate a paging announcement through all loudspeakers.

PAL—see *phase alternate line*.

PAM—see *pulse amplitude modulation*.

parallel transmission—the simultaneous transmission of all the bits making up a character or byte, either over separate channels or on different carrier frequencies on the same channel. Contrast with *serial transmission*.

parity check—addition of noninformation bits to data to make the number of ones in a grouping of bits either always even or always odd. This procedure allows detection of bit groupings that contain single errors. It can be applied to characters, blocks, or any specific bit grouping. Also called *vertical redundancy checking (VRC)*.

part 68—section of FCC rules governing the direct connection of nontelephone company-provided terminal equipment to the telephone network.

party line—subscriber line upon which several subscribers' stations are connected, possibly with selective calling.

party line stations—two-party station service can be expanded to support multiparty service.

PATX—see *private automatic telex exchange*.

PAX—see *private automatic exchange*.

PBX—see *private branch exchange*.

PCB—printed circuit board.

PCM—see *pulse code modulation*.

PDM—pulse duration modulation; see *pulse width modulation*.

PDN—see *public data network*.

PDX—see *private digital exchange*.

peg count—tally of the number of calls made or received over a specified period.

peripheral device or equipment—input or output unit that is not included within the confines of the primary system, e.g., data printer or diskette.

permanent virtual circuit—in packet switching, a connection for which a single dedicated path is chosen for a particular transmission. The network is aware of a fixed association between two stations; permanent logical channel numbers are assigned exclusively to the permanent circuit, and devices do not require permission to transmit to each other. A new connection between the same users may be routed along a different path. Contrast with *virtual circuit*.

SEPTEMBER 1989

Glossary of Telecommunications Technologies

phantom circuit—third voice circuit, which is superimposed on two 2-wire voice circuits.

phase alternate line (PAL)—color television broadcasting system developed in West Germany and the U.K. that uses 650 picture lines and a 50Hz field frequency. See also *National Television Systems Committee (NTSC)* and *Sequential Couleur à Mémoire (SECAM).*

phase jitter—a random distortion of signal lengths caused by the rapid fluctuation of the frequency of the transmitted signal. Phase jitter interferes with interpretation of information by changing the timing.

phase modulation (PM)—a way of modifying a sine wave signal to make it carry information. The sine wave, or carrier, has its phase changed in accordance with the information to be transmitted.

phase-shift keying (PSK)—modulation technique for transmitting digital information whereby that information is conveyed as varying phases of a carrier signal.

physical layer—within the OSI model, the lowest level of network processing, below the link layer, that is concerned with the electrical, mechanical, and handshaking procedures over the interface that connects a device to a transmission medium (i.e., RS-232-C).

pilot tone—test frequency of controlled amplitude transmitted over carrier system for monitoring and control purposes.

plant—general term used to describe the physical equipment of a telephone network that provides communications services.

plotters—devices that convert computer output into drawings on paper or on display-type terminals instead of a printed listing.

plug-in stations—station cabling requirement remains constant for all types of station instruments. For ease in station moves and rearrangements, all stations are provided and installed as plug-in instruments.

PM—see *phase modulation.*

PMBX—see *private manual branch exchange.*

PMOS—P-channel metal oxide semiconductor.

POCSAG—a U.K. standard for radio-paging formulated by the Post Office Coding Standards Advisory Group in 1982, now adopted by the CEPT as a pan-European standard.

point of presence (POP)—since divestiture, the physical access location within a LATA of a long-distance and/or inter-LATA common carrier; the point to which the local telephone company terminates subscribers' circuits for long-distance, dial-up, or leased-line communications.

point-to-point line—describing a circuit that connects two points directly, where there are generally no intermediate processing nodes, although there could be switching facilities. Synonymous with *two-point.* See also *multipoint* and *broadcast.*

polar keying—a technique of current loop signaling in which current flow direction establishes the two-level binary code.

polarization—property of an electromagnetic wave characterized by the direction of the electric field.

polling—a method of controlling the sequence of transmission by terminals on a multipoint line by requiring each terminal to wait until the controlling processor requests it to transmit. Contrast with *contention.*

port—a point of access into a communications switch, a computer, a network, or other electronic device; the physical or electrical interface through which one gains access; the interface between a process and a communications or transmission facility.

position—part of a switchboard normally controlled by an operator or attendant.

Postal, Telegraph, and Telephone Organization (PTT)—usually a governmental department that acts as its nation's common carrier.

POTS—plain old telephone service.

power failure—PBX users attempting to cope with commercial power failures have a number of alternatives available, ranging from setting up alternative power sources to arranging for the system to fail gradually. See *brownout operation, reserve power, uninterruptible power supply (UPS).*

power failure transfer—if the PBX is unable to get enough power, this feature provides service to and/or from the exchange network for a limited number of prearranged stations at the customer location. This feature is not available with DID service. Power failure stations usually have an external button that will establish dial tone during outage.

PPS—pulses per second.

Glossary of Telecommunications Technologies

prefix—digits that must be dialed to indicate that a call is directed outside the local area.

presentation layer—in the OSI model, that layer of processing that provides services to the application layer, allowing it to interpret the data exchanged, as well as to structure data messages to be transmitted in a specific display and control format.

preset call forwarding—incoming calls are rerouted to a predetermined secondary number.

preventive maintenance—the routine checking of components to keep the system functioning.

primary block—see *primary group.*

primary carrier—long-distance carrier selected by a subscriber as the first-choice provider of long-distance service. Calls placed through the primary carrier require no additional digits, while calls placed with other carriers require that a five-digit access code be dialed.

primary center—in international terms, a switching center through which trunk traffic is passed and to which local exchanges are connected; i.e., the equivalent of a Class 4 office in the U.S. However, in the U.S. a primary center is defined as a Class 3 office and is used for toll switching. See *class of exchange.*

primary group—group of basic signals that are combined by multiplexing; the lowest level of the multiplexing hierarchy. The term is also used for the signal obtained by multiplexing these basic signals, or for the transmission channel that carries it. Also called *primary block.*

primary rate interface—in ISDN, the interface to the primary rate CCITT 23B + D, 23 channels + 1 signaling channel. See *integrated services digital network.*

primary station—the data terminal in a network that selects and transmits information to a secondary terminal and has the responsibility to insure information transfer.

priority trunk queuing—through a customer-chosen preferential trunk access level, this feature places any caller with this or higher level in the class of service assignment ahead of all callers with a lower trunk access level in the queue of callers waiting for the same trunk group.

privacy and privacy release—all other extensions of a line are unable to enter a conversation in progress on that line unless the initiating station releases the feature.

privacy lockout—automatically splits the connection whenever an attendant would otherwise be included on a call with more than one person. When privacy is provided, the attendant lockout feature is also supplied. A tone warning is generated when the attendant bridges into a conversation in progress.

privacy override—activation of a special pushbutton allows the station user to gain access to a given busy line, even though the automatic exclusion facility is engaged by the station using that line.

private automatic branch exchange (PABX)—See *private branch exchange.*

private automatic exchange (PAX)—dial telephone exchange that provides private telephone service to an organization and that does not allow calls to be transmitted to or from the public telephone network.

private branch exchange (PBX)—a telephone switch located on a customer's premises that primarily establishes voice-grade circuits, over tie-lines between individual users and the switched telephone network. Typically, the PBX also provides switching within a customer premises' local area and usually offers numerous other enhanced features, such as least-cost routing and call-detail recording.

private digital exchange (PDX)—private exchange employing digital transmission techniques.

private exchange (PX)—exchange serving a particular organization and having no means of connection with a public exchange.

private line—denotes the channel and channel equipment furnished to a customer as a unit for exclusive use, generally with no access to or from the public switched telephone network. Also called *leased line.*

private manual branch exchange (PMBX)—private, manually operated telephone exchange that provides private telephone service to an organization and that allows calls to be transmitted to or from the public telephone network.

private network—network established and operated by a private organization or corporation. Compare with *public switched telephone network.*

presentation layer—in the OSI model, that layer of processing that provides services to the application layer, allowing it to interpret the data exchanged, as well as to structure data messages to be transmitted in a specific display and control format.

SEPTEMBER 1989

Glossary of Telecommunications Technologies

process control—the monitoring and controlling of industrial operations by a computer system.

project planning—a preliminary activity that defines areas to be studied and desired objectives before a formal study begins.

propagation delay—the period between the time when a signal is placed on a circuit and when it is recognized and acknowledged at the other end. Propagation delay is of great importance in satellite channels because of the great distances involved.

protector—interface between inside and outside plant providing against hazardous voltages or currents.

protocol—a set of strict procedures required to establish, maintain, and control communications. Protocols can exist at many levels in one network such as link-by-link, end-to-end, and subscriber-to-switch. Examples include BSC, SDLC, X.25, etc.

protocol conversion—the process of translating the protocol native to an end-user device (e.g., a terminal) into a different protocol (e.g., ASCII to BSC), allowing that end-user device to communicate with another device (e.g., a computer) with which it would otherwise be incompatible. Protocol conversion can be performed by a dedicated device (a protocol converter), by a software package loaded onto an existing system, such as a general-purpose computer, front-end processor, or PBX system, or by a value-added network, such as Telenet.

PRX—program.

PSE—packet switching exchange.

PSK—see *phase-shift keying.*

PSS—British Telecom's Packet Switch Stream network. Packet switched data network operating to X.25 standards.

PSTN—see *public switched telephone network.*

PTAT—private transatlantic fiber optic cables.

PTT—see *Postal, Telegraph, and Telephone Organization.*

public data network (PDN)—digital leased circuits, digital switched circuits, or packet switching network that is designed to provide low error-rate data transmission by using digital rather than analog techniques.

public switched telephone network (PSTN)—the complete public telephone system, including telephones, local and trunk lines, and exchanges.

Public Utilities Commission (PUC)—a state regulatory body.

pulse—a momentary, sharp alteration in the current or voltage produced in a circuit to operate a switch or relay which can be detected by a logic circuit; a sharp rise and fall of finite duration.

pulse amplitude modulation (PAM)—form of modulation in which the amplitude of the pulse carrier is varied in accordance with successive samples of the modulating signal.

pulse carrier—series of identical pulses intended for modulation.

pulse code modulation (PCM)—used to convert an analog signal into a digital bitstream for transmission. Digital transmission technique that involves sampling of an analog information signal at regular time intervals and coding the measured amplitude value into a series of binary values, which are transmitted by modulation of a pulsed, or intermittent, carrier. A common method of speech digitizing using 8-bit code words, or samples, and a sampling rate of 8 KHz.

pulse duration modulation (PDM)—see *pulse width modulation.*

pulse width modulation (PWM)—process of encoding information based on variations of the duration of carrier pulses. Also known as *pulse duration modulation (PDM).*

pushbutton dialing—use of keys or pushbuttons instead of a rotary dial to generate a sequence of digits to establish a circuit connection. The signal form is usually multiple tones. See *dual tone multifrequency signaling (DTMF).*

pushbutton dialing to stations—special attendant console feature in which the switching system is served by rotary dial exchange trunk circuits. A 10-button keyset is provided on the console to allow fast dialing of extension numbers in order to complete incoming outside calls.

PWM—see *pulse width modulation.*

pwr—power.

PX—see *private exchange.*

Glossary of Telecommunications Technologies

PX64—expected to be the videoconferencing standard accepted worldwide by 1990 (for 2-way, full motion video).

Q

QOS—see *quality of service.*

quad—a cable of four separately insulated conductors, twisted together in such a way as to provide two pairs.

quality of service (QOS)—measure of the performance of a telephone system in terms of the quality of the lines and the amount of call blocking experienced.

quantization noise—signal errors caused by the process of digitizing a continuously variable slope.

query—a request for information entered while the computer system is processing.

queue—series of telephone calls, arranged in sequence, the two ends being the head and tail. New calls are added to the tail. Calls can be removed either from the head or tail.

queuing—1) in telephony, a feature that allows calls to be "held" or delayed at the origination switch while waiting for a trunk to become available. 2) sequencing of batch data sessions.

R

RACE—Research and Development Program in Advanced Communications in Europe. European Community sponsored project for the development of broadband technologies.

rack—framework or structure on which apparatus is mounted, usually by means of shelves or mounting plates. Also known as a *bay.*

radio determination—the use of radio waves to determine the position or velocity of, or to collect the information for, a remote object.

radio determination-satellite service (RDSS)—a service using one or more satellites for radio determination.

radio channel—frequency band allocated to a service provider or transmitter.

radio circuit—physical circuit consisting of two unidirectional radio links and connections to terminal exchanges.

radio frequency (RF)—a frequency that is higher than the audio frequencies but below the infrared frequencies, usually above 20 KHz. See also *electromagnetic spectrum.*

radio paging—provides attendant and station user dial access to customer-owned radio paging equipment to selectively tone-alert or voice-page individuals carrying pocket radio receivers. The paged party can answer by dialing an answering code from a station within the PBX system.

radio paging access with answerback—allows access to customer-provided paging systems and provides the capability in PBX to connect the paged party to the paging party when the former answers the radio page by dialing a special code from any PBX telephone.

radio wave—electromagnetic waves of frequencies between 20 KHz and 3 GHz approximately.

RAM—random-access memory.

rate averaging—telephone companies' method for establishing uniform toll rates based on distance rather than on the relative cost and/or volume of telephone traffic on a given route.

rate base—total invested capital on which a regulated company is entitled to a reasonable rate of return.

rate center—defined geographic point used by telephone companies in determining distance measurements for inter-LATA mileage rates.

rate of return—percentage net profit that a telephone company is authorized to earn.

RBHC—see *Regional Bell Holding Company.*

RCL—see *recall.*

RDT—see *recall dial tone.*

realtime—1) pertaining to actual time during which a physical process transpires. 2) pertaining to an application in which response to input is fast enough to effect subsequent input, as when conducting the dialog that take place at terminals in interactive systems.

recall (RCL)—PBX feature allowing a station user engaged in a call to signal the operator, often by a switch hook flash, to enter the conversation.

recall dial tone (RDT)—stutter, or interrupted, dial tone indicating to a station user that the switchook flash has been properly used to gain access to system features.

SEPTEMBER 1989

Glossary of Telecommunications Technologies

receive only (RO)—1) a printer terminal without a keyboard for data entry. 2)'a satellite earth station capable of receiving, but not transmitting, a signal.

recorded information service—special type of access trunk that, when dialed, will connect the caller to a prerecorded message.

recovery—the restoration of a system to full operation after a malfunction has been corrected.

recovery from fallback—when the system has switched to a fallback mode of operation and the cause of the fallback has been removed. This is the process that restores the system to its former condition.

redundancy—1) portion of the total information contained in a message that can be eliminated without loss of essential information. 2) provision of duplicate, backup equipment to immediately take over the function of equipment that fails. 3) in a database, the storage of the same data item or group of items is two or more files.

redundancy check—automatic or programmed check based on the systematic insertion of components or characters used especially for checking purposes.

redundant processor—a computer that duplicates or partially duplicates the operation of another computer to substantially reduce the possibility of system failure.

regenerative repeater—1) repeater utilized in telegraph applications to retime and retransmit the received signal impulses and restore them to their original strength. These repeaters are speed- and code-sensitive and are intended for use with standard telegraph speeds and codes. 2) repeater used in PCM or digital circuits which detects, retimes, and reconstructs the bits transmitted. See also *repeater*.

regenerator—see *regenerative repeater*.

Regional Bell Holding Company (RBHC)—one of the seven holding companies formed by the divestiture of AT&T to oversee the 22 Bell Operating Companies (BOCs), which, provide regulated services, and other subsidiaries, which provide nonregulated services; includes Bell Atlantic, NYNEX, BellSouth, Pacific Telesis, U S WEST, Southwestern Bell Corp., and Ameritech.

regional center—see *class of exchange.*

regional (computer) network—a communications pathway with connections limited to a defined geographic area.

register—first unit in the assembly of common control equipment in an automatic exchange. The register receives address information in the form of dial pulses or dual-tone multifrequency (DTMF) signals and stores it for possible conversion or translation.

regression analysis—a technique for determining the mathematical expression that best describes the functional relationship between two or more variables.

reinitiation time—time required for a device or system to restart (usually after a power outage).

relational database—a method of organizing a file of information to link information contained in separate records.

relay—device, operated electrically, that causes by its operation abrupt changes in an electrical circuit, such as breaking the circuit, changing the circuit connection, or varying the circuit characteristics.

relocatability—a capability that allows programs or information to be transferred to different places in main memory at different times without modifying the program.

remote access—pertaining to communications with a computer or PBX in one location from a device that is physically removed from the location of the computer.

remote-access software—sometimes called remote-control software, this type of program is a superset of the asynchronous communications software market. Remote-access software allows a PC to have complete control over another PC at a different site.

remote data concentration—communications processors used for multiplexing data from low-speed lines or terminals onto one or more high-speed lines; see also *multiplexing.*

remote maintenance—feature or service in which a service technician can dial the PBX and be connected, usually through the attendant, to the system processor to test or modify the program.

remote traffic measurement—traffic and feature usage data transmitted by the system to a distant service technician.

reorder tone (RT)—tone signal placed on a line by the switching equipment to tell the user that an error has been made in dialing the number or selecting a feature, and/or that the call cannot be completed.

repeater—1) in analog transmission, equipment that receives a pulse train, amplifies it, and retimes it for

Glossary of Telecommunications Technologies

retransmission. 2) in digital transmission, equipment that receives a pulse train, reconstructs it, retimes it, and then amplifies the signal for retransmission. 3) in fiber optics, a device that decodes a low-power light signal, converts it to electrical energy, and then retransmits it via an LED or laser-generating light source, often including some form of signal amplification. See also *regenerative repeater*.

request for information (RFI)—general notification of an intended purchase of communications or computer equipment sent to potential suppliers to determine interest and solicit product materials.

request for proposal (RFP)—follow-up to RFI, sent to interested vendors to solicit a configuration proposal, with prices, that meets a user's requirements.

resale carrier—company that redistributes the services of another common carrier and retails the services to the public.

reserve power—in case of the failure of a PBX power supply, rechargeable batteries can be added to the system allowing system operation for some length of time—15 minutes to 12 hours—after commercial power has failed. Two different situations can occur: 1) the system has enough power to maintain memory but is unable to supply talking voltage and ringing signals, leading to an interruption in services but allowing a fast recovery, and 2) a more adequate reserve power supply will keep the system running during the failure or brownout.

residual error rate, undetected error rate—the ratio of the number of bits, unit elements, characters, or blocks incorrectly received but undetected or uncorrected by the error-control equipment, to the total number of bits, unit elements, characters, or blocks sent.

reverse-battery signaling—type of loop signaling in which battery and ground are reversed on the tip and ring of the loop to give an off-hook signal when the called party answers.

reverse channel—a simultaneous data path in the reverse direction over a half-duplex facility. Normally, it is used for positive/negative acknowledgments of previously received data blocks; see also *half-duplex*.

reverse charge call—call in which the caller specifies that the charge should be paid by the called party. Also called *collect call*.

RF—see *radio frequency*.

RFI—see *radio frequency interference*.

RFI—see *request for information*.

RFP—see *request for proposal*.

ring—1) ring-shaped contact of a plug usually positioned between, but insulated from, the tip and sleeve. 2) audible alerting signal on a telephone line. 3) a network topology in which stations are connected to one another in a closed logical circle, with access to the medium passing sequentially from one station to the next by means of polling from a master station or by passing an access token from one station to another; also called a *loop*.

ring again—the PBX remembers the last number called by a station and will redial it when the feature is activated. Also called *last number dialed*.

ringback tone—see *ringing tone*.

ringdown—to gain the attention of an operator, a ringing current is applied to a line to operate a device producing a steady signal.

ringing key—key whose operation causes the sending of a ringing current.

ringing signal—any AC or DC signal transmitted over a line or trunk for the purpose of alerting a party at the distant end of an incoming call. The signal can operate a visual or sound-producing device.

ringing tone—tone received by the calling telephone indicating that the called telephone is being rung.

ringing transfer—provides for designated bells in a group of stations to ring for incoming calls. Ringing transfer allows additional sets of bells to be designated, with the user controlling which set is to ring.

RJE—remote job entry.

RO—receive only.

ROH—receiver off-hook (permanent signal).

rollback—a programmed return to a prior checkpoint or rerun point; see also *checkpoint/restart facility, rerun point*.

ROM—see *read-only memory*.

rotary dial calling—system that will accept dialing from conventional rotary dial sets which generate pulses, although pushbutton (DTMF) dial sets offer faster calling and greater reliability.

SEPTEMBER 1989

Glossary of Telecommunications Technologies

rotary output to exchange—many systems are equipped to provide pushbutton dial service in all areas. In cases where the telephone exchange trunks are not designed to accept tone signaling, the system will translate the number entered by a station in tones into rotary dial pulses that can be processed by the exchange.

route advance—variation of Automatic Route Selection that allows the caller to select the first-choice trunk group. If that group is busy, the system will attempt to place the call over alternate trunk groups. Unlike ARS, translation is not provided. If ARS is available, Route Advance is generally unnecessary.

routing—assignment of the communications path by which a message or telephone call will reach its destination.

routing code—address, or group of characters, in the heading of a message defining the final circuit or terminal to which the message has to be delivered. Also called *routing indicator*.

routing indicator—see *routing code*.

routing table—table associated with a network node that states for each message (or packet) destination the preferred outgoing link that the message should use.

RPQ—request for price quotation.

RS-232-C—a technical specification published by the Electronic Industries Association that establishes mechanical and electrical interface requirements between computers, terminals, modems, and communications lines.

RS-423-A—EIA specification for electrical characteristics of unbalanced-voltage digital interface circuits.

RS-449—EIA specification for general-purpose, 37-position and 9-position interface for data terminal equipment (DTE) and data circuit terminating equipment (DCE) employing serial binary data interchange. RS-449-1 includes Addendum 1.

RT—1) reperforator/transmitter. 2) reorder tone.

RTNR—ringing tone no reply.

S

SAA—see *Systems Application Architecture*.

sampling—a statistical procedure where generalizations are drawn from a relatively small number of observations.

satellite communications—the use of geostationary orbiting satellites to relay transmissions from one earth station to another or multiple earth stations.

scattering—cause of lightwave signal loss in optical fiber transmission. diffusion of a lightbeam caused by microscopic variations in the material density of the transmission medium.

schematic—diagram that details the electrical elements of a circuit or system.

SCPC—see *single channel per carrier*.

scrambler—coding device applied to a digital channel that produces an apparently random bit sequence. A corresponding device is used to decode the channel, i.e., the coding is reversible.

scratch pad memory—a fast, temporary internal storage area used to hold subtotals for various unknowns that are needed for final results.

SDLC—see *synchronous data link control*.

secondary channel—a low-speed channel established on a four-wire circuit over which diagnostic or control information is passed. User data is passed on the primary, high-speed channels of the circuit.

secretarial intercept—call forwarding of executives' telephones to a secretary/receptionist who can take messages.

sectional center—see *class of exchange*.

security—the protection of information against unauthorized access or use.

selective calling—ability of the transmitting station to specify which of several stations on the same line is to receive a message.

selective paging to station—an originating station is able to page to specific individual station instruments.

selective ringing—system designed with the capability of ringing only the desired subscriber's telephone on a multiparty line. Ringers tuned to one of five possible frequencies are used to achieve this effect.

self-test and fault isolation—most systems include a processor-check capability that allows the controlling computer to test itself and the rest of the system. If a

Glossary of Telecommunications Technologies

fault is found an alarm light is lit and a message is given on the system printer teletype, if one is provided. This feature also expedites service since the computer can pinpoint faulty equipment, saving diagnostic time.

sender—device that receives address information from a register or routing information from a translator and then outpulses the proper routing digits to a trunk or to local equipment. Sender and register functions are often combined in a single unit.

sensor-based system—a configuration made up of devices that measure an external phenomenon.

series call—operator arranges for a call to return to the console after the extension it was connected to hangs up. Lets the attendant easily connect a caller with a series of inside extensions without the risk of losing the call. Also called *serial call.*

server—a processor that provides a specific service to the network, e.g., a routing server connects nodes and network of like architectures, a gateway server connects nodes and networks of different architectures, etc.

service bureau—a computer organization that offers timesharing and software services; typical applications include payroll, billing, and bookkeeping; see also *timesharing, online services company.*

service code—one or more digits dialed by a customer to access services such as directory enquiries or operator assistance.

service order—request to a telecommunications vendor or carrier for service or equipment.

service organization—any company that contracts to provide for and operate computers not owned or leased by the company.

serving area—1) region surrounding a broadcasting station where signal strength is at or above a stated minimum. 2) geographic area handled by a telephone exchange, generally equivalent to a LATA.

session— 1) a period of time in which an end user engages in dialog with an interactive computer system. 2) layer 5 of the International Organization for Standardization's (ISO) Open Systems Interconnection (OSI) reference model for network architectures.

SF—single frequency.

shared tenant service—see *tenant service.*

shelf—a device used to mount equipment (such as printed circuit boards, power supplies, etc.) in an equipment rack or bay (see rack). A rack can contain several shelves.

shielded pair—two insulated wires in a cable wrapped with metallic braid or foil to prevent interference and provide noise-free transmission.

short code address (SCA)—those few digits allocated to any frequently dialed number, which when dialed are translated by the exchange into the required full number. See also *abbreviated dialing.*

sideband—the frequency band on either the upper or lower side of the carrier frequency band within which the frequencies produced by the process of modulation fall. Various modulation techniques make use of one or both of the sidebands, some of which also suppress the carrier frequency.

sidetone—reproduction in a telephone receiver of sounds picked up by the associated microphone. The microphone can pick up either the voice of the speaker or the room noise.

signal—a physical, time-dependent energy value used for the purpose of conveying information through a transmission line. Contrast with *noise.*

signal-to-noise ratio (SNR)—the relative power of a signal compared to the power of noise on a line, expressed in decibels (dB). As the ratio decreases, it becomes more difficult to distinguish between information and interference.

signaling—process by which a caller or equipment on the transmitting end of a line informs a particular party or equipment at the receiving end that a message is to be communicated.

simplex—pertaining to the capability to transmit in one direction only. Contrast with *half-duplex and full-duplex.*

simplex circuit—circuit permitting the transmission of signals in one specified direction only.

single channel per carrier (SCPC)—transmission system in which a physical channel is allocated solely to one carrier for the duration of the transmission.

single-digit dialing—provides for single-digit dialing to reach a preselected group of stations. A variation of speed dialing, it also helps reduce the need for key systems by replacing the intercom function.

Glossary of Telecommunications Technologies

single-mode fiber—a fiber with a small core diameter allowing the propagation of a single light path.

single-precision—the number of memory locations used for a number in the computer; each number has one location.

single-sideband transmission—to make efficient use of the frequency band available, the carrier and the unwanted sideband of an amplitude-modulated wave can be filtered out so that only the sideband that contains all the information is transmitted.

single threading—a group of instructions that completes the processing of one message before starting another; see also *multithreading*.

sleeve—1) third contacting part on a telephone plug preceded in the location by the tip and ring. 2) the sleeve wire is the third control wire of each telephone in an automatic switching office.

SMDR—see *station message detail recording*.

smooth scrolling—the continuous vertical movement (up or down) of lines of data displayed on a CRT screen, much in the same manner as a credit roll at the end of a movie.

SNA—see *Systems Network Architecture*.

SNR—see *signal-to-noise ratio*.

software—computer instructions that perform common functions for all users as well as specific applications for particular user needs.

SOGT—Senior Officials Group-Telecommunications. An ad hoc committee composed of senior Member State and Commission Officials in the European Community to define, oversee, and agree on the technical work of the GAP committees in telecommunications.

solidstate device—electronic pathways made of solid materials, e.g., chips and bubble memories.

space-diversity reception—to reduce the effects of fading and attenuation, a radio signal is received at more than one site, the sites being separated by a few wavelengths. The signals are then combined, selected, or both.

space-division switching—method for switching circuits in which each connection through the switch takes a physically separate path.

specialized carrier—a company that provides value-added or limited communications facilities. Contrast with *common carrier*.

speech circuit—circuit designed for the transmission of speech, either analog or encoded, but which can also be used for data transmission or telegraphy.

speed dialing—feature that enables a PBX or PBX station to store certain telephone numbers and dial them automatically when a short (1-, 2- or 3-digit) code is entered. Also called *speed calling*.

split access to outgoing trunks—provides two separate trunk groups for direct outward dialing. The groups can be accessed by dialing the same trunk access code.

split screen—to section the display portion of a video screen into two or more areas, sometimes called windows, so that the different areas can be viewed and compared simultaneously by the user.

splitting—permits the operator to consult privately with one party on a call without the other party hearing.

spooling—temporary storage of batch input and output on disk or tape files until the processor is ready; see also *batch*.

SSB—single sideband; see *single-sideband transmission*.

SSN—see *switched service network*.

standby central control—second control computer that can be provided to direct PBX operations if the primary one fails.

star—a network topology in which each station is connected only to a central station by a point-to-point link and communicates with all other stations through the central station.

StarLAN—a local network design and specification within IEEE 802.3 standards subcommittee, characterized by 1M bps baseband data transmission over two-pair, twisted-pair wiring.

start-stop (signaling)—signaling in which each group of code elements corresponding to an alphabetical signal is preceded by a start signal which serves to prepare the receiving mechanism for the reception and registration of a character and is followed by a stop signal which serves to bring the receiving mechanism to rest in preparation for the reception of the next character. Also known as *asynchronous* or *start-stop transmission*.

Glossary of Telecommunications Technologies

start-stop transmission—see *asynchronous transmission*.

static RAM—random-access memory that requires continuous power but not continuous regeneration to retain its contents.

station—one of the input or output points of a communications system—e.g., the telephone set in the telephone system or the point where the business machine interfaces the channel on a leased private line.

station equipment—hardware located at a network station. Telephone examples include rotary-dial and pushbutton telephones, key telephones, speakerphones, and IVDTs.

station message detail recording (SMDR)—processor-generated records of all calls originated and/or received by a PBX system.

station message registers—message unit information is centrally recorded on a per-station-line basis for each completed outgoing local call made by the station user. Most systems provide for surcharges on station usage and automatically reset the counter after readout.

station monitoring—allows selected stations to monitor any other stations within the system.

station override security—on an individual station line basis, designated stations can be "shielded" against the executive busy override facility that is being used by another station.

station rearrangement and change—allows the customer to move stations, change the features and/or restrictions assigned to a station, administer features associated with electronic telephones, and perform search routines on individual stations in order to identify the services provided for that station. Station rearrangements and changes are made on a per-line basis.

station transfer security—if a trunk call is transferred from one station to another and the second station does not answer within a predetermined time interval, the trunk call is automatically rerouted to the attendant console.

station-to-station dialing—system feature that allows calling between stations by direct dialing without the need for operator assistance in call completion.

status information—data about the logical state of a piece of equipment, e.g., a peripheral device reporting its current state to the computer.

STD—subscriber trunk dialing.

step—one operation in a computer program.

step call—allows the attendant or station user, upon finding that the called station is busy, to call a nearby idle station by dialing a single additional digit when the nearby station number has only a different last digit.

step-by-step switch (SXS)—switch that moves in synchronism with a pulse device such as a rotary telephone dial. Each digit dialed causes the movement of successive selector switches to carry the connection forward until the desired line is reached.

stop bit—in asynchronous transmission, the quiescent state following the transmission of a character; usually required to be at least 1, 1.42, 1.5, or 2 bit times long.

stop element—last bit of a character in asynchronous serial transmission, used to ensure recognition of the next start element.

store-and-forward—the process of accepting a message or packet on a communications pathway, retaining it in memory, and retransmitting it to the next station. Synonymous with *message switching*.

straightforward outward completion—operator can place an outgoing call for the station user, either by dialing "0" or by an intercept arrangement, without requiring the station user to hang up and redial the operator.

strap—a hard-wired connection. A strapping option is one that is implemented by changing wires.

stunt box—1) device to control the nonprinting functions of a teletypewriter terminal, such as a carriage return and line feed. 2) device to recognize line control characters.

STX—start of text (of message).

subscriber line—telephone line connecting the exchange to the subscriber's station. Also called *access line* and *subscriber loop*.

subscriber loop—see *subscriber line*.

subsystem—a secondary or subordinate configuration, generally capable of operating within a controlling configuration.

subvoice grade channel—channel with bandwidth narrower than that of voice grade channels. Such channels are usually subchannels of a voice grade line.

super high frequency (SHF)—denotes frequencies from 3 GHz to 30 GHz

SEPTEMBER 1989

Glossary of Telecommunications Technologies

supergroup—assembly,of five 12-channel groups occupying adjacent bands in the spectrum for the purpose of simultaneous modulation or demodulation, i.e., 60 voice channels.

supervised station release—station that is "off-hook" (that is, the user has not dialed, or is connected to a busy signal for more than a predetermined time interval) is automatically routed to the attendant console.

supervision—process of detecting a change of state between idle and busy conditions on a circuit.

supervisory lamp—lamp illuminated during a call and indicating to an operator the status of the call.

switched line—one of a series of lines that can be interconnected through a switching center; a line on the public telephone network. Contrast with *leased line.*

switched loop operation—attendant position is arranged so that each call requiring attendant assistance is automatically switched to one of several switched loops in position. Normally, the call automatically releases from the position when answered by the called station (released loop operation). Incoming calls are queued in the order of arrival when all attendant positions are busy and are switched to each attendant position automatically to distribute the call load evenly to each attendant. A console lamp indication is normally given to the attendant when calls are waiting in the queue to be served.

switched message network—a network service, such as Telex, providing interconnection of message devices such as teletypewriters.

switched network—a multipoint communications pathway with circuit-switching capabilities; e.g., the telephone network, Telex, and TWX.

switched service network (SSN)—network consisting of terminals, transmission links, and at least one exchange, on which any user can communicate with any other user at any time.

switched virtual call (SVC)—the stream of packets which forms the data flow for a single data message.

switchook—switch on a telephone set, associated with the structure supporting the receiver or handset, often used to signal the switching equipment or an attendant during a call, e.g. to transfer the call.

switching—establishment of transmission path from a particular inlet to a particular outlet of a group of such inlets and outlets.

switching center—location that terminates multiple circuits and is capable of interconnecting circuits or transferring traffic between circuits.

switching matrix—see *matrix.*

switchover—when a failure occurs in the equipment, a switch can occur to an alternative component.

SXS—step-by-step switch.

sync character—a symbol or defined bit pattern that is used by the receiving terminal to adjust its clock and achieve synchronization.

synchronization—the process of adjusting a receiving terminal's clock to match the clock of the transmitting terminal.

synchronous—having a constant time interval between successive bits, characters, or events. Synchronous transmission uses no redundant information (such as the start and stop bits in asynchronous transmission) to identify the beginning and end of characters, and thus is faster and more efficient than asynchronous transmission. The timing is achieved by transmitting sync characters prior to data; usually, synchronization can be achieved in two or three character times.

synchronous communications—high-speed transmission of contiguous groups of characters; the stream of monitored and read bits is using a clock rate.

synchronous data link control (SDLC)—an IBM communications line discipline or protocol associated with SNA. In contrast to BSC, SDLC provides for full-duplex transmission and is more efficient.

synchronous network—network in which all the communications links are synchronized to a common clock.

synchronous transmission—transmission process whereby the information and control characters are sent at regular, clocked intervals so that the sending and receiving terminals are operating continuously in step with each other.

Systems Application Architecture (SAA)—IBM's announced framework for allowing development of consistent applications across six software environments (TSO/E, CICS/MVS, IMS/ESA/TM, VM/CMS, OS400, and OS/2 Extended Edition) running on three hardware computing platforms (System/370 ESA, OS/400, and PS/2).

Glossary of Telecommunications Technologies

systems approach—a general term for reviewing all implications of a condition or group of conditions rather than the narrow implications of a problem at hand.

Systems Network Architecture (SNA)—IBM's standardized relationship between its virtual telecommunication access method (VTAM) and the network control program (NCP/VS).

systems test—a complete simulation of an actual running configuration for purposes of ensuring the adequacy of the configuration.

T

T carrier—a time-division multiplexed, digital transmission facility, operating at an aggregate data rate of 1.544M bps and above. T carrier is a pulse code modulation (PCM) system using 64K bps for a voice channel.

T,R—1) tip, ring. 2) transmit and receive.

T1—a digital carrier facility used to transmit a DS1 formatted digital signal at 1.544M bps; the equivalent of 24 voice channels. The European equivalent transmits at 2.048M bps.

T1C—a digital carrier facility used to transmit a DS1C formatted digital signal at 3.152M bps; the equivalent of 48 voice channels.

T2—a digital carrier facility used to transmit a DS2 formatted digital carrier signal at 6.312M bps; the equivalent of 96 voice channels.

T3—a digital carrier facility used to transmit a DS3 formatted digital carrier signal at 44M bps; the equivalent of 672 voice channels.

T4—a digital carrier facility used to transmit a DS4 formatted digital carrier signal at 273M bps; the equivalent of 4,032 voice channels.

TAAS—trunk answer from any station.

table-driven—describes a logical computer process, widespread in the operation of communications devices and networks, where a user-entered variable is matched against an array of predefined values. Also, a frequently used logical process in network routing, access security, and modem operation.

TACS—total access communications system. A derivative of the AT&T-developed analog cellular radio standard AMPS (advanced mobile phone service) adopted by U.K. Cellnet and Racal-Vodaphone, operating at 900 MHz.

talking battery—dc voltage supplied by the exchange to the subscriber's loop to operate the carbon transmitter in the handset.

talking path—in a telephone circuit, the transmission path consisting of the tip and ring conductors.

tandem—the connection of networks or circuits in series, i.e., the connection of the output of one circuit to the input of another.

tandem data circuit—a data circuit that contains two or more data circuit terminating equipment (DCE) in series.

tandem exchange—exchange that serves to switch traffic between other exchanges when direct trunks are not available. Also called *tandem office*.

tandem office—see *tandem exchange*.

tandem switching—switching of circuits between exchanges only.

tandem tie line switching—PBX permits tie lines to "tandem" through the switch. This means that an incoming tie line call from the distant PBX receives a dial tone instead of automatically connecting with the operator. The outgoing line can be a local trunk or another tie line that links a third system.

tandem trunk—circuit between a tandem exchange and an exchange.

tariff—the published rate for the use of a specific unit of equipment, facility, or type of service provided by a communications common carrier; also, the vehicle by which the regulating agencies approve or disapprove such facilities or services; see also *common carrier*.

TAS—see *telephone answering service*.

TASI—see *time-assignment speech interpolation*.

task—a unit of work.

TAT—transatlantic telephone cable.

TCM—see *traveling class mark*.

TCP/IP—see *transmission control protocol/internet protocol*.

SEPTEMBER 1989

Glossary of Telecommunications Technologies

TDF—trunk distribution frame. See *main distribution frame*.

tdm—tandem.

TDM—see *time-division multiplexing*.

TDMA—see *time-division multiple access*.

Telco—telephone company.

telecommunication lines—telephone and other communication pathways that are used to transmit information from one location to another.

telecommunications—any process that permits the passage of information from a sender to one or more receivers in any usable form (printed copy, fixed or moving pictures, visible or audible signals, etc.) by means of any electromagnetic system (electrical transmission by wire, radio, optical transmission, waveguides, etc.). Includes telegraphy, telephony, video-telephony, data transmission, etc.

telegraphy—branch of telecommunications concerned with processes providing reproduction, at a distance, of written, printed, or pictorial matter or the reproduction at a distance of any kind of information in such form.

teleinformatics/telematics—terms describing non-voice services, particularly those emerging from the integration of computing and telecommunications.

telephone answering service (TAS)—private concern that answers telephone calls and takes messages for a large number of different people and organizations. Because the called number is identified on a special console, the answering service attendant can answer each call as if he or she were actually on the called party's premises.

telephone channel—transmission path designed for the transmission of signals representing human speech or other telephone communication (e.g., facsimile) requiring the same bandwidth. The bandwidth allotted is usually 64K bps, but can be reduced to 32K or even 16K bps with multiplexing techniques.

telephone circuit—electrical connection permitting the establishment of a telephone communication in both directions between two points. Also called *telephone line*.

telephone exchange—switching center for interconnecting the lines that terminate therein. Also called *central office (CO)*.

telephone frequency—any frequency within that part of the audiofrequency range essential for the transmission of speech of commercial quality, i.e., 300 to 3000Hz. Also called *voice frequency (VF)*.

telephone line—see *telephone circuit*.

telephone receiver—device within the handset that converts electrical energy into sound energy and designed to be placed next to the ear.

telephony—generic term describing voice telecommunications.

teleprinter—see *teletypewriter (TTY)*.

teleprocessing—the processing of information that is received from or sent to remote locations by way of telecommunication lines; such systems are necessary to hook up remote terminals or connect geographically separated computers; see also *telecommunications*.

teletext—generically, one-way data transmission designed for widespread broadcasting of graphics and textual information for display on subscriber televisions or low-cost terminals.

teletype—frequently used as a generic name for keyboard/printers and for asynchronous transmission.

teletypewriter (TTY)—start-stop apparatus comprising a keyboard transmitter, together with a printing receiver.

television receive-only—(TVRO) an antenna designed specifically to receive but not transmit a television signal.

telex—1) dial-up telegraph service enabling subscribers to communicate directly and temporarily among themselves by means of start-stop apparatus and of circuits of the public telegraph network. The service operates worldwide using Baudot equipment. 2) Western Union provides such services within the U.S. and abroad under its Telex and Telex II trademarks. Other international record carriers (IRCs) also provide telex services to both the domestic and international markets.

Telex II—see *TWX*.

TELNET (TELetype NETwork)—provides a virtual terminal capability for accessing remote systems as a terminal.

temporary station disconnection—allows the attendant to completely remove selected stations from total service at any time on a temporary basis.

Glossary of Telecommunications Technologies

tenant service—two or more closely located customers can simultaneously be served by the same PBX equipment. Each customer is provided with separate attendant facilities, dedicated trunk facilities, and separate feature and class of service complements. It is often provided by the owner of an office building to the tenants.

terminal—1) a point at which information can enter or leave a communication network. 2) any device capable of sending and/or receiving information over a communication channel.

test center—facility that receives customer trouble reports, tests communications lines and equipment, and can dispatch repair technicians.

test desk—switchboard equipped with testing apparatus, so arranged that connections can be made from it to telephone lines or exchange equipment for testing purposes. Also called *test board*.

TFR—transfer.

TGB—trunk group busy.

TGW—trunk group warning.

third-party maintenance—see *service organizations*.

three-way conference transfer—by depressing the switch hook a user can dial another extension and either 1) hang up and transfer call, 2) get information from the called party and then resume the first call, or 3) bridge all three parties together for a three-way conference.

throughput—the total useful information processed or communicated during a specified time period; expressed in bits per second or packets per second.

tie line (TL)—private-line communications channel of the type provided by communications common carriers for linking two or more points together, typically PBXs. Also called tie trunk; see also *common carrier, trunk line, leased line*.

tie trunk access—allows the system to handle tie lines that can be accessed either by dialing a trunk group access code or through the attendant. Tie lines link a PBX or a distant key system.

time-assignment speech interpolation (TASI)—specialized switching equipment that connects a party to an idle circuit while speech is taking place, but disconnects the party when speech stops, so that a different party can use the same circuit. During periods of heavy traffic, TASI can improve line efficiency from 45 percent to 80 percent.

time-division multiple access (TDMA)—communicating devices at different geographical locations share a multipoint or broadcast channel by means of a technique that allocates different time slots to different users.

time-division multiplexing (TDM)—means of obtaining a number of channels over a single path by dividing the path into a number of time slots and assigning each channel its own intermittently repeated time slot. At the receiving end, each time-separated channel is reassembled. Contrast with *frequency-division multiplexing*.

time-division switching—switching method for a TDM channel requiring the shifting of data from one slot to another in the TDM frame. The slot in question can carry a bit, byte, or, in principle, any other unit of data.

time out—a set time period for waiting before a terminal system performs some action. Typical uses include a poll release (when a terminal is disconnected if the time-out period elapses before keying resumes), or an access time-out (when a terminal on a local area network using a CSMA/CD access method is prevented from transmitting for a specified time period after a collision occurs).

timed recall—the PBX can be instructed to place a call at a designated time. When the time comes, the PBX rings the station. When the station answers, the call is placed automatically.

timed recall on outgoing lines—outgoing trunk calls can be automatically transferred to the attendant after a selected time interval. A warning tone is sent to the calling party several seconds before the transfer takes place.

time series—a sequence of quantitative data assigned to specific intervals.

timesharing—method of operation in which a computer facility is shared by several users for different purposes at (apparently) the same time. Although the computer actually services each user in sequence, the high speed of the computer makes it appear that the users are all handled simultaneously.

time slice—an interval on the central processing unit allocated for performing a task; once the interval is expired, CPU time is allocated to another task; see also *timesharing*.

tip—contacting part at the end of a telephone plug or the top spring of a jack. The conductors associated with these contacts. The other contact is called a ring.

SEPTEMBER 1989

Glossary of Telecommunications Technologies

TL—see *tie line*.

TLF—trunk link frame (crossbar switching).

TLP—see *transmission level point*.

token bus—a local network access mechanism and topology in which all stations actively attached to the bus listen for a broadcast token or supervisory frame. Stations wishing to transmit must receive the token before doing so; however, the next physical station to transmit is not necessarily the next physical station on the bus. Bus access is controlled by preassigned priority algorithms.

token passing—a local area network access technique in which participating stations circulate a special bit pattern that grants access to the communications pathway to any station that holds the sequence; often used in networks with a ring topology.

token ring—a local network access mechanism and topology in which a supervisory frame or token is passed from station to station in sequential order. Stations wishing to gain access to the network must wait for the token to arrive before transmitting data. In a token ring, the next logical station receiving the token is also the next physical station on the ring.

toll call—call outside the local exchange area, charged at toll rates.

toll center—class 4/primary telephone exchange where time- and distance-based toll charge information is collected.

toll-connecting trunk—trunk used to connect a class 5/local exchange to the long-distance network.

toll restriction—blocking a telephone user's access to the toll network.

toll terminal access—allows guest stations to access toll calling trunks. These can be direct dial or operator access depending on the servicing public exchange office.

tone ringing—either a steady or oscillating electronic tone is provided at the station instrument to provide incoming calls with audible signaling. Also called *tone calling*.

tone signaling—transmission of supervisory, address, and alerting signals over a telephone circuit by means of tones.

tone-to-dial-pulse conversion—converts DTMF signals to dial pulse signals when the trunks associated with outgoing trunk calls are not equipped to receive tone signals. Auxiliary dial pulse conversion equipment is not necessary.

TOP—technical office protocols developed by Boeing.

topology—the logical or physical arrangement of stations on a network in relation to one another, e.g., bus, ring, star, and tree.

touch sensitive—refers to the technology that enables a system to identify a point of contact on the screen by coordinates and transmit that information to a program.

Touch-tone—registered AT&T trademark for pushbutton dialing. See also *dual tone multifrequency signaling (DTMF)*.

trace packet—in packet switching, a special kind of packet that functions as a normal packet but causes a report of each stage of its progress to be sent to the network control center.

tracking, telemetry, control, and monitoring (TTC&M)—specialized ground stations used to track and control satellites and to monitor their performance.

traffic—1) messages sent and received over a communications channel. 2) quantitative measurement of the total messages and their length, expressed in hundred call seconds (CCS) or other units.

traffic data to customer—customer can poll switching locations on a daily or hourly basis to obtain traffic measurements, including usage and overflow data. Summary reports, exception reports, and complete traffic register outputs can be obtained.

traffic flow—measure of the density of traffic, expressed in Erlangs.

traffic matrix—matrix of which the X,Y element contains the amount of traffic originated at node X and destined for node Y. The unit of measurement could be calls or packets per second, for example, depending on the kind of network.

traffic monitor—PBX feature that provides basic statistics on the amount of traffic handled by the system.

traffic overflow—occurs when traffic flow exceeds the capacity of a particular trunk group and is automatically switched over to another trunk group.

traffic service position system (TSPS)—stored-program computer with telephone operator

Glossary of Telecommunications Technologies

consoles permitting calls needing operator intervention to be handled as efficiently as possible.

transaction file—a collection of records for any business activity or request that is entered into a computer system.

transaction processing—a procedure in which files are interactively updated and results are generated immediately as a result of data entry.

transceiver—device that can transmit and receive traffic.

transducer—a device for converting signals from one form to another, such as a microphone or a receiver.

transfer rate—the speed at which information can be sent across a bus or communications link.

transistor-transistor logic (TTL)—a type of signaling, in which a nominal +5V is equated with logic 1, and a nominal 0V is equated with logic 0.

translation—when using automatic route selection or trunk-to-trunk connection features, the PBX can add or delete area codes and toll access digits from number codes, and toll access digits from numbers so that the call will be handled properly by the switching network.

translator—device that converts information from one system of representation into equivalent information in another system of representation. In telephone equipment, it is the device that converts dialed digits into call routing information.

transmission—sending information in the form of electrical signals over electric wires, waveguides, or radio.

transmission control protocol/internet protocol (TCP/IP)—ensures packets of data are delivered to their destinations in the sequence in which they were transmitted.

transmission level point (TLP)—any point in a transmission system at which the power level of the signal is measured.

transmission speed—the rate at which information is passed through communications lines, generally measured in bits per second (bps).

transmit—to send information from one location to another.

transport layer—in the OSI model, the network processing entity responsible, in conjunction with the underlying network, data link, and physical layers, for the end-to-end control of transmitted data and the optimized use of network resources.

traveling class mark (TCM)—when automatic route selection (ARS) or uniform numbering/automatic alternative routing selects a tie trunk to a distant tandem PBX, the traveling class mark is sent over the tie trunk. It is then used by the distant system to determine the best available facility consistent with the user's calling privileges. The TCM indicates the restriction level to be used based on the station, trunk, or attendant originating the call or the authorization code, if dialed.

traveling wave tube amplifier (TWTA)— a satellite component used to amplify an incoming signal before transmitting to receiving earth stations.

tree—a type of bus network topology in which the medium branches at certain points along its length to connect stations or clusters of stations; also called a *branching bus*.

truncation—the removal of one or more digits, characters, or bits from an item of data when a string length or precision of a target variable has been exceeded; also, to cut off a computational procedure at a specified spot.

trunk—transmission paths that are used to interconnect exchanges in the main telephone network. Also, a telephone exchange line that terminates in a PBX.

trunk code—code consisting of one or more digits used to designate a called numbering plan area.

trunk exchange—exchange to which trunk circuits are connected, but not subscribers' lines.

trunk group—those trunks that connect two points, both of which are exchanges and/or individual message distribution points and both of which employ the same multiplex terminal equipment. Also, a discrete group of trunk lines with a specific function in a PBX.

trunk prefix—one or more digits to be dialed before the trunk code when making a call to a subscriber outside the local area but in the same country.

trunk reservation—attendant can hold a single trunk in a group and then extend it to a specific station.

trunk-to-tie trunk connections—ability of the switching system to provide the attendant with the capability of extending an incoming exchange call to a tie trunk that terminates within the system.

SEPTEMBER 1989

Glossary of Telecommunications Technologies

trunk-to-trunk connections, station—station already in connection with an incoming or outgoing trunk is able to utilize the consultation hold and add-on conference circuitry to effect a conference with another circuit or with another trunk.

trunk-to-trunk consultation—allows a station connected to an outside trunk circuit to gain access to a second outside trunk circuit for "outside" consultation; however, no conference capability is available with this feature.

trunks, direct termination—each incoming or combination trunk appears on a key at a console or at a jack position on a cord switchboard and the attendant is always in full visual supervision of the status of all such circuits.

TSPS—see *traffic service position system*.

tst—test.

TTC&M—tracking, telemetry, control, and monitoring.

TTL—see *transistor-transistor logic*.

TTTN—tandem tie trunk network. Private network making use of tandem switching.

TTY—teletype.

tuning—the process of adjusting computer system control variables to make a system divide its resources most efficiently for a workload.

turnkey system—complete communications system, including hardware and software, assembled and installed by a vendor and sold as a total package.

TVRO—see *television receive-only*.

TVS—trunk verification by station. See *trunk verification by customer*.

twisted pair—two insulated wires twisted together but not covered with an outer sheath.

two-party station service—PBX system with two internal stations and with selective ringing to each.

two-way alternate operation—see *half-duplex*.

two-way simultaneous operation—see *full-duplex*.

two-way splitting—attendant is able to consult privately with either party (internal or external) on a call.

two-wire channel—a circuit containing two single wires in pair (or logical equivalent) for nonsimultaneous (i.e., half-duplex) two-way transmission. Contrast with *four-wire channel*.

two-wire circuit—a circuit formed by two conductors insulated from each other that can be used as either a one-way transmission path, a half-duplex path, or a duplex path. Contrast with *four-wire circuit*.

TWTA—see *traveling wave tube amplifier*.

TWX—an obsolete but still popular term for Western Union's switched teleprinter exchange service, now called Telex II. Interconnection is provided with the Telex network through Western Union's computers even though speeds and codes are different.

TXD—telephone exchange, digital. Exchange making use of digital transmission techniques.

TXE—telephone exchange, electronic. Exchange making use of electronic switching techniques, as opposed to electromechanical means such as crossbar or step-by-step switches.

TXC—telephone exchange, crossbar. See *crossbar switch*.

TXK—telephone exchange, crossbar. See *crossbar switch*.

TXS—telephone exchange, stronger. See *step-by-step switch (SXS)*.

U

UCD—see *uniform call distribution*.

UDLC—Universal Data Link Control; a bit-oriented communications standard developed by Sperry. (Sperry is now Unisys.)

UG—underground.

UHF—see *ultra high frequency*.

ultra high frequency (UHF)—portion of the electromagnetic spectrum ranging from about 300 MHz to about 3 GHz. The frequency band includes television and cellular radio frequencies.

unattended operation—transmission and/or reception that is controlled automatically and does not require a human operator.

Glossary of Telecommunications Technologies

unbalanced-to-ground—with a two-wire circuit, condition in which the impedance-to-ground on one wire is measurably different from that on the other. Compare with *balanced-to-ground*.

uniform call distribution (UCD)—allows calls coming in on a group of lines to be assigned stations as smoothly as possible so that all stations can handle similar loads. Most call distribution systems also provide for a queuing of incoming calls with the longest holding time presented for service first.

uniform numbering plan—permits station users at a PBX to place calls over tie trunks using a uniform dialing plan.

uniform-spectrum random noise—noise distributed over the spectrum in such a way that the power-per-unit bandwidth is constant. Also called *white noise*.

uninterruptible power supply (UPS)—usually includes an inverter, drawing its power from batteries, which generates an extremely "well-behaved" AC power signal for a PBX or other equipment. The UPS cost is related to the amount of power needed and the length of time it must operate during a failure. If a particularly heavy demand is anticipated, the system can be coupled with an auxiliary generator that is started when commercial power is interrupted.

universe—the population from which samples are drawn.

UNIX—a multiuser operating system developed by Bell Laboratories.

unlisted number—telephone number that is not listed in the telephone directory and not provided by directory-assistance operators. There is usually an additional charge to the subscriber for the deletion from the directory.

unrestricted extension—PBX extension permitted to make exchange line calls without the assistance of the PBX operator.

UPS—see *uninterruptible power supply*.

USASCII—see *American Standard Code for Information Interchange (ASCII)*.

USITA—United States Independent Telephone Association.

USOC—Uniform Service Order Code.

V

VA—volt ampere.

VAC—volts alternating current.

vacant code intercept—routes all calls made to an unassigned "level" (first digit dialed) to the attendant, a busy signal, a "reorder" signal, or a recorded announcement.

vacant number intercept—usually routes all calls of unassigned numbers to the attendant or a recorded announcement.

VADS—value added data network.

validation—an attempt to find errors by executing a program in a given environment.

validity check—verification that each element of information is an actual character of a code in use.

value-added carriers—vendors that design and enhance the phone network and resell the service.

value-added common carrier—company that sells services of a value-added network. It can be a PTT or subsidiary or an independent company.

value-added network (VAN)—a public data communications network that provides basic transmission facilities (generally leased by the VAN vendor from a common carrier) plus additional, "enhanced" services such as computerized switching, temporary data storage, protocol conversion, error detection and correction, electronic mail service, etc.

value-added service—a communications facility utilizing communications common carrier networks for transmission and providing enhanced extra data features with separate equipment; such extra features, including store-and-forward message switching, terminal interfacing, and host interfacing, are common.

VAN—see *value-added network*.

variance—the difference between the expected (or planned) and the actual.

VDC—volts direct current.

VDT—1) video display terminal. 2) visual display terminal.

VDU—visual display unit.

SEPTEMBER 1989

Glossary of Telecommunications Technologies

vendor—a company that supplies products and services.

verification—an attempt to find errors by executing a program in a test or simulated environment.

vertical redundancy check—an error-checking method that uses a parity bit for each character; see also *parity*.

very high frequency (VHF)—portion of the electromagnetic spectrum with frequencies between about 30 MHz and 300 MHz. Operating band for radio and television channels.

very low frequency (VLF)—refers to frequencies below 30 KHz.

very small aperture terminal (VSAT)—an earth station with a small antenna, usually 6 meters or less. VSATs are typically used in point-to-multipoint data networks.

VF—see *voice frequency*.

VHF—see *very high frequency*.

videoconferencing—an electronic meeting in which geographically separated groups communicate using interactive audio and video technology.

video signal—signal comprising frequencies normally required to transmit pictorial information (1 MHz to 6 MHz).

videotex—an interactive data communications application broadcast over the public switched telephone network and received on adapted television receivers. It is designed to allow unsophisticated users to converse with remote databases, enter data for transactions, and retrieve textual and graphics information. Also called *viewdata*.

viewdata—see *videotex*.

virtual circuit—proposed CCITT definition for a data transmission service in which the user presents a data message for delivery with a header of a specified format. The system delivers the message as though a circuit existed to the specified destination. One of many different routes and techniques could be used to deliver the message, but the user doesn't know which is employed. Contrast with *permanent virtual circuit*.

visually impaired attendant service—achieved by augmenting the normal visual signals provided on a standard attendant position with special tactile devices and/or audible signals that enable a visually impaired person to operate the position.

VLF—see *very low frequency*.

VLSI—very large scale integration. See *large scale integration*.

voice band—see *voice grade channel*.

voice digitization—conversion of an analog voice into digital symbols for storage or transmission.

voice frequency (VF)—any frequency within that part of the audio frequency range essential for the transmission of speech of commercial quality; i.e., 300 to 3000Hz. Also called *telephone frequency*.

voice grade—telecommunications link with a bandwidth (about 3 KHz) appropriate to an audio telephone line.

voice grade channel—a channel with a frequency range from 300 to 3000 Hz and suitable for the transmission of speech, data, or facsimile.

voice message service—provides the ability for a station user to access an optional voice message recording facility and leave a message for a particular station user.

voice-operated device—a piece of equipment that permits telephone currents to control a task.

voice paging access—allows attendants and station users the ability to have dial access to customer-provided loudspeaker paging equipment.

voice print—a technique for verifying an individual's identity through the pattern produced by his or her voice.

voice recognition—a system of sound sensors that translates the tones of human sounds into computer commands.

voice response unit—a computer-connected device that can selectively link sentences of stored words to create a spoken message.

voice store-and-forward system (VSF)—processor-controlled system that allows voice messages to be created, edited, sent, stored, and forwarded. Users access and operate the system by means of any 12-button dialpad, in response to voice prompts from the system.

voice synthesis—computer-generated sounds that simulate the human voice.

volatile storage—memory that loses its contents when electrical power is removed.

VRC—see *vertical redundancy check*.

VSF—voice store-and-forward.

Glossary of Telecommunications Technologies

W

WARC—World Administative Radio Conference. ITU conference for deciding the allocation of international radio frequencies and satellite geostationary orbit locations.

WATS—see *wide area telephone service.*

WATTC—World Administative Telegraph and Telephone Conference. ITU conference for deciding provisions for the interconnecting and interworking of the world's telecommunications networks.

waveguide—transmission path in which a system of boundaries guides electromagnetic energy. The most common of these are hollow metallic conducting tubes (microwave communications) or rods of dielectric material. See also *fiber-optic waveguides.*

white noise—see *uniform-spectrum random noise.*

wide area telephone service (WATS)—telephone company service allowing reduced costs for certain telephone call arrangements. This can be in-wats, or 800-number service, where calls can be placed to a location from anywhere at no cost to the calling party, or out-wats, where calls are placed out from a central location. Cost is generally based on hourly usage per wats circuit and on distance based on zones, or bands, to which or from which calls are placed.

wide frequency tolerant power plant—provides PBX power facilities that will operate from ac energy sources that are not as closely regulated as commercial ac power. The wide tolerant plant will tolerate average frequency deviations of up to ± 3 Hz or voltage variations of – 15 percent to + 10 percent as long as both of the conditions do not occur simultaneously. This feature permits operation with customer-provided emergency power generating equipment.

wideband—see *broadband.*

wideband channel—channel wider in bandwidth than a voice grade channel. See also *broadband.*

world numbering plan—CCITT numbering plan that divides the world into nine zones. Each zone is allocated a number that forms the first digit of the country code for every country in that zone. The zones are as follows:

- (1) North America,
- (2) Africa,
- (3 and 4) Europe,
- (5) South America,
- (6) Australia,
- (7) USSR,
- (8) North Pacific (Eastern Asia), and
- (9) Far East and Middle East.

X

X.21—CCITT recommendation that describes the interface used in the CCITT X.25 packet switching protocol and in certain types of circuit switched data transmissions.

X.25—CCITT recommendation that specifies the interface between user data terminal equipment (DTE) and packet-switching data circuit-terminating equipment (DCE).

X.400—a standard for electronic mail exchange; developed by the CCITT.

X.75—a standard for connecting X.25 networks; developed by the CCITT.

xbar—crossbar.

XD—ex-directory. XD refers to a subscriber number that is not listed in a printed directory. Also called *unlisted number.*

xfr—transfer.

X-on/X-off—a method of controlling data communications; it essentially allows a terminal to activate a line when it is ready to receive information and suspend line activity when the terminal is overloaded; see *asynchronous.*

Z

zip tone—short burst of dial tone to the headset of an ACD agent, indicating that a call is being connected to the agent console. □

SEPTEMBER 1989

Index

Index note: The *f.* after a page number refers to a figure; the *n.* to a note; and the *t.* to a table.

AC (air conditioning), **5.**16, **5.**18
ACD (automatic call distribution), **6.**12, **6.**16
Act to Regulate Commerce (1887), **1.**10
Advanced Mobile Phone Service (AMPS), **6.**26, **6.**27
Air conditioning (AC), **5.**16, **5.**18
American Bell, **1.**16
American National Standards Institute (ANSI), **1.**8
   Subcommittee on Information Processing Systems, **1.**7, **1.**8
American Radio Telephone Service (ARTS), **6.**26, **6.**27
American Telephone and Telegraph Co. (AT&T), **1.**10 to **1.**13, **3.**10, **6.**13, **6.**26
   antitrust suit, **1.**14 to **1.**17
   cable placement, **10.**5
   planning guidelines, **9.**5*n.*, **9.**23*n.*, **9.**25*n.*, **9.**26*n.*
   reorganization, **1.**17 to **1.**19, **2.**4, **6.**20
Ameritech, **6.**12
Amplitude-frequency distortion, **5.**10
AMPS (Advanced Mobile Phone Service), **6.**26, **6.**27
ANSI (American National Standards Institute), **1.**7, **1.**8
Apparatus closets, **4.**1 to **4.**3, **5.**17
   layout, **4.**1, **4.**2*f.*
   management and control systems, **8.**13, **8.**15*f.*, **8.**16
   specifications, **4.**2 to **4.**3
Archer, Rowland, **7.**1*n.*
ARS (automatic route selection), **6.**13, **6.**15 to **6.**16
ARTS (American Radio Telephone Service), **6.**26, **6.**27
Asynchronous communications, **7.**2

AT&T (*see* American Telephone and Telegraph Co.)
AT&T Information Systems, **1.**12
AT&T Long Lines, **1.**12, **1.**15
AT&T Technologies, **6.**23
Attenuation distortion, **5.**10
Automatic call distribution (ACD), **6.**12, **6.**16
Automatic route selection (ARS), **6.**13, **6.**15 to **6.**16

Baker, Donald C., **5.**12*n.*
Bell Atlantic, **6.**12, **6.**14
Bell Communications Research, **1.**15
Bell Laboratories, **1.**12, **1.**14, **6.**26
Bell operating company (BOC), **1.**12, **1.**17 to **1.**19, **3.**3
   telecommunications systems and services, **6.**1, **6.**2, **6.**5, **6.**13, **6.**14, **6.**16, **6.**18 to **6.**20
Bell System, **1.**3 to **1.**5, **1.**10, **1.**12 to **1.**14, **2.**4, **5.**11
BellSouth, **6.**12
Bias distortion, **5.**11 to **5.**12
   negative (spacing), **5.**11
   positive (marking), **5.**11
BOC (*see* Bell operating company)
British Independent Authority, **6.**44
Business television, **6.**44

Cable (*see* Wire, cable and coax applications)
California, **2.**7, **2.**10
California Public Utility Commission, **2.**10
California Telecommunications Division
   bid specifications, **9.**30 to **9.**128*f.*

Campus distribution systems, **4.**19 to **4.**24
  aerial construction, **4.**21 to **4.**22
    cable dancing, **4.**21
  buried cable, **4.**22
  conformance tests, **10.**9
  layout, **4.**19, **4.**20*f.*
  location records, **8.**21
  outside plant, **4.**19
  underground conduit, **4.**22 to **4.**24
Carter Electronics Corp., **1.**11
Carterfone, **1.**11
CATV (community antenna television),
  2.3
CB (citizens' band), **6.**27
CCIR (International Radio Consultive
  Committee), **1.**6 to **1.**8
CCITT (Consultative Committee on
  International Telegraph and
  Telephone), **1.**6 to **1.**8, **5.**11, **5.**12, **6.**7,
  6.14, **6.**49, **7.**1, **7.**4 to **7.**7
CCSA (common-control switching
  arrangement), **6.**13
Cellular radio (telephone), **6.**26 to **6.**30
  cells, **6.**26
  components, **6.**27 to **6.**29
    cell site, **6.**27, **6.**28
    mobile and portable units, **6.**27, **6.**29
    switching office, **6.**27
    system interconnections, **6.**27 to **6.**29
  frequency reuse, **6.**26
  services, **6.**29 to **6.**30
    data transmission, **6.**30
      microcom networking protocol
        (MNP), **6.**30
    roaming, **6.**29
  technology, **6.**27
Central office (exchange), **1.**4
Central office-local-area network (CO-
  LAN), **6.**14, **6.**15, **6.**18 to **6.**20, **6.**24
Central services organization (CSO), **1.**15
Centrex, **6.**8, **6.**12 to **6.**13
  data communications, **6.**14
    CO-LAN, **6.**14, **6.**15, **6.**18 to **6.**20, **6.**24
  digital, **6.**13
  features, **6.**9 to **6.**12*t.*, **6.**14 to **6.**16
    automatic route selection, **6.**13, **6.**15
      to **6.**16
    enhanced, **6.**12 to **6.**13
    SMDR, **6.**14, **6.**15, **6.**22
    station rearrangement (CSR), **6.**15
    uniform call distribution, **6.**13, **6.**15
    voice messaging, **6.**14 to **6.**15

Centrex (*Cont.*):
  versus PBX as customer premises
    equipment, **6.**16 to **6.**25
    capital investment and recurring
      costs, **6.**17 to **6.**19
    floor space, **6.**17
    maintenance, **6.**18 to **6.**19
    mileage charges, **6.**18
    power considerations, **6.**17 to **6.**18,
      6.21
    price comparison, **6.**19
    subscriber line charge, **6.**18
    government regulations, **6.**19
  key and hybrid systems, **6.**20 to **6.**25
    centrex-compatible equipment,
      6.23
    electromechanical, **6.**20 to **6.**21
    electronic, **6.**21 to **6.**23
    hybrid electronic, **6.**22 to **6.**23
    installation charges, **6.**24
    key service unit (KSU), **6.**20 to **6.**22
    key telephone unit (KTU), **6.**20 to
      6.22
    main distribution frame (MDF),
      6.20 to **6.**21
    non-KSU, **6.**23
    reconfiguration, **6.**24
    selection guidelines, **6.**24 to **6.**25
    spare parts, **6.**23
    technology, **6.**19 to **6.**20
Centrex station rearrangement (CSR),
  6.15
CEPT (European Conference of Postal
  and Telecommunications
  Administrations), **1.**7
Characteristic distortion, **5.**12
Cincinnati Bell, **6.**12
Citizens' band (CB), **6.**27
Cluster One, **7.**13, **7.**14
CO-LAN (central office-local-area net-
  work), **6.**14, **6.**15, **6.**18 to **6.**20,
  6.24
Coax (*see* Wire, cable and coax applica-
  tions)
Codec, **6.**4
Common-control switching arrangement
  (CCSA), **6.**13
Communication, **1.**1 to **1.**3
Communications Act (1934), **1.**10, **1.**12,
  1.14, **1.**15
Communications Satellite Act (1962),
  1.11

Community antenna television (CATV), **2.3**

Conditioning, **5.12**

Connectivity, **1.19**

Consent Decree, **3.3**

Consultative Committee on International Telegraph and Telephone (CCITT), **1.6** to **1.8**, **5.11**, **5.12**, **6.7**, **6.14**, **6.49**, **7.1**, **7.4** to **7.7**

Corporation for Open Systems (COS), **1.7**, **1.8**

CPE (*see* Customer premises equipment)

Cross talk, **5.5** to **5.7**

CSO (central services organization), **1.15**

CSR (centrex station rearrangement), **6.15**

Customer premises equipment (CPE), **1.12**, **1.14** to **1.19**, **6.1**, **6.2**, **6.13**, **6.15** to **6.25**

capital investment and recurring costs, **6.17** to **6.19**

floor space, **6.17**

maintenance, **6.18** to **6.19**

mileage charges, **6.18**

power considerations, **6.17** to **6.18**, **6.21**

price comparison, **6.19**

subscriber line charge, **6.18**

government regulations, **6.19**

key and hybrid systems, **6.20** to **6.25**

centrex-compatible equipment, **6.23**

electromechanical, **6.20** to **6.21**

electronic, **6.21** to **6.23**

hybrid electronic, **6.22** to **6.23**

installation charges, **6.24**

key service unit (KSU), **6.20** to **6.22**

key telephone unit (KTU), **6.20** to **6.22**

main distribution frame (MDF), **6.20** to **6.21**

non-KSU, **6.23**

reconfiguration, **6.24**

selection guidelines, **6.24** to **6.25**

central equipment, **6.25**

contract, **6.25**

features, **6.24** to **6.25**

request for proposal (RFP), **6.25**

system examination, **6.25**

spare parts, **6.23**

technology, **6.19** to **6.20**

Data processing applications, **1.20**, **7.1** to **7.17**

data speed, **7.7** to **7.8**

bandwidth, **7.7**

baud, **7.7** to **7.8**

802 Committee on Local-Area Networks, **1.8**, **7.1**, **7.3** to **7.4**

industry overview, **7.1**

open system interconnection, **1.3**, **1.7**, **1.8**, **7.1** to **7.3**

layer 1 (physical), **7.2**

layer 2 (data link), **7.2**

layer 3 (network), **7.2**

layer 4 (transport), **7.2**, **7.3**

layer 5 (session), **7.2**, **7.3**

layer 6 (presentation), **7.2**, **7.6**

layer 7 (application), **7.2**, **7.3**

protocol converters, **7.14** to **7.16**

cathode-ray tube (CRT), **7.15**

protocols, **7.11** to **7.16**

cable access, **7.12** to **7.15**

carrier-sending multiple access with collision avoidance (CSMA/CA), **7.13**

carrier-sending multiple access with collision detection (CSMA/CD), **7.13** to **7.14**

contention, **7.13** to **7.14**

exhaustive service, **7.13**

multilevel multiple access (MLMA), **7.14**

nonexhaustive service, **7.13**

ordered access bus, **7.14** to **7.15**

slot, **7.12**

token passing, **7.12** to **7.13**

topology (architecture), **7.8** to **7.11**

branching tree, **7.10** to **7.11**

repeater, **7.10**

splitters, **7.10**

bus, **7.9**

ring, **7.10**

star, **7.8** to **7.9**

vendor connectivity specifications, **7.16** to **7.17**

wide-area network, **7.4** to **7.7**

Datapro Research Corp., **1.4***n.*, **1.5***n.*, **1.8***n.*, **1.9***n.*, **1.18***n.*

on power considerations, **5.13***n.*, **5.16***n.*

on telecommunications systems and services, **6.3***n.*, **6.8***n.*, **6.16***n.*, **6.25***n.*, **6.26***n.*, **6.34***n.*, **6.39***n.*, **6.46***n.*, **6.49***n.*

deButts, John B., **1.15**

Delay (phase-frequency) **5.**10 to **5.**11
Demarcation point, **2.**7, **2.**9*f.*, **2.**10, **5.**17
  campus distribution systems, **4.**19
  environmental specifications, **2.**10 to
      2.11
  equipment room, **5.**17
  horizontal building systems, **2.**5
  management and control systems, **8.**1,
      8.12 to **8.**14
  riser cable, **2.**14
  service entrance, **2.**11
  support structure and connectivity
      hardware, **4.**4, **4.**6 to **4.**8
    hardware, **4.**4, **4.**6
      blue, **4.**6
      gray, **4.**6
      green, **4.**6, **4.**7*f.*
      orange, **4.**6
      purple, **4.**6
      red, **4.**6
      white, **4.**6
      yellow, **4.**6
    physical security, **4.**6, **4.**8
  vertical building systems, **2.**6, **2.**7
Department of Commerce, **1.**10
Department of Justice, **1.**14 to **1.**18
Department of State, **1.**7, **1.**8
Digital switching systems, **6.**13

Echo, **5.**7
ECMA (European Computer
    Manufacturers Association), **1.**7
Economic evaluation (*see* Planning guide-
    lines)
ECSA (Exchange Carriers Standards
    Association), **1.**7, **1.**8
EIA (Electronic Industries Association),
    1.7, **1.**8
802 Committee on Local-Area Networks
    (IEEE), **1.**8, **7.**1, **7.**3 to **7.**4
EKTS (Electronic Key Telephone
    System), **5.**18
Electrical power (*see* Power considera-
    tions)
Electromechanical interference (EMI), **5.**4
Electronic directory, **6.**13
Electronic Industries Association (EIA),
    1.7, **1.**8
Electronic Key Telephone System
    (EKTS), **5.**18
Electronic switching system (ESS), **6.**18

Electronic tandem switching (ETS), **6.**13
EMI (electromechanical interference), **5.**4
Equipment room, **5.**17
ESS (electronic switching system), **6.**18
Ethernet, **7.**3, **7.**9, **7.**13, **7.**14
ETS (electronic tandem switching), **6.**13
European Computer Manufacturers
    Association (ECMA), **1.**7
European Conference of Postal and
    Telecommunications Administrations
    (CEPT), **1.**7
Exchange (central office), **1.**4
Exchange Carriers Standards Association
    (ECSA), **1.**7, **1.**8

Fax, **6.**30
Federal Communications Commission
    (FCC), **1.**7 to **1.**15, **1.**17, **1.**18
  telecommunication systems and ser-
      vices and, **6.**1, **6.**19, **6.**23, **6.**24,
      6.26, **6.**31, **6.**33, **6.**34, **6.**38, **6.**41
Federal Telecommunications System
    (FTS), **6.**13
Fiber optics (*see* Optical-fiber applica-
    tions)
Fiber superhighway (fiber rings), **6.**1
Flexible route selection, **6.**13
Foreign exchange (FX), **6.**13
Four-wire, **6.**6, **6.**7
FTS (Federal Telecommunications
    System), **6.**13
FX (foreign exchange), **6.**13

Generic program, **6.**18
Government agency use of centrex ser-
    vices, **6.**8
Government regulations, **1.**9 to **1.**19, **3.**3
  Act to Regulate Commerce (1887), **1.**10
  AT&T antitrust suit, **1.**14 to **1.**19, **2.**4
  Communications Act (1934), **1.**10, **1.**12,
      1.14, **1.**15
  Communications Satellite Act (1962),
      1.11
  Computer Inquiry I, **1.**11
  Computer Inquiry II, **1.**12, **6.**16
  customer premises equipment, **6.**19
  Fully Distributed Cost Method 7, **1.**13
  jurisdictional separations, **1.**13 to **1.**14
  mobile communications, **1.**11, **1.**13, **1.**14
  natural monopoly, **1.**9, **2.**3

Government regulations(*Cont.*):
  Post Roads Act (1866), **1.**9 to **1.**10
  Radio Act (1912), **1.**10
Green, Harold H., **1.**16
GTE, **6.**12

Harmonic distortion, **5.**10
Hilton, **6.**44
Holiday Inn, **6.**44

IBM, **3.**7, **7.**15
ICC (Interstate Commerce Commission),
  1.10
IEEE (*see* Institute of Electrical and
  Electronics Engineers)
Impulse noise, **5.**3 to **5.**5
Information exchange process, **1.**1 to
  1.3
Information industry, **1.**3
Infrared communications, **6.**34 to **6.**39
  costs, **6.**39
  hardware components, **6.**36 to **6.**37
    very large scale integration (VLSI),
      6.37
  system architecture, **6.**36
    time-division multiple-access
      (TDMA), **6.**36
  system components, **6.**37 to **6.**39
    coherent light, **6.**37
    light-emitting diode (LED), **6.**37
    noncoherent (random-phase) light,
      6.37
    receivers, **6.**38 to **6.**39
    transmitters, **6.**37 to **6.**38
      terrestrial atmospheric laser com-
        munication, **6.**38
  technology, **6.**34 to **6.**36
    skin effect, **6.**34 to **6.**35
    transmission, **6.**35 to **6.**36
      bandwidth, **6.**35
      modulation, **6.**35
Institute of Electrical and Electronics
  Engineers (IEEE), **1.**7, **1.**8
  802 Committee on Local-Area
    Networks, **1.**8, **7.**1, **7.**3 to **7.**4
Instruments, **2.**1
Integrated services digital network
  (ISDN), **1.**8, **6.**1, **6.**12 to **6.**14, **6.**20,
  6.47 to **6.**48
  protocols, **6.**14

Interexchange carrier termination, **6.**13,
  6.14
Interexchange company (IXC), **1.**18
Intermodulation noise, **5.**6 to **5.**7
International Radio Consultive
  Committee (CCIR), **1.**6 to **1.**8
International Standards Organization
  (ISO), **1.**6, **6.**49
  open system interconnection, **1.**3, **1.**7,
    1.8, **7.**1 to **7.**3
    layer 1 (physical), **7.**2
    layer 2 (data link), **7.**2
    layer 3 (network), **7.**2
    layer 4 (transport), **7.**2, **7.**3
    layer 5 (session), **7.**2, **7.**3
    layer 6 (presentation), **7.**2, **7.**6
    layer 7 (application), **7.**2, **7.**3
International Telecommunications Union
  (ITU), **1.**6, **6.**6 to **6.**7, **7.**1
Interstate Commerce Commission (ICC),
  1.10
ISDN (*see* Integrated services digital net-
  work)
ISDN user part (ISUP), **6.**14
ISO (*see* International Standards
  Organization)
ISUP (ISDN user part), **6.**14
ITU (International Telecommunications
  Union), **1.**6, **6.**6 to **6.**7, **7.**1
IXC (interexchange company), **1.**18

Job scheduling, **1.**20, **10.**1 to **10.**9
  conformance testing, **10.**7 to **10.**9
    data cable tests, **10.**8
    defect types, **10.**7 to **10.**8
      crosses, **10.**8
      ground, **10.**8
      open, **10.**8
      short, **10.**8
      splits, **10.**8
    interbuilding or campus distribution
      systems, **10.**9
  coordination responsibility, **10.**1 to **10.**2
  inspection requirements, **10.**6 to **10.**7
  performance control system, **10.**5 to
    10.6
    cable placement, **10.**5 to **10.**6
    cable termination, **10.**5, **10.**6
  purpose of, **10.**1
  scheduler authority and responsibility,
    10.2

Job scheduling (*Cont.*):
 scheduling techniques, **10.**2 to **10.**5
  equipment and service order organization, **10.**2, **10.**4
  facilities design, **10.**2, **10.**4
  office/building planning, **10.**2, **10.**3
  procurement services, **10.**3, **10.**4
  public sector utility companies, **10.**3, **10.**4
  vendor and contractor schedules, **10.**3, **10.**5
 system acceptance, **10.**9

LAN (*see* Local-area network)
LATA (local access and transport area), **1.**16 to **1.**18
LEC (local exchange carrier), **1.**18
Lighting, **5.**16
Local access and transport area (LATA), **1.**16 to **1.**18
Local-area network (LAN), **1.**8, **3.**5, **3.**9, **3.**14, **6.**31
 cable access protocols, **7.**12 to **7.**15
 standards, **7.**1 to **7.**4
 topology (architecture), **7.**8 to **7.**11
  branching tree, **7.**10 to **7.**11
   repeater, **7.**10
   splitters, **7.**10
  bus, **7.**9
  star, **7.**8 to **7.**9
 vendor connectivity specifications, **7.**16 to **7.**17
Local exchange carrier (LEC), **1.**18
Local loops, **2.**1 to **2.**3

Main building terminal (*see* Demarcation point)
Management and control systems, **1.**20, **8.**1 to **8.**33
 building terminal records, **8.**1, **8.**12 to **8.**16
  apparatus and satellite closet, **8.**13, **8.**15*f.*, **8.**16
  data patch panel, **8.**16
  demarcation point, **8.**1, **8.**12 to **8.**14
  equipment room, **8.**16, **8.**17*f.*
 cable assignment records, **8.**16, **8.**18 to **8.**21
  cable type, **8.**16
  count, **8.**16

Management and control systems, cable assignment records (*Cont.*):
  number, **8.**16
  place of origin, **8.**16
  size, **8.**16
  termination location, **8.**16
 facility and equipment diagnostic system, **8.**24 to **8.**28
  data troubleshooting procedure, **8.**24, **8.**26, **8.**27*f.*
  diagnostic equipment, **8.**26, **8.**28
   ohmmeter, **8.**28
   tone generator, **8.**28
   hand-held digital multimeter, **8.**28
  line troubleshooting procedure, **8.**24, **8.**26*f.*
  lineman's test set (butt-in), **8.**28
  modular and nonmodular set flowchart, **8.**24, **8.**25*f.*
  modular phone jack tester, **8.**28
  optical time-domain reflectometer (OTDR), **8.**28
  time-domain reflectometer (TDR), **8.**28
 facility and equipment inventory system, **8.**1 to **8.**10
  data equipment, **8.**3 to **8.**5
  key telephone system, **8.**3, **8.**8 to **8.**9*f.*
  PBX equipment, **8.**2, **8.**3, **8.**10
  single-line telephone set, **8.**3, **8.**6 to **8.**7*f.*
 facility modifications, **8.**22 to **8.**24
  date service requested, **8.**22
 location records, **8.**21 to **8.**22
  aerial facilities, **8.**21 to **8.**22
  buried cable, **8.**22
  cable, **2.**5*f.*, **2.**6*f.*, **4.**20*f.*, **8.**21
  campus, **8.**21
  conduit, **8.**22
 numbering plan, **8.**2
  cable, **8.**2, **8.**10
   cable type, **8.**10
   numbering, **8.**10
   terminating location, **8.**10
  serving (building) closet, **8.**2, **8.**11
   closet identification, **8.**11
   code, **8.**11
   floor number, **8.**11
  universal information outlet (UIO), **8.**2, **8.**11 to **8.**12
   floor number, **8.**12
   number, **8.**12

Management and control systems, numbering plan, universal information outlet (*Cont.*):
  serving closet type, **8.12**
  service request, **8.2**
  system administration requirements, **8.1** to **8.2**
  trouble reporting and repair analysis procedure, **8.2**, **8.28** to **8.33**, **9.**18
    daily report, **8.29**, **8.30**f.
    monthly report, **8.29**, **8.31** to **8.33**
Management information service (MIS), **6.2**
Martin, James, **2.1**n., **2.2**n., **2.4**n., **3.14**n., **5.1**n., **5.12**n., **7.7**n.
MCI, **1.15**
Microwave communications (*see* Short-haul microwave communication systems)
Minimum point of presence (*see* Demarcation point)
MIS (management information service), **6.2**
Mobile radio (telephone) [*see* Cellular radio (telephone)]
Mobile telephone service (MTS), **1.11**, **1.13**, **1.14**
Modem, **5.7**, **6.4**, **6.7** to **6.8**, **6.30**
Modem pooling, **6.13**
Motorola, **6.26**
MPOP (*see* Demarcation point)
MTS (mobile telephone service), **1.11**, **1.13**, **1.14**

NAP (*see* Demarcation point)
National Association of Regulatory Utility Commissioners (NARUC), **1.13**, **1.16** to **1.17**
National Center for Devices and Radiological Health, **6.38**
National Electric Code® (NEC®), **4.24** to **4.25**, **4.29**, **4.35**, **5.13**, **5.18**
National Electrical Safety Code, **4.24**
National Television Standards Committee (NTSC), **6.43**
NEC® (National Electric Code®), **4.24** to **4.25**, **4.29**, **4.35**, **5.13**, **5.18**
NetView (IBM), **6.48**
Network access point (*see* Demarcation point)

Noise suppression, **1.20**, **5.1** to **5.12**
  data transmission, **5.1** to **5.2**
  fortuitous distortion, **5.2** to **5.9**
    amplitude change, **5.7** to **5.8**
    cross talk, **5.5** to **5.7**
    echoes, **5.7**
    impulse noise, **5.3** to **5.5**
      electromechanical interference, **5.4**
      reducing, **5.5**
    intermodulation noise, **5.6** to **5.7**
      data tone, **5.6** to **5.7**
    line outage, **5.8**
    phase change, **5.8** to **5.9**
    radio fading, **5.8**
    white noise, **5.2** to **5.3**, **5.6**
  systematic distortion, **5.2**, **5.9** to **5.12**
    attenuation, **5.10**
      amplitude-frequency, **5.10**
    bias, **5.11** to **5.12**
      negative (spacing), **5.11**
      positive (marking), **5.11**
    characteristic, **5.12**
    conditioning, **5.12**
    delay (phase-frequency), **5.10** to **5.11**
    frequency offset, **5.11**
    harmonic, **5.10**
    loss, **5.9**
    quantum noise, **5.12**
  voice transmission, **5.1** to **5.2**
Northern Telecom, **6.13**
NTSC (National Television Standards Committee), **6.43**
Nynex, **6.12**

Omninet, **7.14**
1.1 terminal (*see* Demarcation point)
Open system interconnection (OSI), **1.3**, **1.7**, **1.8**, **7.1** to **7.3**
  layer 1 (physical), **7.2**
  layer 2 (data link), **7.2**
  layer 3 (network), **7.2**
  layer 4 (transport), **7.2**, **7.3**
  layer 5 (session), **7.2**, **7.3**
  layer 6 (presentation), **7.2**, **7.6**
  layer 7 (application), **7.2**, **7.3**
Optical-fiber applications, **1.19**, **3.1** to **3.14**
  cable systems, **3.4** to **3.6**
    operation method with FITL and HSDL capabilities, **3.4** to **3.5**
    transitional operation method, **3.5**

Optical-fiber applications (*Cont.*):
  communication links, **3.**2 to **3.**4
    guided propagation, **3.**2 to **3.**3
    light sources and system components,
      3.7 to **3.**8
    metropolitan fiber rings, **3.**3, **3.**4*f.*
    superhighway, **3.**3
    unguided propagation, **3.**2 to **3.**3
  integration, **3.**14
  light sources and system components,
    3.7 to **3.**12
    cable connectors and splices, **3.**9 to
      3.11
      biconic, **3.**10
      focusing system, **3.**11
      groove-block, **3.**10 to **3.**11
      SMA, **3.**10
      smooth, straight rods, **3.**11
      ST, **3.**10
    communication links, **3.**7 to **3.**8
      injection laser, **3.**7, **3.**8
      light-emitting diode (LED), **3.**7 to
        3.8
    directional couplers, **3.**9, **3.**12
    fiber-optic switches, **3.**9
    optical detectors, **3.**8 to **3.**9
      avalanche diode, **3.**9
      PIN diode, **3.**9
    wavelength multiplexers, **3.**9, **3.**11 to
      3.12
  modes, **3.**1 to **3.**2
    multimode graded index, **3.**2
    multimode step index, **3.**2
    single, **3.**2
  optical loop calculations, **3.**12 to **3.**13
  optical transmission origins, **3.**1, **3.**3
  quantum noise, **5.**12
  types, specifications and capabilities,
    3.6 to **3.**7
    graded-index fiber, **3.**7, **3.**10 to
      3.12
    multimode fiber, **3.**7, **3.**10 to **3.**12
    single-mode fiber, **3.**6 to **3.**7, **3.**10
OSI (*see* Open system interconnection)

Pacific Telesis, **6.**12
PBX (*see* Private branch exchange)
Phase-frequency (delay) distortion, **5.**10
  to **5.**11
Planning guidelines, **1.**20, **9.**1 to **9.**128
  bid evaluation, **9.**29

Planning guidelines (*Cont.*):
  bid specifications, **9.**26 to **9.**27
    California Telecommunications
      Division, **9.**30 to **9.**128*f.*
    wire specifications, **9.**109 to **9.**126*f.*
  bid strategies, **9.**26
  contractor strategies, **9.**27 to **9.**29
    bid development, **9.**28
    knowledge collection, **9.**28
  economic variables, **9.**6
  information gathering, **9.**6 to **9.**15
    additional costs, **9.**8
    cost applications, **9.**7 to **9.**8
    per-unit costs, **9.**7 to **9.**15
    estimating process, **9.**7
  maintenance expense study technique
    (status quo versus rehabilitation),
    9.16 to **9.**20
    alternatives selection, **9.**17
      basic alternatives, **9.**17
      incremental computations, **9.**17
    assumptions, **9.**16 to **9.**17
    cable and support structure lives,
      9.19
    considerations, **9.**19 to **9.**20
    data collection, **9.**17 to **9.**18
      capital expense, **9.**18
      one-time noncapital expense, **9.**18
      operating expenses, **9.**18
      revenue income, **9.**18
    program time conversion, **9.**18 to
      9.19
  study analysis and budget process, **9.**23
    to **9.**26
    project presentation, **9.**25 to **9.**26
    sample, **9.**20 to **9.**23
  study techniques, **9.**2 to **9.**5
    benefit-to-cost ratio, **9.**4
    break-even study, **9.**4 to **9.**5
    capital expenditure, **9.**16
    internal rate of return, **9.**3
      incremental, **9.**3
    payback period, **9.**4
    present worth of annual costs
      (PWAC), **9.**3 to **9.**4, **9.**16
    present worth of expenditure, **9.**3
    summary, **9.**5
  system approach, **9.**1 to **9.**2
    alternatives considered, **9.**2
    study objective, **9.**2
Post Office Department, **1.**10
Post Roads Act (1866), **1.**9 to **1.**10

Power considerations, **1.**20, **5.**13 to **5.**18
AC requirements, **5.**16, **5.**18
apparatus closet, **5.**17
customer premises equipment, **6.**17 to
6.18, **6.**21
dedicated ground circuit, **5.**13, **5.**16
equipment and power clearance, **5.**16 to
5.17
equipment room, **5.**17
lighting, **5.**16
lightning protection, **5.**13
line filter, **5.**14
line voltage regulator, **5.**14
main terminal room, **5.**17
motor-generator (MG), **5.**13 to **5.**15
planning, **5.**15 to **5.**16
power line conditioner, **5.**14, **5.**15
satellite closet, **5.**17
station locations, **5.**17 to **5.**18
accessories, **5.**18
data, **5.**18
voice telephone equipment, **5.**18
ultraisolation transformer, **5.**14
uninterruptible power system (UPS),
5.13 to **5.**15
voltage, **5.**13 to **5.**14
fluctuations, **5.**14
line noise, **5.**14
Premises equipment (*see* Customer
premises equipment)
Privacy issues, **5.**6
Private branch exchange (PBX), **1.**4 to
1.5, **2.**3, **2.**4, **6.**8, **6.**12, **6.**15, **7.**8, **7.**15
versus centrex as customer premises
equipment, **6.**16 to **6.**25
capital investment and recurring
costs, **6.**17 to **6.**19
floor space, **6.**17
maintenance, **6.**18 to **6.**19
mileage charges, **6.**18
power considerations, **6.**17 to **6.**18,
6.21
price comparison, **6.**19
subscriber line charge, **6.**18
government regulations, **6.**19
key and hybrid systems, **6.**20 to **6.**25
centrex-compatible equipment,
6.23
electromechanical, **6.**20 to **6.**21
electronic, **6.**21 to **6.**23
hybrid electronic, **6.**22 to **6.**23
installation charges, **6.**24

Private branch exchange (PBX), versus
centrex as customer premises equipment, key and hybrid systems
(*Cont.*):
key service unit (KSU), **6.**20 to
6.22
key telephone unit (KTU), **6.**20 to
6.22
main distribution frame (MDF),
6.20 to **6.**21
non-KSU, **6.**23
reconfiguration, **6.**24
selection guidelines, **6.**24 to **6.**25
spare parts, **6.**23
technology, **6.**19 to **6.**20
electrical protection, **4.**27
management and control systems, **8.**2,
8.3, **8.**10
Protocol conversion, **6.**13
Public network services, **6.**3*f.*, **6.**8 to
6.16
centrex, **6.**8, **6.**12 to **6.**13
data communications, **6.**14
CO-LAN, **6.**14, **6.**15, **6.**18 to **6.**20,
6.24
features, **6.**9 to **6.**12*t.*, **6.**14 to **6.**16
automatic route selection, **6.**13,
6.15 to **6.**16
enhanced, **6.**12 to **6.**13
SMDR, **6.**14, **6.**15, **6.**22
station rearrangement (CSR), **6.**15
uniform call distribution, **6.**13, **6.**15
voice messaging, **6.**14 to **6.**15
versus PBX as customer premises
equipment, **6.**16 to **6.**25
capital investment and recurring
costs, **6.**17 to **6.**19
government regulations, **6.**19
key and hybrid systems, **6.**20 to
6.25
technology, **6.**19 to **6.**20
digital switching systems, **6.**13
full service, **6.**8*f.*
integrated services digital network, **1.**8,
6.1, **6.**12 to **6.**14, **6.**20, **6.**47 to **6.**48
protocols, **6.**14
Public Satellite Network, **6.**44
Public utility, **1.**9
Public utility commission (PUC), **1.**18, **6.**1

Quantum noise, **5.**12

Radio Act (1912), **1.**10
Radio determination satellite service
(RDSS), **6.**45
Radio fading, **5.**8
Rate stability, **6.**19
RBHC (regional Bell holding company),
1.17, **1.**18, **3.**3, **6.**12, **6.**14, **6.**15
RDSS (radio determination satellite ser-
vice), **6.**45
Regional Bell holding company (RBHC),
1.17, **1.**18, **3.**3, **6.**12, **6.**14, **6.**15
Ring, **1.**5
Roaming, **6.**29

Satellite closet, **4.**3 to **4.**4, **5.**17, **6.**1
management and control systems, **8.**13,
8.15*f.*, **8.**16
satellite cabinets, **4.**3 to **4.**4
specifications, **4.**1, **4.**3 to **4.**4
Satellite communications, **6.**39 to
6.46
advantages, **6.**41
high bandwidth, **6.**41
low error rates, **6.**41
stable costs, **6.**41
applications, **6.**44 to **6.**46
business television, **6.**44
education and training, **6.**44
hotel/motel programming,
6.44
mobile communications, **6.**45
radio determination satellite service,
6.45
satellite news gathering, **6.**44 to
6.45
teleports, **6.**45 to **6.**46
disadvantages, **6.**41 to **6.**42
earth station size, **6.**41
interference, **6.**41 to **6.**42
signal delay, **6.**41
ground element, **6.**40 to **6.**41
signal element, **6.**40 to **6.**41
space segment, **6.**40 to **6.**41
video transmission, **6.**43 to **6.**44
high-definition television (HDTV),
6.43
multiples analog component (MAC),
6.44
phase alteration by line (PAL),
6.43
SECAM, **6.**43

Satellite communications (*Cont.*):
voice and data transmission, **6.**42 to
6.43
frequency-division multiplexing
(FDM), **6.**42
group, **6.**42
supergroup, **6.**42
single-channel-per-carrier frequency
modulation (SCPC/FM), **6.**42
time-division multiplex (TDM), **6.**42
to **6.**43
Satellite news gathering (SNG), **6.**44 to
6.45
SDLC/SNA (Synchronous Data Link
Control/Systems Network
Architecture), **7.**15
Serial port, **7.**2
Sherman Antitrust Act, **1.**14
Short-haul microwave communication
systems, **6.**30 to **6.**34
advantages, **6.**32 to **6.**33
components, **6.**31 to **6.**32
antenna, **6.**31 to **6.**32
parabolic, **6.**32
waveguides, **6.**32
maintenance and diagnostics, **6.**32
repeaters, **6.**32
disadvantages, **6.**33 to **6.**34
technology, **6.**30 to **6.**31
analog electronic transmission, **6.**30
digital, **6.**31
SLC (subscriber line charge), **6.**18
SMDR, **6.**14, **6.**15, **6.**22
SMSA (standard metropolitan statistical
area), **1.**18
SNG (satellite news gathering), **6.**44 to
6.45
SPF (subscriber plant factor), **1.**13 to **1.**14
Standard metropolitan statistical area
(SMSA), **1.**18
Standards, **1.**5 to **1.**8
European, **1.**6 to **1.**7
international, **1.**6
local-area network, **7.**1 to **7.**4
satellite communications, **6.**43 to **6.**44
United States, **1.**7 to **1.**8
wide-area network, **7.**4 to **7.**7
Station rearrangement, **6.**15
Strowger, Almond B., **1.**5
Subscriber line charge (SLC), **6.**18
Subscriber plant factor (SPF), **1.**13 to **1.**14
Superhighway, **3.**3

Support structure and connectivity hardware, **1.**19 to **1.**20, **4.**1 to **4.**35
apparatus closet, **4.**1 to **4.**3
layout, **4.**1, **4.**2*f.*
specifications, **4.**2 to **4.**3
building environment, **4.**8 to **4.**18
horizontal distribution in older buildings, **4.**16 to **4.**18
baseboard raceways, **4.**16, **4.**17
flat cable, **4.**16, **4.**18
molding raceways, **4.**16 to **4.**18
overfloor ducts, **4.**16, **4.**17
horizontal floor distribution methods,
4.11 to **4.**16
cellular floor, **4.**11, **4.**12, **4.**14
raised floor, **4.**11, **4.**14, **4.**15*f.*
underfloor channel, **4.**11 to **4.**13
underfloor conduit, **4.**11, **4.**15 to
4.16
horizontal overhead distribution
methods, **4.**8 to **4.**11
advantages and disadvantages, **4.**8
home-run, **4.**8 to **4.**10
poke-through, **4.**8, **4.**11
raceway, **4.**8 to **4.**11
zone, **4.**8 to **4.**9
campus distribution systems, **4.**19 to
4.24
aerial construction, **4.**21 to **4.**22
cable dancing, **4.**21
buried cable, **4.**22
layout, **4.**19, **4.**20*f.*
outside plant, **4.**19
underground conduit, **4.**22 to **4.**24
component definition, **4.**1
demarcation point, **4.**4, **4.**6 to **4.**8
hardware, **4.**4, **4.**6
blue, **4.**6
gray, **4.**6
green **4.**6, **4.**7*f.*
orange, **4.**6
purple, **4.**6
red, **4.**6
white, **4.**6
yellow, **4.**6
physical security, **4.**6, **4.**8
electrical protection, **4.**26 to **4.**35
bonding, **4.**26, **4.**27
grounding, **4.**26 to **4.**27, **4.**29, **4.**31 to
4.35
aboveground cable sheaths and
cable messengers, **4.**34

Support structure and connectivity hardware, electrical protection, grounding
(Cont.):
AC power branch circuits, **4.**35
building entrances, **4.**34
building steel, **4.**31
cable shields at splice cases, **4.**33 to
4.34
cold-water pipes, **4.**31
concrete-encased electrodes, **4.**31
data communication systems
wiring closets, **4.**33
electrode system, **4.**31 to **4.**32
gas-pipe system, **4.**31
ground pipes, **4.**32
ground ring, **4.**31
ground rods, **4.**32
manholes, **4.**35
metallic power service conduit,
4.31
node electrodes, **4.**32
plate electrodes, **4.**32
telephone switching, **4.**32 to **4.**33
underground cable, **4.**34
voice and data system wiring closet, **4.**33
protectors, **4.**26 to **4.**29
current limiters, **4.**27 to **4.**29
protector panels, **4.**28 to **4.**30
voltage-surge limiters, **4.**27, **4.**28*f.*
electrical separation requirements, **4.**24
to **4.**26, **5.**5
National Electrical Code®, **4.**24 to
4.25
utility company, **4.**25 to **4.**26
satellite closet, **4.**1, **4.**3 to **4.**4
satellite cabinets, **4.**3 to **4.**4
specifications, **4.**1, **4.**3 to **4.**4
testing and patch panels, **4.**4 to **4.**5*f.*
Switching facilities, **2.**2
Synchronous Data Link Control/Systems
Network Architecture (SDLC/SNA),
7.15

Tandem office, **1.**4
TCAP (transaction capabilities applica-
tion part), **6.**14
Technical Committee 32 (ECMA), **1.**7
Technical Committee 97 (ISO), **1.**6
Telecommunications systems and ser-
vices, **1.**20, **6.**1 to **6.**49

Telecommunications systems and services (*Cont.*):

cellular radio (telephone), **6.**26 to **6.**30
cells, **6.**26
components, **6.**27 to **6.**29
cell site, **6.**27, **6.**28
mobile and portable units, **6.**27, **6.**29
switching office, **6.**27
system interconnections, **6.**27 to **6.**29
frequency reuse, **6.**26
services, **6.**29 to **6.**30
data transmission, **6.**30
roaming, **6.**29
technology, **6.**27
customer premises equipment, **1.**12, **1.**14 to **1.**19, **6.**1, **6.**2, **6.**13, **6.**15 to **6.**25
capital investment and recurring costs, **6.**17 to **6.**19
floor space, **6.**17
maintenance, **6.**18 to **6.**19
mileage charges, **6.**18
power considerations, **6.**17 to **6.**18, **6.**21
price comparison, **6.**19
subscriber line charge, **6.**18
government regulations, **6.**19
key and hybrid systems, **6.**20 to **6.**25
centrex-compatible equipment, **6.**23
electromechanical, **6.**20 to **6.**21
electronic, **6.**21 to **6.**23
hybrid electronic, **6.**22 to **6.**23
installation charges, **6.**24
key service unit (KSU), **6.**20 to **6.**22
key telephone unit (KTU), **6.**20 to **6.**22
main distribution frame (MDF), **6.**20 to **6.**21
non-KSU, **6.**23
reconfiguration, **6.**24
selection guidelines, **6.**24 to **6.**25
spare parts, **6.**23
technology, **6.**19 to **6.**20
infrared communications, **6.**34 to **6.**39
costs, **6.**39
hardware components, **6.**36 to **6.**37
very large scale integration (VLSI), **6.**37
system architecture, **6.**36

Telecommunications systems and services, infrared communications, system archictecture (*Cont.*):

time-division multiple-access (TDMA), **6.**36
system components, **6.**37 to **6.**39
coherent light, **6.**37
light-emitting diode (LED), **6.**37
noncoherent (random-phase) light, **6.**37
receivers, **6.**38 to **6.**39
transmitters, **6.**37 to **6.**38
technology, **6.**34 to **6.**36
skin effect, **6.**34 to **6.**35
transmission, **6.**35 to **6.**36
instruments, **2.**1
local loops, **2.**1 to **2.**3
private languages and marketing, **6.**1 to **6.**3
demand creation, **6.**2
embedded obsolete features, **6.**2
pushing the envelope, **6.**2
public network services, **6.**8 to **6.**16
centrex, **6.**8, **6.**12 to **6.**13
data communications, **6.**14
features, **6.**9 to **6.**12*t.*, **6.**14 to **6.**16
digital switching systems, **6.**13
full service, **6.**8*f.*
integrated services digital network, **1.**8, **6.**1, **6.**12 to **6.**14, **6.**20, **6.**47 to **6.**48
protocols, **6.**14
satellite communications, **6.**39 to **6.**46
advantages, **6.**41
high bandwidth, **6.**41
low error rates, **6.**41
stable costs, **6.**41
applications, **6.**44 to **6.**46
business television, **6.**44
education and training, **6.**44
hotel/motel programming, **6.**44
mobile communications, **6.**45
radio determination satellite service, **6.**45
satellite news gathering, **6.**44 to **6.**45
teleports, **6.**45 to **6.**46
disadvantages, **6.**41 to **6.**42
earth station size, **6.**41
interference, **6.**41 to **6.**42
signal delay, **6.**41
ground segment, **6.**40 to **6.**41
signal element, **6.**40 to **6.**41

Telecommunications systems and services, satellite communications, ground segment (*Cont.*):
space segment, **6.**40 to **6.**41
voice and data transmission, **6.**42 to **6.**43
video transmission, **6.**43 to **6.**44
frequency-division multiplexing (FDM), **6.**42
high-definition television (HDTV), **6.**43
multiples analog component (MAC), **6.**44
phase alteration by line (PAL), **6.**43
SECAM, **6.**43
single-channel-per-carrier frequency modulation (SCPC/FM), **6.**42
time-division multiplex (TDM), **6.**42 to **6.**43
short-haul microwave communication systems, **6.**30 to **6.**34
advantages, **6.**32 to **6.**33
components, **6.**31 to **6.**32
antenna, **6.**31 to **6.**32
maintenance and diagnostics, **6.**32
repeaters, **6.**32
disadvantages, **6.**33 to **6.**34
technology, **6.**30 to **6.**31
analog electronic transmission, **6.**30
digital, **6.**31
switching facilities, **2.**2
system integration, **6.**46 to **6.**49
bandwidth partitioning, **6.**46 to **6.**47
multivendor interface options, **6.**48 to **6.**49
applications program interface (API), **6.**49
asynchronous ASCII terminal interface, **6.**49
reverse engineering, **6.**49
software bridging, **6.**49
standard message protocols, **6.**49
third-party "umbrella" systems, **6.**49
wrap boxes, **6.**49
public/private interaction, **6.**47 to **6.**48
software-defined network (SDN), **6.**47, **6.**48
transmission methods, **6.**3 to **6.**8
analog, **6.**3 to **6.**5, **6.**7

Telecommunications systems and services, transmission methods (*Cont.*):
codec, **6.**4
data transmission lines, **6.**6 to **6.**8
error signals, **6.**7
full-duplex, **6.**6, **6.**8
half-duplex, **6.**6, **6.**7
simplex, **6.**6, **6.**7
digital, **6.**3 to **6.**5, **6.**8
bits, **6.**4
modem, **6.**4, **6.**7 to **6.**8
nonvoice, **6.**5
public networks, **6.**3*f.*
telephone connections, **6.**5 to **6.**6
four-wire, **6.**6, **6.**7
two-wire, **6.**7
transmission path, **6.**4
trunk network, **6.**5 to **6.**6
trunk circuits, **2.**2
Telecommunications technology origins, **1.**4 to **1.**5
central office (exchange), **1.**4
private branch exchange, **1.**4 to **1.**5, **2.**3, **2.**4
ring, **1.**5
tandem office, **1.**4
tip, **1.**5
toll office, **1.**4
trunks, **1.**4
Telefunken, **6.**43
Teleports, **6.**45 to **6.**46
Tie (trunk) line, **2.**3 to **2.**4, **6.**13
Tip, **1.**5
Toll office, **1.**4
T1 Committee on Communications (ECSA), **1.**8
Transaction capabilities application part (TCAP), **6.**14
Trunk circuits, **2.**2
Trunk (tie) line, **2.**3 to **2.**4, **6.**13
Trunk network, **6.**5 to **6.**6
Trunks, **1.**4

UCD (uniform call directing), **6.**13, **6.**15
Uniform Building Code, **4.**24
Uniform call directing (UCD), **6.**13, **6.**15
United Telephone, **6.**12
U.S. Organization for the CCIR (USC-CIR), **1.**7, **1.**8
U.S. Organization for the CCITT (USC-CITT), **1.**7, **1.**8

USCCIR (U.S. Organization for the CCIR), **1.**7, **1.**8
USCCITT (U.S. Organization for the CCITT), **1.**7, **1.**8

Voice messaging, **6.**14 to **6.**15

WATS (wide-area telephone service), **1.**13, **1.**14, **6.**13
Western Electric Co., **1.**10 to **1.**12, **1.**14
White noise, **5.**2 to **5.**3, **5.**6
Wide-area network, **7.**4 to **7.**7
Wide-area telephone service (WATS), **1.**13, **1.**14, **6.**13
Wire, cable and coax applications, **1.**19, **2.**1 to **2.**29
  cable performance, **2.**21
    insulating and jacketing compounds nominal temperature range, **2.**24*t.*
    loop resistance, **2.**24*t.*
    manufacturers, **2.**27 to **2.**28*t.*
    100 Mb/s-over copper products, **2.**29*t.*
    plastic insulations, **2.**23*t.*
    sheath and shield resistance, **2.**26*t.*
    solid bare copper wire (American Wire Gage) diameter, weight and resistance, **2.**22*t.*
    standard exchange cables nominal attenuation, **2.**25*t.*
    wiring options, **2.**28*t.*
  environment and applications, **2.**4 to **2.**18
    campus distribution systems, **2.**7, **2.**8*f.*, **2.**11

Wire, cable and coax applications, environment and applications (*Cont.*):
    demarcation point, **2.**7, **2.**9*f.*, **2.**10
      specifications, **2.**10 to **2.**11
    horizontal building systems, **2.**5
    house (distribution; gray) cable, **2.**14
    riser cable, **2.**14
      sizing, **2.**20 to **2.**21
      termination, **2.**18
    service entrance, **2.**11 to **2.**14
      aerial, **2.**13
      buried, **2.**11 to **2.**13
      sizing, **2.**18
      underground conduit, **2.**11 to **2.**14
    specifications, **2.**14 to **2.**17
      data, **2.**17
      high-speed data, **2.**15
      voice, **2.**16
      voice and low-speed data, **2.**15
    station wire, **2.**16
      sizing, **2.**18, **2.**20
      termination, **2.**18
    terminations, **2.**18
      data wire, **2.**18, **2.**20*f.*
      riser cable, **2.**18
      voice station wire, **2.**18, **2.**19*f.*
    vertical building systems, **2.**6*f.*, **2.**7
  intraagency tie cable, **2.**21
  local distribution, **2.**2 to **2.**4
  public and private networks, **2.**1 to **2.**2
  riser/distribution cable sizing, **2.**20 to **2.**21
  service entrance/interbuilding cable sizing, **2.**18
  station wiring sizing, **2.**18, **2.**20
Wire Plan Manual, **2.**16
World Administrative Telegraph and Telephone Conference, **1.**8

ABOUT THE AUTHOR

Harry B. Maybin has been involved in information flow and
telecommunications technologies for more than 25 years.
He was the founding managing director, and CEO of the
Pacific Rim Group, an international consulting firm
specializing in information systems design and implementation. Mr. Maybin has developed connectivity strategies for
the United States Environmental Protection Agency, the
United States Army Corps of Engineers, the State of
California Telecommunications Division, Pacific Bell, and
many other government and private sector entities. He is
also credited with developing the State of California Low
Voltage Wiring Standard and is an accredited instructor for
the California Community College System on information
technology subjects.